Evolutionary Linguistics

How did the biological, brain and behavioural structures underlying human language evolve? When, why and where did our ancestors become linguistic animals, and what has happened since? This book provides a clear, comprehensive but lively introduction to these interdisciplinary debates. Written in an approachable style, it cuts through the complex, sometimes contradictory and often obscure technical languages used in the different scientific disciplines involved in the study of linguistic evolution. Assuming no background knowledge in these disciplines, the book outlines the physical and neurological structures underlying language systems, and the limits of our knowledge concerning their evolution. Discussion questions and further reading lists encourage students to explore the primary literature further, and the final chapter demonstrates that, while many questions still remain unanswered, there is a growing consensus as to how modern human languages have arisen as systems by the interplay of evolved structures and cultural transmission.

APRIL MCMAHON is Vice-Chancellor of Aberystwyth University.

ROBERT MCMAHON is a Research Associate in the Institute of Biological, Environmental & Rural Sciences at Aberystwyth University

CAMBRIDGE TEXTBOOKS IN LINGUISTICS

General editors: P. AUSTIN, J. BRESNAN, B. COMRIE, S. CRAIN,
W. DRESSLER, C. EWEN, R. LASS, D. LIGHTFOOT, K. RICE, I. ROBERTS,
S. ROMAINE, N. V. SMITH.

Evolutionary Linguistics

Evolutionary Linguistics

APRIL MCMAHON

AND

ROBERT MCMAHON

Aberystwyth University

CAMBRIDGE UNIVERSITY PRESS
Cambridge, New York, Melbourne, Madrid, Cape Town,
Singapore, São Paulo, Delhi, Mexico City

Cambridge University Press
The Edinburgh Building, Cambridge CB2 8RU, UK

Published in the United States of America by Cambridge University Press, New York

www.cambridge.org
Information on this title: www.cambridge.org/9780521891394

First published 2013

Printed and bound in the United Kingdom by the MPG Books Group

A catalogue record for this publication is available from the British Library

Library of Congress Cataloguing in Publication data
McMahon, April M. S.
    Evolutionary linguistics / April McMahon, Robert McMahon.
        p.   cm. – (Cambridge textbooks in linguistics)
    Includes bibliographical references and index.
    ISBN 978-0-521-81450-8 (hardback) – ISBN 978-0-521-89139-4 (paperback)
    1. Language and languages–Origin.   2. Anthropological linguistics.   3. Human
    evolution.   I. McMahon, Robert, 1964–   II. Title.
    P116.M45 2012
    401–dc23
    2012014211

ISBN 978-0-521-81450-8 Hardback
ISBN 978-0-521-89139-4 Paperback

Contents

Figures

Preface

This book is for anyone who has ever wondered why humans are linguistic animals. There can be no doubt at all that this is exactly what we are: we use language constantly, creatively and almost compulsively. We talk to each other, our pets, and ourselves. We talk to our babies, and are quite unreasonably delighted when they talk back to us (at least for a while). We use language to get to know people; to parade our skills for prospective employers; to share our views; to express our emotions; to negotiate and establish our identities; to argue, lie and mislead; and even, sometimes, to exchange information. We invent new words, which become normal currency within families and social groups; we struggle with which of our various accents to get out of the wardrobe for particular occasions, or which of our languages to use in certain circumstances if we are bilingual; and we get white-hot furious when we feel others are taking liberties with 'our' language. One person's clever neologism is another person's linguistic mangling.

Language, then, is natural. It is part of our lives from the very start: Mehler *et al.* (1988), for example, have demonstrated that newborns are capable of differentiating their mother's language from other languages, even on the basis of a signal filtered so only suprasegmental information remains. Children learn language, as we shall see in Chapter 2, remarkably quickly and efficiently. Language forms part of our identity as individuals, as social groups – and as a species, because one of the remarkable things about language is that no other species seems to do it quite like us. To be sure, other species make noises (some of them quite persistently); and some use vocal or gestural signs, or a combination of both, to communicate. What marks out human language as special is the extraordinary structural complexity all languages display, and the inventiveness with which humans use these systems, applying them regularly to new situations and using new utterances we have neither said nor heard before. Many species use their systems of communication to establish group membership, to issue warnings about dangers in the environment, to keep track of other group members, and to share information; but in general they are restricted in the amount of information they can convey (Hauser 1996). Humans also use language as an art form, and as a tool; and we talk freely about people, objects and situations which are not right here in the physical context for the conversation, and which might indeed be imaginary, hypothetical or impossible.

Given this view of the absolute centrality of language in human life and interactions, it is scarcely surprising that humans also use language, as we are doing
now in this introduction, to talk about language. We might in particular expect
that linguists and non-linguists alike would be intrigued by the question of where
human language came from. If we are linguistic animals, and if this makes us at
least quantitatively but probably also qualitatively different from other species
(including the other primates who are our reasonably close relatives), how has
this situation come about? How did the unique system that is human language
arise, and why in particular are we its beneficiaries?

Until relatively recently, the origin and evolution of language have arguably not
received the attention they deserve in linguistics or its precursors. There is a long
and distinguished history of linguistic analysis of synchronic structure, which
has given rise to the many and varied linguistic theories we find today for the
description and possible explanation of aspects of phonology, morphology, syntax and semantics. Similarly, there is a long-standing commitment to diachronic,
or historical linguistics, with emphasis on the tracking and, again, explanation of
changes in these same structural domains, and on the establishment of language
families. But in contrast, the origin of language is one of the very few disciplines
to have been outlawed by its own practitioners, when the Société Linguistique de
Paris announced in 1866 that 'The Society does not accept papers on either the
origin of language or the invention of a universal language'.

This prohibition was a response to a nineteenth-century proliferation of theories, many based on the flimsiest of evidence, and some frankly fanciful, which
had brought the topic of the origin of language into disrepute among linguists.
We shall review some of these ideas briefly in Chapter 1; but though we might
concede that the ban on discussion of our topic was amply justified in 1866, it is
important to note that the same conditions no longer obtain now. The intellectual
climate has changed, so that unsubstantiated theorising is no longer generally
accepted; and perhaps even more importantly, we now have vastly more information than the average Victorian gentleman-scientist could have dreamed of.
As we shall see in later chapters, the development of medicine, computer technology, archaeology, palaeontology, linguistic theory and typological description, neuroimaging and primatology together mean that we are able to take at
least some tottering steps towards answering questions which our predecessors
in 1866 could not even have formulated. We do not know everything we need
to know to account for the development of language in our species; but we are
beginning to understand where we need to look, and to construct reasoned and
well-supported hypotheses about what we might find there. These hypotheses are
for the most part still plural, and there is still healthy debate over many aspects of
the field. This is to be welcomed: it is only by engaging in such interdisciplinary
debates that we can expect to move forward.

Although the Paris Linguistic Society's ban makes a good story, it is unfortunate that mentioning it was, until fairly recently, virtually the only element of
coverage of the origin and evolution of language to be found in many introductory

books on linguistics. In the early 1990s, however, there began to appear unmistakeable signs that the evolution of language was ready to come in from the intellectual cold. Interest in this domain came from two very different directions. First, researchers in a range of scientific disciplines outside linguistics began to focus on issues like the organisation and function of the brain, and on similarities and differences between spoken human language and the signed systems which are either used natively or can be taught artificially to other primates. The Human Genome Project, with its target of sequencing the whole human genetic system, attracted a great deal of initial publicity on the grounds that it was hopelessly over-ambitious, and even more as it began to achieve its goals; and as it did so, philosophical as well as genetic attention was focused on which aspects of human behaviour might be 'in our genes', and which not. These developments began to converge on the extent to which human language is species specific, and how it might be genetically and neurologically encoded. At the same time, the development of a range of powerful linguistic theories, ranging from Chomsky's various syntactic models to Optimality Theory in phonology, refocused linguistic attention on the question of innateness, so that both those seeking to defend an innate component underlying language and those keen to attack it found they had to engage with aspects of evolutionary theory. The key issue here is that if we are going to claim that any aspect of language or of any other human physical, mental or behavioural system is innate, and therefore genetically prespecified, we also have a responsibility to explain how it got there. Inevitably, then, questions of origin began to be asked again, both inside and beyond linguistics.

Aims

The origin and evolution of language is inherently interdisciplinary, both because it appeals to such a wide range of intellectual interests, and because it requires contributions from so many fields. This makes the area fascinating: even established researchers are always learning something new. But it also brings its own challenges, not least in terms of providing accessible materials for students. Contributions to the field come from genetics and neurology, psychology and anthropology, archaeology and primatology, sociolinguistics and the sociology of language, computational linguistics (especially simulation), and historical linguistics – and this list is not exhaustive. Sometimes these findings are complementary, and we find emerging syntheses; but in other cases they seem flatly contradictory, and it can be hard to know whether this is an impression created by our relative ignorance of some of the contributory fields, or whether the contradictions are real. True, the Evolang conference series which began in 1996 has provided a welcome and hugely valuable forum for interaction among practitioners of these varied disciplines, and the volumes arising from these (Hurford, Studdert-Kennedy and Knight 1998, Knight, Studdert-Kennedy and Hurford 2000, Wray 2002b, and several more to which we shall refer as we progress through the book) have disseminated and encouraged scholarship from a wide range of fields. But with the exception of a few introductory textbooks (notably Aitchison 1996, Fitch 2010), and some books on specialist topics (like de Boer 2001), such edited volumes of individual essays by specialists seem to have become the norm in the field (see also Christiansen and Kirby 2003, Tallerman 2005). These are undoubtedly high-quality collections, and are invaluable for colleagues and students seeking to develop a specialism in the area; but they may not meet the needs of students beginning work on the origin and evolution of language, or taking a single course in the area while specialising elsewhere. This is increasingly the case, since this area interacts with historical linguistics; the history of linguistics; language and mind; language acquisition; biological linguistics and neurolinguistics; and the philosophy of language. Indeed, as we have seen, students working in either formal or functional approaches to syntax or phonology may also find they need to confront the issue of innateness, and why it is so controversial; and this clearly involves dealing with evolutionary questions.

In all these cases, students and colleagues investigating the origin and evolution of language would benefit from an accessible treatment of the issues in a single volume, and this is what we seek to provide here. As a historical linguist and a human geneticist, both with research interests in evolution, we approach the topic from different but related perspectives which allow us to interpret and summarise the primary literature, and outline and evaluate the arguments, in order to encourage and equip those new to the field to participate in the many ongoing debates.

These aims mean that we will inevitably be starting from a relatively introductory level in many of the topics we discuss. There will, where necessary, be introductions to human genetics; to the anatomy and functioning of the brain and the vocal tract; to the history and prehistory of our species; to historical as opposed to evolutionary linguistics; and to relevant aspects of language acquisition. Some of these topics will only be dealt with at this introductory level, since they are important as part of the context, to allow us to progress to a consideration of evolutionary linguistics in its own right; our readers do not need to be completely au fait with all the individual complexities of each area at the highest level of expertise. Given the mixed readership we anticipate for this book, this means everyone should realistically expect to find some sections introductory, and others more challenging; some readers will also feel frustrated that we have not dealt with certain issues in greater depth. We will, however, indicate throughout in the text and at the ends of chapters where further information is to be found. Furthermore, although part of our job is to introduce and to summarise what researchers have already established in a range of background areas, we assume that this does not mean we must remain studiedly neutral throughout the book. When we think something is a poor argument, or a strong one, we will tell you. And we will warn you when we are pulling arguments together to make our own evolutionary story, which makes good sense to us; the point of proceeding in the way outlined below is that, by that stage, you should be able to decide for yourself whether you agree.

Organisation, and how to use this book

Much of the introductory contextualisation above has inevitably involved sweeping statements and unsubstantiated claims. The following chapters will develop the arguments underlying these, and help readers to understand why we approach the issues from the perspectives we do, and to decide whether they agree with us or not. The book is structured to start within linguistics, work outwards to concerns mainly within adjoining disciplines (though with clear implications for language), and finish squarely within linguistics again. Chapters are organised according to topics, but are not wholly self-contained, though Chapters 4, 5 and 6 are more so than the others. There is considerable interaction between the topics of Chapters 1–3, and 7–9, in particular.

In the first chapter, we introduce the background issues; review some of the problematic early ideas about language origins; and deal with questions of terminology as we contrast evolution with history, noting a number of recent cases where linguists say 'evolution' without meaning it. Chapter 2 goes on to consider why we might find it necessary to invoke evolution to account for language in the first place, outlining aspects of the relevant evidence from such varied areas as language acquisition, genetic disorders, creolisation and ape language experiments which suggest the development of some genetically specified underpinnings for language in our species. In Chapter 3, we ask the obvious question of whether linguists can access information about the earliest human language by taking the linguistic systems we have today and simply extrapolating backwards, and show why this cannot in practice be achieved. At the end of this first section of the book, we will have established that it is not language systems or linguistic behaviour that have evolved; rather, it must be the genetic instructions and biological structures underlying these.

To investigate the genetic and biological characteristics of a species, we must ask where that species has come from, and how it has evolved compared to its ancestors and relatives; and this is the topic of Chapter 4. We shall consider fossil evidence, and evaluate different theories on where and when *Homo sapiens sapiens* evolved. In Chapter 5, we consider the physical system which seems most clearly specialised for language in modern humans, namely the vocal tract; while in Chapters 6 and 7, we turn to an outline of the interacting roles of the brain and genes in the acquisition, use and evolution of language. Chapter 8 is central to the book, addressing the question of whether the genetics underlying language evolved slowly and gradually, via incremental pathways of mutation and natural selection, or whether a more abrupt and drastic mechanism might have been involved; it is here that we become most closely involved with questions of evolutionary theory. Chapter 9 begins by discussing how we might model what happens in populations after something linguistic has originated, but during its development to a more complex and structured set of systems, and continues to consider possible social explanations for these innovations, stressing the possibility of selective advantages provided by improved communication. Here, we face a lack of concrete evidence, since social alliances do not fossilise, and we cannot rerun human evolutionary history to assess the dynamics involved; but we are in no worse a position than many synchronic or historical linguists, who must often assess the most likely hypothesis in the light of limited facts, and by considering the best fit of a number of disparate pieces of the puzzle.

Each chapter will end with a list of primary reading, and because exercises tend not to be appropriate to this kind of field (imagine: 'Exercise 2: Rerun the last 7 million years of primate evolution and assess what conditions are needed for primates other than humans to speak'), we have provided points for discussion. These can be used in conjunction with the primary reading for reflection on the issues; as essay topics; or as pivots for discussion in tutorials or workshops. In a field like evolutionary linguistics, there are many things we do not yet know, and debate and discussion can only be of benefit in tackling these problems; likewise, the existence of many unanswered questions should not discourage beginners in the field, but should encourage them to see that they are just as able to make a contribution as any of the rest of us. With this in mind, we encourage readers to use their evolved linguistic systems well and reflexively in considering their development in our species.

Acknowledgements

This book has been a long time in the making; one of the problems with rapidly moving interdisciplinary fields is that nobody (even if there are two of you) can keep up with all that is going on, so we have been very reliant on talking things over. Our thinking has been shaped by discussions with colleagues in Cambridge, Sheffield and Edinburgh, and through the AHRC Centre for the Evolution of Cultural Diversity based at University College London, and while mentioning individuals by name always seems invidious when so many influences are indirect and cumulative, we would like to acknowledge Simon Kirby, Rob Foley, Marta Lahr, Jim Hurford, David Chivers, Colin Renfrew, Peter Matthews, Francis Nolan, James Steele, Fiona Jordan, Ruth Mace, Jamie Tehrani, Guy Deutscher, Roger Lass, David Barton, Hannah Cornish, Thom Scott-Phillips, Gareth Roberts, Paul Heggarty, Maggie Tallerman, Dan Dediu, Bob Ladd, Alison Wray, Andrew Smith and Kenny Smith, among many others. We would like to pay particular tribute to colleagues and students in the Language Evolution and Computation Unit in Linguistics and English Language at the University of Edinburgh, who have done so much to put the study of evolutionary linguistics on the map, and indeed to establish it as a discipline.

There are two sets of people we are very happy to thank for their patience with this project, which has been held up partly by the swift development of the many intellectual areas involved, and partly by the fact that at least one of us always seems to be either moving to a new job or adjusting to one. So, thank you to Andrew Winnard and Sarah Green at Cambridge University Press; but second and much more importantly, thank you to our children, Aidan, Fergus and Flora, not only for their tolerance of our days working on the book, but also for their increasingly excellent company as the years go by. The teenage years would not be what they are if humans didn't have language.

1   Evolution and history

1.1   Overview

In this chapter, we shall outline exactly what we mean by 'evolutionary linguistics'. We consider some early theories and speculations about the origin of language, and make an important distinction between evolution and history. In discussing the evolution of language (or strictly, the evolution of the human capacity for language), we need to understand the terminology of evolutionary theory, and we therefore introduce the basics of the theory of natural selection. It is worth noting, however, that many linguists do use terms like 'evolution' and 'evolve' to refer to developments in historical rather than evolutionary time, adopting evolutionary metaphors to help us understand language change. Finally, we discuss the notion that humans might now be beyond the forces of evolution, but conclude that evolution is indeed continuing, though humans may be more capable of altering our environment than other species.

1.2   Evolutionary linguistics

Evolutionary linguistics has two interacting subjects of study, namely the issue of how, why and from what earlier systems the human capacity for language originated; and how, when and why it has subsequently developed, through descent with modification, to allow us to acquire and use the language systems we have now. The first aspect requires us to investigate where the systems we call language have come from: as Shakespeare puts it, 'Nothing will come of nothing' (*King Lear*, I:i), so it would be scientifically improper as well as completely unenlightening to invoke a mysterious blinding flash somewhere in the early history of our species before which we were silent and agrammatical beings, and after which we had fricatives, diphthongs and subordinate clauses. On the other hand, what we know of the processes of biological evolution indicates clearly that something very highly structured and apparently very specialised can indeed come from not very much: the eye, for example, has evolved independently a number of times in different lineages, beginning on each occasion from a much simpler source (Nilsson and Pelger 1994). Moreover, at least when we are dealing with simpler organisms and shorter generation intervals, this process of development of complexity can happen rather quickly: Nilsson and Pelger make a pessimistic best estimate of approximately 364,000 years for a light-sensitive patch to develop into a camera eye corresponding to roughly the structure of the modern human eye.

This means that there are likely to have been precursors to human language as we know it; and perhaps aspects of these might still be discernible either in the communication systems of our primate relatives (which are sometimes vocal, sometimes gestural, and sometimes both), or in other human behaviours which we have maintained since before we had language. However, the second part of our enterprise involves looking at how and why these pre-existing systems have developed into the capacity for language and the linguistic systems we have now. As we shall see, this does not mean we will be focusing on how languages today differ from one another: that is a matter for historical linguistics and typology. Evolutionary linguistics looks at possible sources for human language in other systems and its development up to the point when it achieved the level of structural complexity common to languages today.

This preamble may explain why we are not calling the enterprise in which we are engaged here the origin of language, or the evolution of language: each of these terms focuses on only one of the two key aspects of the whole process. It is also important that we include the term 'linguistics' in the title, because the 'of language' part is either misleading or at best ambiguous, stressing as it does the superficial and behavioural aspects of languages themselves. As we shall see, we are primarily concerned with the genetic developments which have shaped the physical and neurological structures responsible for predisposing us to acquire these rather highly variable behavioural patterns.

1.3   Early ideas about the origin of language

1.3.1   The first language

Early theories tended to focus only on one aspect of what we are calling evolutionary linguistics: they were much more concerned with the origin question than with any subsequent development. The very earliest discussions in this area, in fact, tended to make the assumption that human language must have originated in a system just like some of the languages known to us in historical time: the question was not how human language originated in some other system, but which human language came first. This model is familiar from the biblical account (in Genesis 11) of the Tower of Babel, which tells how God created a single language for humans. Those humans, in turn, used their language to develop a plan to build a tower high enough to reach heaven; whereupon God, increasingly concerned by this outbreak of cooperation, 'confounded' their languages, scattering different groups of people around the earth, and presumably creating the family-tree model of language families at a stroke.

What the Bible does not tell us is which language was the first; the implication seems to be that this initial system no longer exists. However, there are various records of attempts to identify the first human language through rudimentary (and typically inhumane) 'experiments'. Pharaoh Psamtik, or Psametticus I (663–610 BC), for instance, left two small children to be brought up by a shepherd, who was permitted to feed them but not talk to them, and who was to report to the Pharaoh on what language they had learned. Unsurprisingly, they were not particularly verbose; but in due course, after about two years, the shepherd reported to the Pharaoh that they had approached him with hands outstretched, saying *becos*, which on investigation turned out to be the Phrygian word for 'bread', both a reasonable request and an apparent demonstration that Phrygian (an Indo-European language of Asia Minor which died out around the sixth century AD) was the world's first language. A similar approach was adopted by King James IV of Scotland (1473–1513), who sent babies to be raised on the rather inhospitable island of Inchkeith in the Firth of Forth, with the added sophistication of preventing cheating by employing a nursemaid who could not speak. Robert Lindesay of Pitscottie, a contemporary commentator, comments laconically that 'Sum sayis they spak goode hebrew bot as to my self I knew it not bot be the authoris reherse' (Mackay 1899: 237).

However, neither Phrygian nor Hebrew is a realistic candidate for the role of first human language. This is no reflection on these particular languages, but on the simple fact that the role of 'first human language' is not open to be filled, certainly not with any developed, structurally complex system such as either of these. Nominating any language of the type we know now is simply invalid: in doing so, we would be confusing Language with languages, and evolution with history.

1.3.2 Early theories of origin

Before going on to develop this distinction between evolution and history, we should look at some further early theories which do seek a source for human language beyond languages we know now. Jespersen (1922: Ch. XXI) provides an outline of a number of these, referring to them all rather disparagingly as having 'been advanced by followers of the speculative or a priori method', by which

> those who have written about our subject have conjured up in their imagination a primitive era, and then asked themselves: How would it be possible for men or manlike beings, hitherto unfurnished with speech, to acquire speech as a means of communication of thought? (1922: 413)

These early theories tend also to rejoice in rather expressive, cartoon shorthand names: the three we shall consider briefly here are the bow-wow, pooh-pooh, and yo-he-ho theories.

All these early ideas rest on the key – and reasonable – assumption that a sound or sequence of sounds originally uttered for some non-linguistic purpose might in time come to have linguistic meaning. First, the bow-wow theory is based on onomatopoeia: if cats meow, or birds tweet, early humans (or rather, pre-human hominids; see further Chapter 4) may well have used relatively close imitations of those sounds to refer to the creatures making them. Second and similarly, the pooh-pooh theory suggests that language has developed from instinctive, emotional cries, like the *ouch*es and *ow*s and *ooh*s and *aah*s of pain or pleasure or anger. Third, the yo-he-ho theory notes that strenuous physical activity can lead to repeated and sometimes rhythmic exhalation of breath, often accompanied by grunts and groans of various kinds; again, elementary words for actions might develop from the sounds typically encountered along with the actions, so that 'the first words would accordingly mean something like "heave" or "haul"' (Jespersen 1922: 415–16).

However, the main and intrinsic limitation of all these early theories lies precisely in their restriction to and preoccupation with the first words of human language, when words in many ways are the very least of our worries as evolutionary linguists. It is true, but not particularly illuminating, that some early words might derive from imitation of sounds heard in the natural environment of early hominids; or that cooperative grunts might turn into a few early verbs; or that noises triggered by pain, for example, might change into language-specific signals: Jespersen (1922: 415) notes that 'in pain, a German and a Seelander will exclaim *au*, a Jutlander *aus*, a Frenchman *ahi* and an Englishman *oh*, or perhaps *ow*'. But there are many more objects, let alone concepts, which have no characteristic noises of their own for us to imitate, than those that do; many activities are not strenuous and would not be expected to provoke cooperative noises which might in time come to stand for the event; and it is questionable whether Jespersen's *au*, *ahi* and *ow* are words at all, or something else, a difficulty he notes himself in recognising that such interjections frequently contain sounds, like voiceless vowels or clicks, not found in the normal segmental phonological inventory of the language in question. These theories might tell us a plausible story about the origin of a few early words, but they cast no light on the subsequent development of whole lexicons, full as these are of totally conventionalised, arbitrary and unmotivated associations of sound and meaning, or of the more mysterious but arguably far more central development of order and structure above the level of the word.

Jespersen's own contribution is to reject the 'speculative method', and instead to consider evidence from child language, 'primitive' languages, and the histories of languages. He has rather little to say about the first of these, only making the point that any insight from language acquisition should really come from the very earliest, babbling stage, since it is erroneous to attempt to develop a theory of language origins based on what children are doing while learning a language that already exists. His observations on 'the languages of contemporary savages' (1922: 417) are clearly best left on one side as a product of the thinking of the time; they are difficult to reconcile even with Jespersen's own earlier and more enlightened statement that 'no race of mankind is without a language which in everything essential is identical in character with our own' (1922: 412–13). For the most part, the features he highlights here are simply reflections of typological differences and consequently, as we shall see below, products of historical rather than evolutionary processes and time.

Finally, however, Jespersen suggests that we might productively consider evidence from earlier stages of present-day languages and from linguistic reconstruction, and 'attempt from that starting-point step by step to trace the backward path. Perhaps in this way we may reach the very first beginnings of speech' (1922: 418). We shall test this method of backwards extrapolation in Chapter 3 and find it wanting, but for the moment can note that Jespersen nonetheless comes to some interesting conclusions on the basis of his retracings of change. On the basis that language change typically involves simplification of pronunciation, 'We may perhaps draw the conclusion that primitive languages in general were rich in all kinds of difficult sounds' (1922: 419); the same kind of argument is taken to suggest that early language had long words which have subsequently become progressively shorter. Grammatically, Jespersen suggests that the historical trend towards analysis, with the development of small and separable units, may have its roots in an extreme form of synthesis, where units with the shape of single words and no discernible internal structure may have expressed the meaning of whole sentences. Since tone and pitch-accent often disappear in documented histories, we might assign tone to the first language(s), which Jespersen also envisages as highly musical in its intonation, on the grounds that intonation is strongly linked with emotion, and that (much more dubiously) 'it is a consequence of advancing civilization that passion, or at least, the expression of passion, is moderated, and we must therefore conclude that the speech of uncivilized and primitive men was more passionately agitated than ours, more like music or song' (1922: 420).

This link with song is perhaps one of the most striking elements of Jespersen's own suggestions about early language. Conceptually at least, his ideas have some connection with the pooh-pooh theory, since he argues explicitly against the suggestion that language developed primarily to allow us to express and communicate our thoughts: 'Thoughts were not the first things to press forward and crave for expression; emotions and instincts were more primitive and far more powerful' (1922: 433). From here, however, Jespersen departs on more of a flight of fancy, developing his theme in a claim that (1922: 433–4):

> the genesis of language is not to be sought in the prosaic, but in the poetic side of life; the source of speech is not gloomy seriousness, but merry play and youthful hilarity ... In primitive speech I hear the laughing cries of exultation when lads and lasses vied with one another to attract the attention of the other sex, when everybody sang his merriest and danced his bravest to lure a pair of eyes to throw admiring glances in his direction.

This Camelot-like preoccupation with love and spontaneous outbreaks of song as a key to the early development of language has attracted its share of derision:

Aitchison (1996: 9) quotes Diamond (1959), who observes acidly that 'As for courtship, if we are to judge by the habits of the bulk of mankind, it has always been a singularly silent occupation'. Jespersen himself seems aware of such reactions, and in a rather hurt footnote suggests that critics should not get too preoccupied with his 'remarks on primitive love-songs, etc.', which some have sought to demolish 'by simply representing it as a romantic dream of a primitive golden age in which men had no occupation but courting and singing' (1922: 434). Jespersen claims that the real utility of his view is the suggestion that we should not simply speculate on language origins, but use tendencies of linguistic change to provide evidence. As we shall see in Chapters 3, 8 and 9 below, this methodology is seriously problematic; but some of the suggestions Jespersen makes about the importance of prosody in the development of language from pre-language will resurface again later, while his hypothesis that the earliest language is likely to have contained long, unanalysable utterances which were subsequently divided to produce smaller, individually meaningful units, is strikingly similar to some aspects of Alison Wray's (2000, 2002a) account of the evolutionary linguistic importance of formulaic language (though without, in this case, the singing).

1.4   Evolution and history

1.4.1   Schleicher's distinction

All the languages we know about today, whether they are still spoken now or attested in written records, are at approximately the same level of structural complexity: we shall essentially take this for granted, as any linguist would, but what we mean by it is that complexity, roughly speaking, evens out across the grammar, so that a language with a highly complex morphology might have a rather more simple syntax, for example. For one recent and extensively documented example, consider Everett (2005), on Pirahã, a language spoken in a number of villages along the Maici River in Brazil (see also Nevins, Pesetsky and Rodrigues 2009, Everett 2009). Everett argues, controversially, that Pirahã lacks a whole range of features which are often considered to be vital to, or even definitional of, human language: these include number words and numerals; colour terms; and embedding of one syntactic structure in another, giving rise, for instance, to subordinate clauses. Pirahã likewise has one of the smallest segmental phoneme inventories recorded, and a remarkably small pronoun system (which is moreover likely to have been borrowed in its entirety). Nonetheless, Everett emphasises that 'No one should draw the conclusion from this paper that the Pirahã language is in any way "primitive". It has the most complex verbal morphology I am aware of and a strikingly complex prosodic system' (2005: 62, note 1). Everett argues that there is radical reduction in some aspects of the structure of Pirahã as compared with many or even perhaps all other human languages; but equally, some areas of the grammar are highly complex. As Everett also notes, in comparing languages we are not concerned with particular vocabulary items, or coverage of particular semantic fields: not having words for vaccines or satellite television does not make one language less structurally complex than others, just as not having the things these words refer to does not make a human group less highly evolved, reflecting instead simply the nature of its society.

In fact, we should not say that properties of individual human societies, or of individual human languages, have evolved at all: behavioural differences of this kind have arisen in historical rather than evolutionary time, and through historical rather than evolutionary processes. This distinction between evolution and history has certainly been made in linguistics before now, though not always in the way we envisage here. Invoking evolution, though not necessarily with its Darwinian biological meaning, was quite popular in late nineteenth-century linguistics, and August Schleicher in particular used biological terminology plentifully in his development of the family-tree mode of representation for language affiliations. 'Philosophically, however, Schleicher was a nineteenth-century German Romantic progressivist, influenced more by Hegel than by Darwin' (McMahon 1994: 319); and although he knew Darwin's work, his concept of evolution was an earlier, pre-Darwinian interpretation couched in an overall model of progress and decay. Schleicher saw the evolution of language as an essential aspect of the evolution of our species, but regarded evolution as a progressive expansion of earlier simple forms into later complex ones Evolution in this sense is a period of novelty and progress, culminating in the considerable morphological complexity of inflecting languages like Latin, Greek and Sanskrit. However, as soon as any languages reached this peak of perfection, evolution stopped and history began. History, in Schleicherian terms, is bad news: if evolution is progress and the development of complexity, history involves a return to simplicity via a lengthy period of decay and loss, with no prospect of creating anything new.

This rather bleak view of history is not one we are going to follow here; nor will we be adopting these negative attitudes to change, which tends in any case to follow cycles from more complex to less and back again rather than the linear and directed (and downhill) developments envisaged by Schleicher. However, although we do not accept Schleicher's definitions of evolution and history, we do need some distinction between the two.

1.4.2   Evolution by natural selection

Perhaps the easiest way to conceptualise the difference between evolution and history is precisely to consider the timescales involved. A well-worn but still highly pertinent expression of this comes from Maynard Smith (1993: 327):

> About 400 million years ago the first aquatic vertebrates evolved; at least two million years ago man's ancestors first chipped stones to make simple tools. Less than ten thousand years ago, in the neolithic revolution, animals and plants were first domesticated. If a film, greatly speeded up, were to be made

of vertebrate evolution, to run for a total of two hours, tool-making man would appear only in the last minute. If another two-hour film were made of the history of tool-making man, the domestication of animals and plants would be shown only during the last half minute, and the period between the invention of the steam engine and the discovery of atomic energy would be only one second. These figures show how rapid are historical changes when compared to evolutionary ones.

When we talk about evolution, we mean a slow, incremental process across many generations, which we shall explore in more detail in Chapter 7, but which essentially consists of three interrelated stages. First, there is mutation. Accept for the moment that humans, just like all other living organisms, are composed of individual cells, which work together to make larger organs. Each cell contains a nucleus, and each nucleus in turn contains genetic material, or DNA. Strings of adjacent elements of DNA (which are conventionally written as sequences of four distinct units, A, C, G and T) make up genes. Each gene contains the coded information to construct a protein, which in turn may have a particular function, either alone or as part of a group. If any element of such a sequence is changed, we have mutation: thus, a sequence of AACGTTCGC may become AACATTCGC during the process of reproduction, so that the parent organism has the former sequence, and its offspring the latter. Processes of mutation may be physically conditioned by environmental mutagens such as ultraviolet light, but are much more commonly random 'errors' occurring during replication, which create differences between earlier and later versions of 'the same' DNA sequence. This is why evolution is often described as persistence with modification: the sequence as a whole is maintained, and the structure or behaviour for which it codes is also highly likely to be inherited by the next generation with only minor modification, since the sequence is virtually intact. This assumes that the genetic sequence has some kind of structural or behavioural significance in the first place; but that does not appear to be true for absolutely every 'letter' in the genetic sequence. In many cases, sequences which do code for the construction of a particular protein are interspersed along a stretch of DNA with sequences which do not appear to code for anything at all; hence the frequent reference to 'junk DNA', which either has no surface or phenotypic meaning, or none we have discovered yet. In fact, the Human Genome Project (Human Genome *Nature* issue 2001) has established that only approximately 1.5 per cent of the 3,200,000,000 (that's three thousand two hundred million) bases (those are the As, Cs, Gs and Ts) in the human sequence actually do code for proteins.

Regardless of any physical or behavioural relevance, however, any mutation will cause a small but measurable change at the genetic level to take place. However, although mutation necessarily comes first, it is only an enabling step in the evolutionary story that follows. Without mutation, there could be no evolution; but there may be mutation which stops without developing any evolutionary meaning, if, for instance, the effects of the new, mutated sequence are so disastrous that the organism in which the mutation arises just drops dead. On the other hand, if the organism survives, then we have a situation of variation, such that individuals with the old and new sequences co-exist in the population. What happens next will depend on what effect that new variant has, if any. Crucially, if its effect is positive, then organisms with it are likely, all other things being equal, to prosper, and to have more viable offspring than those with the old-fashioned variant. Consequently, in subsequent generations the proportion of the population with the new variant will increase until it is the norm. Moreover, any other mutations affecting the same trait will tend to be increasingly favoured, so that a simple and small change from the previous norm can become exaggerated over evolutionary time into an extreme or a completely novel structure or behaviour. This would be the usual evolutionary account for the development of long necks in giraffes, or humps in camels, for example, since long necks are beneficial in terms of reaching high foliage, and humps are superb fat-retention (and therefore indirectly water-retention) devices in deserts.

Nonetheless, there are trade-offs to take into account, too, since long necks and humps have their cumbersome aspects and might restrict movement or flexibility in other ways; we would not expect to find a lot of giraffe-type creatures in situations where foliage is low-growing, or humpy beasts among rivers and lakes. Furthermore, all this takes a very long time indeed: necks may become progressively longer, but the increase has to be gradual and incremental, especially bearing in mind that the original mutation will typically have had only a small effect, and that subsequent development of the trait in question will have to wait for further relevant mutations to arise. There is, after all, no point in developing such an enviably long neck in the womb that you can't be born successfully at all, or growing the same long neck in infancy without the associated development of a vascular system capable of pumping blood to the brain at the end of it. Evolution illustrates a constant compromise between developing a more extreme form of some trait, and overshooting, which is frequently fatal. Moreover, adapting too radically for a specific environment leaves a species unable to venture anywhere else, while environmental changes of even a fairly restricted type will leave it seriously vulnerable. Prudence, it seems, is the order of the day in evolutionary terms; and the whole process will typically happen gradually.

This, in short, is the Darwinian (or more accurately, neo-Darwinian; see Chapter 7 below for a much fuller development) account of evolution by natural selection. True, some mutations have such an immediately and catastrophically deleterious effect that organisms carrying them cannot survive: understandably, these do not spread, though they may survive at relatively low levels in populations because the mutation responsible recurs from time to time in different families. In the case of some inherited human diseases, the picture can also be more complex: severe forms of osteogenesis imperfecta (brittle-bone disease) are found in all human populations as a result of just such recurrent mutation, while recessive diseases may be maintained in part because the genetic defects responsible also increase resistance to other diseases. For example, a specific mutation leads to sickle cell anaemia when it is inherited from both parents; but when present in a single copy (as in those parents themselves), has only very mild anaemic effects and simultaneously protects against malaria. However, for the most part, mutations will be either neutral, and drift along unremarked in populations until someone develops an analogue of the Human Genome Project to sequence them; or they will be positive – in a particular environment.

Natural selection is often profoundly misunderstood and misrepresented as suggesting that particular changes and the resulting physical traits are naturally and globally superior and would spread and become the norm in populations wherever they arose. But this is not the case: what is selected depends completely on the environment where a species finds itself, and more specifically on the fit between what its genes force or allow it to do in behavioural terms, and what resources the environment offers for exploitation. This also addresses the confusion often encountered between the description of mutation as a random process, and the apparently directed nature of evolution. While mutation creates a pool of variants, it is natural selection that acts on these variants, picking out those which represent positive developments in, or adaptations for, a particular environment. Mutation is therefore initially at least a development within the individual genome; but natural selection, and the consequent spread of particular highly adapted variants, is active within and between populations.

1.4.3   Genes, structures and behaviour

Although we tend to talk about the evolution of particular behavioural or physical traits, like long necks or humps or flying or burying nuts under trees in autumn and digging them up again in winter, there is in fact a risk of fundamental confusion here which we must seek to avoid at all costs if we are to keep evolution and history separate. There is a three-way distinction we must make and maintain between *genes*, *structures*, and *systems* or *behaviour*.

Mutation takes place at the level of genes. Over long expanses of time, measured in terms of many generations, these mutations may spread through species and become the norm; as this happens, it will tend to have structural consequences, creating physical differences between parents and offspring which will become progressively more marked insofar as their results are still environmentally beneficial and do not become unwieldy. These physical or neurological structures in turn will permit, or condition, or predispose individuals within the species to particular behavioural patterns. But neither the behaviour nor the physical traits evolve in themselves; they are underpinned by the processes of evolution, but those processes operate at the genetic level. Behaviour and structures are phenotypic and represent the results of an interaction between genes and environment; mutation and selection take place in the genotype.

However, things are just a little more complicated than this. Sometimes our genetic systems and the physical structures to which these give rise allow a certain amount of flexibility at the behavioural level. Long necks mean food can be taken from higher trees; but relatively few species (giant pandas and koalas notwithstanding – and look how rare those are) are completely restricted in terms of the specific plants they can eat, and if a new tall food plant evolves or is introduced into an environment, animals with long necks will often be able to use that to supplement their existing diet. We might say that such creatures have evolved genetically specified physical structures, in the shape of long necks, which allow them to eat High Foliage, but not just a particular tree or plant which produces high foliage. Creatures like moles are predisposed to dig holes, because genetic evolution in their species has produced strong, spade-like feet with strong claws; but they are not necessarily restricted to digging only through clay. They are adapted for Digging, not for digging through soil with only particular chemical or physical properties.

This is where we return to language. Here again, we will throughout this book be making precisely the same three-way distinction between evolution at the genetic level, physical phenotypic consequences at the structural level, and the variable behavioural systems these structures enable. In particular, we will be interested in the evolution of the genetic instructions which mean humans grow vocal tracts and brains with particular, consistent, language-related and language-enabling properties across our species. Such vocal tracts and brains in turn predispose us to acquire and learn Language; but they do not require that we learn any particular language. The language system we will learn depends completely on the society in which we are brought up; indeed, in many societies citizens will normally learn more than one language. Humans are linguistic animals because of the consequences of genetic evolution, which means each individual grows particular physical and neurological structures; within certain levels of tolerance, these can be said to be the same across our whole species, as the giraffe's long neck is 'the same', within a few centimetres. What we do with those structures is to a certain degree absolutely preconditioned and determined by our genetic heritage: barring any inherited defect or physical or neurological injury, and provided that he or she receives some environmental stimulus in the form of exposure to linguistic data to set the whole process in motion, every human will grow up to speak at least one language. However, the properties that distinguish one language from another have not developed by the processes of genetic mutation, variation and selection; whether we have velar fricatives or not; subjects before verbs or verbs before subjects; agreement for case and number between adjectives and nouns, or no inflectional morphology at all; or noun classes as opposed to grammatical gender, is absolutely not determined genetically. Linguistic options of this kind are permitted by the plasticity of our genetic and physical systems; they are determined only at the level of the individual, by what he or she hears from his or her immediate family (and later, his or her peer group); and they need not stay constant in historical time, but can all be affected by linguistic change, whether caused by language-internal systemic factors, or induced by contact.

What this means, more generally, is that differences between languages must have developed in historical time. They have not evolved, and they do not require or reflect any difference at the genetic level between populations of humans. This is not simply an assertion: the best evidence for it is the observation that humans will learn different languages depending on the linguistic environment in which they find themselves, so that if a child born to English-speaking parents then moves at an early age with her family to Japan, she is highly likely to learn both Japanese and English. It would therefore be straightforwardly counterfactual to claim that individual human populations are genetically specialised for the learning and use of any particular language, or of languages only with particular characteristics.

But even if we suspend disbelief for a moment and ask whether in principle it would be possible for a genetic specialisation for a particular language to have arisen and subsequently spread by natural selection to all and only the members of a particular population, such that all and only the speakers of Yoruba, or Scots Gaelic, or Hungarian would have that genetic predisposition, the answer again has to be no, largely for reasons connected with the distinction between evolution and history which we have already made. To begin with, what any two human populations share in genetic terms vastly outweighs what differentiates them, since on average 84 per cent of the total human genetic variation is found within any given population (Barbujani *et al.* 1997). There is much discussion of so-called population-specific polymorphisms in the genetics literature (see Cavalli-Sforza, Luca and Feldman 2003), but the actual number of such variants is rather small, and typically the variants concerned are found only in a small number of individuals within the population and are not shared by all its members. They are arguably therefore better conceptualised as highly localised individual-specific variants than as reliable signals of population membership. This in large part reflects the sheer amount of evolutionary time which would be required for a variant to arise and spread to fixation through a human population. In our earlier discussion of the evolution of the eye (1.2 above), we noted Nilsson and Pelger's (1994) convincing argument that only around 364,000 years would be required for a complex, lens-type eye to evolve; but this assumes evolution over approximately 363,992 generations. A generation interval of around one year (that is, the average period between when an organism is born and when all its own offspring are born), which Nilsson and Pelger note would be completely appropriate for small and medium-sized aquatic animals, is impossibly low not only for modern humans (where the generation interval is 20–25), but also for our primate relatives, where the generation interval is of the same order as our own (approximately 25 years for gorillas, 16–20 years for chimpanzees). Given this more extended generation interval, Maynard Smith (1989: 43) estimates that between 5,000 and 275,000 years (200 to 1,100 generations) would be required even for one simple genetic variant to spread through a whole population; for a more complex structure like the eye, the equivalent number of years would be $363,993 \times 25$, which is 9,099,800, or nearly 10 million years. While again this is not vastly long in evolutionary time, it is well beyond human history or even prehistory, since the split from the human–chimpanzee common ancestor is often estimated at 5–7 million years before the present day. As we shall see in later chapters, given the various specialisations for language we can discern in the brain and the vocal tract, for example, we are likely to be dealing with a large number of interacting genetic variants, not a single genetic variant, so that the timescale involved for the evolution of a capacity for Language will be closer to estimates for the eye than to Maynard Smith's spread of a single variant to fixation. Evolving capacities for individual languages after this initial development, which would mean waiting for individual mutations to arise and then to spread for each language population, would imply timescales far longer than the period of existence usually claimed even for language families (leaving aside the associated point that languages are constantly changing anyway, even during the lifetime of individual speakers).

This has two immediate consequences. First, assuming as we do that evolutionary linguistics is about evolution rather than history, our focus in this book must be firmly on the genetic processes which have given rise to the physical and neurological structures and systems responsible for our multifarious linguistic behaviours, but not on that linguistic behaviour itself – except as evidence for the structures we need to posit. We will be interested in what behavioural systems might have preceded human language; what the necessary characteristics of Language might be; and how these differ from the communication systems of other species. We will also be concerned with the general question of how the original human linguistic system(s), starting from more rudimentary structures and limited variability, developed greater complexity to become what we would recognise as human language. However, we will have nothing to say about how these early, full linguistic systems subsequently developed in historical time into Hebrew and Phrygian, Welsh and Tamil, Yoruba and Quechua, or how earlier stages of languages, like Old English or Old Welsh, have changed into today's versions. This is the concern of historical linguistics, as is the study of ongoing variation and change in languages today. The second consequence is that we will have to learn to be very wary indeed of evolutionary metaphors, which have been extraordinarily popular in recent linguistics. In the next section, we shall consider some of these metaphorical uses of the terms 'evolution' and 'evolutionary', and show how these can often be encountered referring to developments completely within historical time, with no evolutionary meaning in our sense whatsoever.

1.5   Saying 'evolution' without meaning it

Historical linguists and variationists frequently observe that something akin to the three-step pathway of mutation, variation and selection can be discerned in language variation and change. We might consider the initial appearance of a new form as mutation: let us say that a word, or several words with initial [h] at an earlier stage of the language in question develop variants without the [h]. This creates variation in the system, such that words with initial [h] will come to have alternative pronunciations without it, and indeed this reflects the situation in most varieties of English today, where certain words, especially more grammatical ones like the pronouns *he* and *his*, will tend to be pronounced without initial [h] when they are in positions of low stress. At this point, however, we are dealing with variation rather than change; the change is a later possibility if the different variants develop different meanings (either linguistically or, more commonly, socially), and one is adopted to the detriment and perhaps ultimately to the exclusion of the other. If a group using the pronunciations without [h] is particularly attractive or fashionable on other, non-linguistic grounds, other speakers may gravitate towards their pronunciation, ultimately dropping [h]s from more words and in more contexts in a (typically subconscious) attempt to emulate the trend-setting group. Variants with [h] may ultimately be lost from this variety, with consequences for the system in that the distribution of the /h/ phoneme may become restricted, or the phoneme itself may be lost.

Seeing linguistic variation and change as a cycle of mutation, variation and selection in this way can be extremely helpful: not only does it provide a framework for the stages of language change, but it can also assist linguists in their understanding of biological evolution by providing a familiar linguistic analogy. However, we must constantly remind ourselves that we are dealing here with evolution as a metaphor: variation and change can be interpreted as analogous to genetic mutation, variation and selection, but in our terms the rise of linguistic variants and their possible embedding in language systems through change are crucially historical rather than evolutionary processes, which therefore do not in fact involve any *genetic* mutation, variation or selection. The genetically prespecified physical and neural structures humans have equip us equally well to learn and use languages which have initial [h], and those which do not; and we need not and in fact absolutely must not invoke any genetic difference to account for the loss or gain of a segment or linguistic structure in this way. Such differences in overt human behaviour are precisely not evolutionary, and are very clearly happening in historical time: if there had to be a genetic change to produce the pronunciations without [h], it would take a very great deal longer than the relatively few human generations between Old English and Modern English to develop and spread. In any case, such a suggestion is obviously nonsensical: speakers of different varieties of English may well be more closely related in genetic terms than those speaking a single dialect (think of families where members of one generation move to different parts of the country, and their children therefore acquire different accents); and invoking a genetic difference to account for varieties with and without initial [h] does not allow for the continuum in between, where many speakers will produce neither categorical [h] nor categorical absence of [h], but rather a variable proportion of [h]s depending on the social and linguistic circumstances.

There is, of course, no intrinsic difficulty with using the steps and ideas of evolutionary theory to provide useful and enlightening metaphors for historical developments in language. There are, however, two requirements we must place on such metaphorical uses of evolution: first, it is crucial that we recall they are indeed metaphors, and do not get tempted into interpreting them literally; and second, even metaphors can be dangerous if they are based on a faulty knowledge of the domain from which they come, in this case biological evolution. Furthermore, linguists have to learn to stop and think about every situation where they find terms like 'evolution' and 'evolve': are we really in an appropriate context to be discussing genetic differences, or are we dealing here with a metaphorical framework to help us understand language change? To illustrate this necessity, let us consider four cases where linguists use the term 'evolution' without, in our sense of evolutionary linguistics, actually meaning it.

Chronologically, the earliest of the four is Samuels (1972), whose very title, *Linguistic Evolution: With Special Reference to English*, would be enough to disqualify it as a contribution to evolutionary linguistics, since, as we have seen, behavioural characteristics of individual languages cannot appropriately be ascribed to genetic differentiation. Samuels himself is completely clear about his intentions, noting at the very outset that 'The title of this book ("Linguistic Evolution") was chosen in preference to "Linguistic Change" although it is about linguistic change', and emphasising that 'We are not concerned here with the prehistoric origins of human language' (1972: 1). Samuels' use of 'evolution' is intended to invoke a particular framework for understanding language change, which 'in its barest essentials ... consists of the two levels, spoken chain and system, linked by the process of selection' (1972: 139). 'Spoken chain' and 'system' here are intended to be close counterparts of Saussure's ideas of *parole* and *langue* respectively; and Samuels stresses selection as the vital ingredient of change, arguing that:

> Every change is, at least in its beginnings, present in the variants of the spoken chain; it is the process of continuous selection that ensures its imitation, spread, and ultimate acceptance into one or more systems.

Depending on this process of selection, variants may be rejected or selected and integrated into the system; or a variant may become so numerous it is effectively imposed on the system, which may thereby change more radically.

This connection of variation and change, and the invocation of selection as a means of transforming one into the other, is fairly commonplace now, but would have been novel and challenging in 1972. However, Samuels defines his use of the terms 'evolution' and 'selection' very clearly in terms of these linguistic applications, and does not attempt to link them to evolutionary theory: other aspects of biological evolution, such as mutation, are not mentioned, and Darwin himself does not figure in the index. More recently, we do find attempts to apply further aspects of evolutionary thinking to language change, for instance in Mufwene's (2001) development of an evolutionary model for creoles. Here,

> language evolution ... is used ... to cover long-term changes observable in the structures and pragmatics of a language, as well as the not-so-unusual cases where a language speciates into daughter varieties identified at times

as new dialects and at others as new languages. It also covers questions of
language endangerment and death. (Mufwene 2001: xi)

Mufwene uses a range of concepts from ecology and population genetics in a novel
approach to creolisation and change. His main innovation is to consider a language,
not as an organism, but as a species, composed of a collection of idiolects, or individuals' language systems: however, languages are not composed entirely like biological species, and do not behave in entirely the same way. Mufwene explicitly
adopts a model of competition and selection, focusing on the notion of ecology for
a language both in terms of the external environment, and the internal interrelations
of elements within a single system, whose neighbouring elements form part of
their immediate ecology. Although this provides a range of enlightening perspectives on the causes and consequences of language change, Mufwene is completely
clear that his focus is on 'evolution' from one language state to another, not from
something in some sense unlike language to language as we know it; again, his
approach is therefore fundamentally historical and metaphorical.

   A second more recent approach, again more explicitly keyed into concepts
from biological evolution, is that of Croft (2000, 2006); though again, like
Samuels and Mufwene, Croft suggests that he is developing 'a framework for
understanding language change as a fundamentally evolutionary phenomenon'
(2000: xiii). However, he is at pains to emphasise that he is not attempting to
visualise language change simply through the metaphorical lens provided by biological evolution:

> The use of biological metaphors or analogies, while valuable for illuminating
> the character of some linguistic phenomena, is ultimately limited. Metaphors
> and analogies cannot clearly identify what biological evolution and language
> change should be expected to have in common. Only a systematic evolu-
> tionary framework can offer a genuinely evolutionary theory of language
> change. (Croft 2006: 92)

By 'a systematic evolutionary framework', Croft means the essentials of evolution, abstracted away from any individual application – including the application in biology which is often seen as primary or privileged. In articulating
these essentials, Croft draws on David Hull's (1988, 2001) generalised analysis
of selection, which is fundamentally based on the notion of copying, or replication: where such copying is imperfect, change may result. Copying produces a
set of replicators, including variants; in biology, the replicator is the gene. The
replicators will be carried or used by what Hull calls interactors (for genes, these
will be biological organisms), which form larger historical entities in the form of
populations. Depending on interaction with the environment, certain replicators
will then be advantaged and others disadvantaged, a picture familiar to us from
biological selection.

   This generalised analysis of selection can, as we have seen, be applied at the
level of biological evolution; Croft seeks to apply it to language change. He defines
languages as populations of utterances; languages are consequently historical

entities, localised in space and time in the same way as biological populations. Each language will incorporate variation of different types, at different levels, and with a range of different motivations. Croft argues that the replicator in language is fundamentally the utterance, since utterances are copied by being produced in language use; but he coins the term *lingueme* to refer to the structures inherent in utterances, which can themselves be produced as different variants depending on a range of linguistic and sociolinguistic factors. Speakers, the interactors of this model, also form part of a population in the form of the speech community. As speakers produce different variants of linguemes in different contexts, frequencies of variants will alter over time, causing change; there can be many motivations for such shifts in frequency and therefore for change itself, though these will be domain specific and do not form part of the generalised analysis of selection.

It would be possible to develop a much lengthier and more detailed account of Croft's terminology and framework, with applications to particular changes in particular communities. However, the necessarily very brief overview here should be sufficient to show that this book is not the appropriate place for that more detailed discussion, precisely because Croft's approach is concerned with language change and with history, and consequently in our terms does not deal with evolutionary linguistics. Croft (2000: 9) is in complete agreement with this, noting that 'this book is concerned with language change itself, not the evolution of a certain biological capacity of human beings'.

There are two major differences between Croft's approach and ours, then. First, Croft is focusing on historical time and historical processes, whereas we are concerned with longer-term evolutionary processes. Croft accepts, as we do, that there must be a difference between the evolution of biological capacities, and the shorter-term development of changes in languages; but his work involves the latter, and ours involves the former. Second, Croft sees both biological evolution and language change as subtypes of a superordinate theory of change through replication, involving units and mechanisms derived from Hull's generalised analysis of selection. Our approach does not address language change at all, and is not intended to. Fundamentally, though of course evolution and history are related and continuous, and we return in Chapter 9 to the linking concept of cultural evolution, we are considering those aspects of language that fall within the purview of biological evolution, not seen as a framework or as a metaphor, but as part of the genetic development of our species. While Croft differentiates biological evolution and language change, then, our evolutionary linguistics precisely addresses the evolution of the human capacity for language, so that our work falls squarely within biological evolution and leaves language change aside. Croft's work is particularly helpful in highlighting these distinctions.

Finally, we turn to Juliette Blevins's (2004) book, *Evolutionary Phonology*. As may by now be familiar, she includes in her preface a disclaimer to the effect that:

> Contrary to what the title may imply, this is not a book about the evolution
> of language in the human species. Though I do draw parallels between the
> evolution of sound patterns and Darwin's theory of natural selection, these

parallels are largely metaphorical. This book does not deal with the biological or neurological foundations of language and should not be read as a
neo-Darwinian treatise. The period of study is roughly the past 7,000 years –
extremely short by biological standards. (2004: xi)

Blevins's main concern is the development of sound patterns in historical time, and
in particular, the fact that many patterns recur even in languages which belong to
entirely different families. Notably, she does not simply present evolutionary linguistics, in our sense, and evolutionary phonology in hers, as unconnected fields:
rather, she suggests that the latter accounts for the continued development in historical time of the initial linguistic systems which arose in evolutionary time: 'Once a
system of categories and contrasts has been established, what are the forces which
continue to shape sound systems, and which have led to similar sound patterns in so
many of the world's languages?' (2004: xiv). Whereas a number of current models
of phonology ascribe recurrent features of phonological systems to innate aspects
of our human capacity for Language, Blevins instead stresses the role of recurrent
sound change in building frequently occurring and regular patterns. In turn, these
sound changes arise variously from misperception, or from inherent ambiguities
in the phonetic signal, or from consequences of listeners' choices from the range
of variants a speaker may produce. This has obvious implications, not just for our
view of the human language capacity, but also for the way we conceive of and write
phonologies of particular languages, since 'once historical explanations are found
for common sound patterns, the same phonetic explanations need not, and indeed
should not, be encoded in synchronic grammars' (2004: 300). Selection, though
at the behavioural rather than the genetic level, is for Blevins a key mechanism
in explaining sound change; but again her concern is with what is encouraged or
facilitated as opposed to what is discouraged by the 'envelope' of physical and
mental structures humans have for Language. In this book, we will be focusing on
the actual, genetic evolution of that envelope itself.

   All these approaches are, of course, both legitimate and valuable in helping
us understand language change. It is in no sense our remit to argue against metaphorical uses of evolutionary concepts or terms; but it is our job to point out that
they are, in the dichotomy we have been developing here, fundamentally concerned with history and not with evolution. At the same time, we must be careful
not to suggest that the division between evolution and history is an absolute one,
or that evolution has stopped; we explore this issue in the next section.

1.6   Beyond evolution?

   To summarise the sections above, we have argued that evolution is not
directly relevant to any aspects of the superficial behavioural systems we call languages. These have not in themselves evolved, at least not in the literal biological
sense we are employing, and they are not evolving now. They are products of

particular structures, some neurological and others physiological, which are sufficiently plastic and permissive to allow a range of such linguistic behaviours to be learned and used, though they will predispose us under normal developmental circumstances to acquire and use at least one such behavioural system. In turn, these phenotypic neurological and physical structures have developed by prior evolution in our species or ancestors at the genetic level, through the processes of mutation, variation and selection. It follows that our topic of evolutionary linguistics is about the development of these capacities for Language, but not about the historical changes we can still observe in languages themselves.

This might mean that we appear to be drawing an absolute line between evolution, which is something that happened in the past and therefore has logically stopped now, and history, which is still going on; and yet earlier in the chapter, we suggested that Schleicher's absolute division of evolution from history was deeply problematic. In fact, evolution is certainly still a continuing process, in our species as in others. Take, for example, wisdom teeth. As a species, humans seem to be losing them progressively: fewer of us have wisdom teeth now than in previous generations. It is possible that this might have some future impact on language (say, by changing the spacing of the other teeth, or the shape of the jaw), though evolutionary processes operate so slowly that this is very difficult to gauge. On the other hand, it might be a consequence of previous evolutionary adaptations that have been essential in allowing us to produce spoken language in the way we do, which, as we shall see later, involve among other things shortening of our muzzles, so we have fairly flat faces with a serious amount of resonating space inside. Add to this the non-Language-specific fact that compacted wisdom teeth would sometimes have proved fatal for our human and hominid ancestors, since infection could well have set in and could not have been counteracted, and it becomes obvious that not having wisdom teeth might be advantageous, so that selection may well be favouring the spread of genetic variants which lack wisdom teeth.

However, there is another interacting factor here, which is that humans have arguably, through many thousands of years of cultural development, altered their own environment, removing or neutralising selection pressures which would have been strongly operative at an earlier stage. For instance, the development of modern dentistry means that humans who would have died and perhaps not left viable offspring in the past because their compacted wisdom teeth would have killed them, are now able to live long and reproductively active (if relatively toothless) lives. This may mean that the variant responsible for lack of wisdom teeth is less likely to spread to fixation in all human populations in the way that might have occurred had dentistry not been available. In other words, deliberately altering the environment may take the pressure off the spread of particular variants. Evolution has not stopped, and mutations continue to enter the gene pool; some will drift along, unremarked; some will be lost; some will have unforeseeable and hugely deleterious effects which will be tragic for individual families but occur too infrequently for medical interventions to be developed successfully;

and yet others, which are problematic but relatively common, will effectively be selected out or reversed by human behavioural intervention, including the development of genetic testing for inherited diseases and selective termination of pregnancy. Evolution goes on, but in individual cases we may have the capacity to block its effects; this becomes a matter of ethics as well as biology.

Evolution, then, is continuing, in our species as in others. The immediate question now is how and why we are seeking to invoke it in a non-metaphorical sense to account for our human capacity for Language. So far, we have laid out the essential ingredients of a model that accounts for a set of behavioural traits, indirectly, using evolutionary forces and processes; but we have not yet justified the adoption of such a model. In the next chapter, we shall explain why our human capacity for Language requires genetic underpinnings at all.

1.7   Summary

In this chapter we have laid out the basis for our investigation of evolutionary linguistics, a term we explained in 1.2, and in particular have developed two vitally important distinctions. The first of these opposes evolution to history: this book is concerned with biological developments in our species, over potentially great expanses of evolutionary time, and with the behavioural consequences of the neurological and physical structures these developments have produced. It is not concerned, as the field of evolutionary linguistics in general is not concerned, with language variation and change in historical time, though, as we saw in 1.5, it is possible and often enlightening to use concepts and terms from evolutionary theory in a metaphorical way to help us understand historical change. Distinguishing history from evolution does not, as we explained in 1.4 and 1.6, imply that evolution stopped when history started; as a species, humans are still subject to the forces of evolution, though we may be more adept in the modern era at modifying or avoiding some of its less palatable consequences. We return to this distinction in the context of cultural evolution, in Chapter 9.

The second major distinction we have argued for opposes genes, structures and behaviours. Evolutionary forces and processes operate at the genetic level, and in 1.4 we outlined the cycle of mutation, variation and natural selection involved. Genetic instructions then build physical or neurological structures; and in combination with environmental factors, these then produce behavioural systems like language. It will be extremely important in the chapters that follow to bear these distinctions in mind. In 1.3, we also outlined some early approaches to the origin of language, such as 'experiments' designed to see which language isolated children might speak, and speculations like the bow-wow, yo-he-ho and pooh-pooh theories, which tried to account for the shape of the first words, but do not successfully generalise to the first grammars. In the next chapter, we consider features of language acquisition, use and breakdown which support our hypothesis that human language does indeed have evolutionary underpinnings.

Further reading

If you would like to read more about the early 'experiments' designed to find the first language, the Psamtik story appears in Herodotus's *Histories*, and the James IV one in Lindesay of Pitscottie's *The Histories and Chronicles of Scotland* (1899, edited by Aeneas J. G. Mackay, Volume I: 237). There is mention of some similar 'experiments', and a general discussion, in Sulek (1989).

Schleicher's distinction of evolution from history is discussed further in Chapter 12 of McMahon (1994), and you can read about his ideas first-hand in Koerner (1983).

There is some discussion of the early 'theories' of language origins in Aitchison (1996); but it is certainly worth reading Jespersen's own account of these in Jespersen (1922, Ch. 12).

There will be more detailed discussion of the theory of evolution, including aspects of natural selection, in later chapters. For the moment, however, anyone finding the outline discussion of genetics difficult or perplexing is recommended to consult Gonick and Wheelis (1991). Accessible popular science accounts are to be found in Jones (2000) and Dennett (1995). A useful set of short readings on evolution (including a reprint of the Nilsson and Pelger paper about the evolution of the eye) is Ridley (1997). We recommend Lewin and Foley (2004) as a comprehensive textbook treatment of all aspects of human evolution.

When it comes to metaphorical uses of evolutionary terminology, and extensions of evolutionary thinking beyond genetics and biology, the evolutionary story, with its sequence of mutation, variation and selection, has proved equally appealing to colleagues in disciplines outside linguistics. Laland and Brown (2002) discuss a range of attempts, some more successful than others, to apply evolutionary approaches to human behaviour and society. The topics covered include sociobiology, evolutionary psychology and memetics.

Finally, on the interplay between genetic and cultural evolution, and the question of whether modern humans are in any sense beyond evolution, there is an interesting perspective in Diamond (1991) – Diamond's theme, printed before his contents page, is 'How the human species changed, within a short time, from just another species of big mammal to a world conqueror; and how we acquired the capacity to reverse all that progress overnight'. The final sections of Jones, Martin and Pilbeam (1992) would be relevant here; and again, this is a useful general sourcebook.

Points for discussion

1. In Schleicher's discussion of evolution versus history, he notes that languages like Greek, Latin and Sanskrit marked the high point of evolution, before the degeneration of history set in. Why might he have identified these particular languages as the most highly developed? What sort of languages would be the opposite type, the ones most seriously affected by historical processes, or

least affected by evolutionary ones? What arguments can you find against the history–evolution dichotomy understood in this way?

2.   As well as the bow-wow, pooh-pooh and yo-he-ho 'theories' of language origins, Jespersen (1922) mentions the ding-dong theory. Find out about this, and about any one other early theory of this kind. Do you find any aspects of these theories convincing? Are they all subject to the same kinds of counter-arguments and limitations?

3.   Throughout this book we will be asking you to plot events (or at least, rough estimates of general time-spans for processes – let's get used to talking in evo-lutionary rather than historical time) on a chart. The first step is to start making your chart. You should draw a line a metre long, and divide it into the conven-tional geological ages, say the Archean, Proterozoic and Phanerozoic. Within the Phanerozoic, plot the start of the Cambrian era, at about 600 million years Before Present (myBP); and then sequentially the Devonian beginning at 400 myBP, and the Carboniferous, Permian, Triassic, Jurassic, Cretaceous and Cenozoic eras. Then draw another line a metre long, which this time is an expansion of the part of this first chart covering just the last 65 million years – in geological terms, this is the end of the Cretaceous and start of the Cenozoic era. Where on your charts would the evolution and extinction of the dinosaurs be? Where would the major expansion of mammals be? What about the origin of our own species, *Homo sapiens*? We will fill in some further details on the chart later, and it may help you envisage the evolutionary timescales involved, and the relationships between certain developments. You can find background information in Jones, Martin and Pilbeam (1992), Lewin and Foley (2004), or other handbooks on evolution or biology.

4.   Have a look through some textbooks or papers on historical linguistics – useful journals might be *Diachronica*, *Language Variation and Change*, *Transactions of the Philological Society*, or the *Journal of Indo-European Studies*, though there are sometimes papers on historical topics in more general journals like *Language*, *Lingua*, or the *Journal of Linguistics*. Look out for uses of the term 'evolution', or of associated terms like 'selection' and 'mutation'. Are these being used in a truly evolutionary, or in a metaphorical, historical sense? Are there cases where this might be misleading, or is it generally easy to tell what is intended from the context?

2   Evidence for evolution

2.1   Overview

Evolutionary linguistics presupposes, by its very name, that it is appropriate to think about language in evolutionary and not just historical terms, and that some structure or structures underlying modern human languages must therefore have evolved somewhere in the family tree of our species. In this chapter, however, we unpack that presupposition, considering evidence from a range of areas to support the suggestion that some aspects of this capacity for language have indeed evolved in humans. This evidence involves the biological argument from design; the hypothesis that there is a critical period for language acquisition; Chomsky's famous argument from the poverty of the stimulus; evidence from children's creativity, especially in the context of creolisation; the notion of specialised language areas in the human brain; and finally, the identification and tracking of genes where defects inherited in families appear to correlate with language (and often other) difficulties.

2.2   The argument from design

There are two logically possible but extreme views on the origin and evolution of language (we return to the more habitable but debatable land between them below). The first of these would claim that language is genetically specified and inherited in every particular; the other would argue that language is learned from what is available in the environment, again in every particular. Like most diametrically opposed and absolutist positions, both are demonstrably wrong.

The first of these extreme hypotheses we have already ruled out in our discussion of Language as opposed to languages. Straightforward observations about family relocations and cross-cultural adoption prove that children learn the languages spoken in the context where they find themselves, not any biologically determined mother (or father) tongue. Should any further refutation be required, our knowledge of the rate of evolutionary change and of human genetic variation shows that no individual language system could be inherited on a population-specific basis. On the other hand, this does not rule out any evolved basis for Language: humans may well be predisposed to learn some linguistic system or systems (the precise details being determined by what is heard in each child's environment) because of certain evolved structures in the brain, or in our physical conformation, or both. Not, of course, that such evolved structures need necessarily be completely specialised for language: it goes without saying that we use elements of the vocal tract for non-linguistic functions (most obviously breathing), and language may well also involve more generally applicable cognitive systems. We are simply suggesting at this stage that there may be physical and neurological structures which have a partial specialisation for language, perhaps among other things. Looking at this from an evolutionary viewpoint, the development of a capacity to analyse patterns might initially have been beneficial in hunting or predicting the availability of water in different places at different seasons; but it might subsequently have facilitated aspects of language and therefore have been co-opted for these purposes too. We return to the question of whether certain capacities are specific to language in later chapters.

In this chapter, we will be arguing against the alternative and diametrically opposed suggestion that absolutely nothing about language has to be 'hard wired', or reliant on evolved structures. In this case, however, there is no single piece of evidence which proves absolutely that human languages are *not* just learned on the basis of observation and repetition. The simplest ways of demonstrating some genetic influence on a particular trait or behaviour involve either observing its inheritance patterns, or disrupting the normal function of the genes involved. In other words, geneticists will typically examine or create mutations which affect the behaviour or characteristic in question. In the former case, they will look across generations and lineages until they are confronted with a major phenotypic change: a naturally occurring mutation may have caused fruit flies, for instance, to develop without eyes. Crossing experiments can then be conducted to track inheritance through the offspring, using a reference map of known variants to help locate the gene or genes responsible. Sequencing at the molecular level and subsequent comparison across species may help to determine when particular changes have occurred. Clearly, ethical considerations prohibit much of this work, except where it involves naturally occurring variants, in humans. Furthermore, a highly complex set of behavioural systems like human language are likely to be influenced by many genes, not by only one; this, plus the indubitable influence of the environment, means that mutations will typically be observable only rarely, when their effects are unusually large, and when systems other than language are also disrupted.

Environmental variation is so broad for partly genetic, partly environmental systems and behaviours that all but the most extremely discrepant variants are typically seen as falling within normal bounds. This means that we cannot rely, at least for the present, on language-relevant mutations even being observed systematically, let alone on their thorough sequencing and cross-generational or cross-species comparison: such studies would be of enormous interest and relevance to work on evolutionary linguistics, but do not appeal so immediately to medical concerns (though see 2.7, and Chapter 7 below). Since we cannot rely on direct genetic confirmation of the evolutionary underpinnings of language, we

will spend the rest of this chapter piecing together a mosaic of less direct, sometimes circumstantial evidence from a whole range of different areas of research.
Taken together, however, we believe that these individual pieces of evidence have
only one cumulative interpretation, namely that humans cannot learn languages
the way we do, at the speed we do, and with the input we have, unless there is
some prespecified drive for us to do so and some assistance in the shape of an
internal framework which supports that learning experience and helps us organise what we then know.

   Biologists, of course, would wonder why we are spending so much time here
on something so obvious, since they are completely accustomed to interpreting
improbably complex structures and behaviours as the result of the long-term
operation of evolutionary processes and principles. Indeed, Richard Dawkins
(1986: 1) begins his book *The Blind Watchmaker* by defining biology as 'the
study of complicated things that give the appearance of having been designed
for a purpose'. Dawkins goes on to quote one of the most famous passages
in nineteenth-century theology, William Paley's (1847) articulation of this
'Argument from Design'. Paley is trying to demonstrate that the only possible
explanation for highly complex systems is the existence of a Creator, and he sets
this argument out in the form of a thought experiment (1847: 49):

> In crossing a heath, suppose I pitched my foot against a *stone*, and were
> asked how the stone came to be there; I might possibly answer, that, for any-
> thing I knew to the contrary, it had lain there for ever: nor would it perhaps
> be very easy to show the absurdity of this answer. But suppose I had found a
> *watch* upon the ground, and it should be enquired how the watch happened to
> be in that place; I should hardly think of the answer which I had before given,
> that for anything I knew, the watch might have always been there … [T]he
> watch must have had a maker … there must have existed, at some time, and
> at some place or other, an artificer or artificers, who formed it for the purpose
> which we find it actually to answer; who comprehended its construction and
> designed its use.

Paley extends these same arguments to the systems of nature, arguing that just
as some human agent must have created and designed the watch, so the inherent
complexities of natural systems like the eye must in turn show the action of a
divine agent. Dawkins, in turn, is keen to stress the clarity of Paley's descriptions
and argumentation, and agrees fully with Paley's observation that the complex
nature of biological systems and structures requires an explanation. However,
Dawkins's explanation is different, being fundamentally physical and biological
rather than theological: complex biological systems which give the appearance
of design and are certainly functional have arisen through the operation of mutation and natural selection on pre-existing physical structures. In Dawkins's terms,
'Natural selection is the blind watchmaker, blind because it does not see ahead,
does not plan consequences, has no purpose in view. Yet the living results of natural selection overwhelmingly impress us with the appearance of design as if by
a master watchmaker' (1986: 21).

In addition to the mechanisms of mutation, variation and selection which we have already encountered, we now have the slightly more controversial ingredients of gradual development of complexity, and of functionality or, more loosely, usefulness. Dennett (1995: 68) talks not just about design, but about the Principle of the Accumulation of Design: if some behaviour or feature is useful, even marginally, then it will confer an advantage on its carriers, meaning that their genes will spread through the population in subsequent generations. Furthermore, if a little bit of the feature or behaviour is good, then probably more will be better, so that favourable mutations which increase the distance between the haves and the have-nots in the population will also be advantaged; hence not just design, but the Accumulation of Design. When we turn to Language, adopting this kind of scenario means accepting that linguistic capabilities and behaviours might have been useful to our ancestors, an issue over which there is rather more controversy than one might anticipate (see Chapters 8 and 9 below), and also that languages as we know them are highly likely to have developed from much simpler precursors. We return to these suggestions later, in discussions of the mechanism of natural selection, the requirement for functionality, and adaptation and other mechanisms; for the moment, however, we require only the suggestive presence of usefulness and apparent design to set us searching for the evolved structures which underlie today's language systems.

2.3   The critical period hypothesis

It is a truth universally acknowledged that children are better at learning languages than adults. Those of us who have tried to learn second and subsequent languages as adults will all too readily confirm that progress can be slow and frustrating: there is an almost palpable sense of rusty machinery cranking along with all too little discernible effect. Jackendoff (2002: 95) develops this 'commonsense observation', noting that 'Any child, taken to any linguistic community at an early age, will come to speak the community's language like a native, while the parents may struggle for years and never achieve fluency.' It follows, he suggests, that 'adult language learning is more like playing chess or the stock market or a musical instrument, domains in which individuals differ widely in talent' (2002: 96). Newport (1990) summarises a number of studies of L2 speakers of English who had attained different levels of grammatical ability (both in their own performance and in judgements of grammaticality) depending on the age when they arrived in the USA. In the case of one particular test, moreover, 'Multiple regression analyses showed that these effects were not attributable to formal instruction in English, length of experience with English, amount of initial exposure to English, reported motivation to learn English, self-consciousness in English, or identification with the American culture' (Newport 1990: 19–20).

In terms of comparisons with other linguistic behaviours, these results make adult second-language learning more like the way children learn to read and write. If they receive appropriate tuition, virtually all of them will man-age to learn some reading and writing; but they certainly have to work at it, and their degree of success and the extent to which they achieve fluency in this other linguistic modality is dependent partly on how hard they work and partly on how good they naturally are at this kind of thing. However, it makes adult second-language learning strikingly unlike children's acquisition of their native language(s): note that 'acquisition' rather than 'learning' is normally used here precisely to highlight this difference. The vast literature on language acquisition, starting from case studies like Bloom (1970), Brown (1973) and Smith (1973), has demonstrated that in acquiring a spoken or signed language (Newport 1990) from infancy, talent does not seem to be a factor, and normally developing chil-dren follow a series of developmental milestones at approximately similar ages (Stromswold 2000), as they do in learning to walk, for instance. Only in cases like the amount of vocabulary used, especially the learned vocabulary developed primarily through literacy and formal education, do we find a continuum for chil-dren of the sort familiar from adults learning a second language. Newport (1990: 12) spells out the resulting paradox:

> in much of developmental psychology, insofar as there are maturational effects, an uncontroversial generalization is typically that big kids are better than little kids. In language acquisition … however, the child, and not the adult, appears to be especially privileged as a learner. The correct account of such a phenomenon must therefore explain not only why children are suc-cessful language learners, but also why adults, who have better capabilities than children at most things, are not.

Behaviours that are more readily acquired early in life than later, or which can only be acquired successfully during a specific phase of life, are seen by biolo-gists as having a critical period for development. Sometimes the term 'sensitive period' is preferred to 'critical period' because such periods very rarely have absolute cut-off points, before and after which no acquisition of the behaviour is possible at all. It is generally more a question of development happening opti-mally and preferentially during the sensitive period, and being more difficult or partial later. A good illustration involves birdsong (as opposed to bird calls, which are typically shorter, and do not appear to have a significant learned component, since they will emerge even in birds which have been isolated or deafened and therefore have no access to environmental auditory input). Birdsong (for further details see Doupe and Kuhl 1999, Hauser 1996) is found in many species, but each species will have a characteristic song of its own, complete with local 'dia-lects'. In order to learn the species-specific song in its local variant, birds have to be exposed to adult birdsong from their conspecifics during a relatively short period; this also varies from species to species, but white-crowned sparrows, for example, appear to require such exposure at between 10 and 50 days of age (Marler 1991). If young birds are isolated during this critical period, they will nonetheless acquire a simplified version of their species' song, but without local dialect features.

The notion of a critical period has also been proposed for human language, with an early, extended and well-known discussion in Lenneberg (1967). In fact, birdsong is more generally a particularly apposite comparison for child language acquisition: as Doupe and Kuhl (1999: 567) suggest,

> Both humans and songbirds learn their complex vocalizations early in life, exhibiting a strong dependence on hearing the adults they will imitate, as well as themselves as they practice, and a waning of this dependence as they mature.

In both cases, input from the environment seems to be required if acquisition of the appropriate local variant is to proceed successfully. We have already seen that the results for second-language learning are materially affected by the age at which such exposure begins. The question now is whether this is also true for first-language acquisition; furthermore, there is clearly a genetic predisposition for birdsong, but we must ask whether this is also the case for language.

In human children who are developing normally and who have been exposed to adult speech in their language(s) through normal social interaction, there are a number of generally agreed, though fairly flexible milestones for language acqui-sition (see Doupe and Kuhl 1999: 577, Stromswold 2000). For approximately the first three months after birth, babies produce non-speech sounds (Stark 1980), though they can be shown to have the capacity to discriminate virtually all the phonetic contrasts found in any language (Streeter 1976, Gerken 1994). They then produce speech sounds during normal babbling, though up to approximately 10 months these sounds are again essentially universal, demonstrating many if not all of the sounds found in human speech. It seems that babies at this stage are partly exercising their vocal organs, though increasingly through this period they show the beginnings of a preference for sounds they hear around them, and, interestingly, from 6–12 months they lose the capacity to perceive sounds which are not contrastive in their own language (Werker and Lalonde 1988, Kuhl 2004). Certainly, older babies show a preference for sounds they hear in their own envir-onment, and are thus perhaps developing an internal representation of speech sounds which is specific to the language they are acquiring. For instance, Jusczyk et al. (1993) used a technique where babies hear the same stimulus while they continue to look towards a particular loudspeaker; the stimulus changes when they look away, and the experimenters then observe any changes in behaviour to assess whether the babies have noticed that the stimulus is different. Their findings showed that English-learning babies could distinguish English from Dutch word lists at 9 months old, but not at 6 months old. This phase of devel-opment goes along with a development of babble towards behavioural norms of the speech community: as Gerken (1994: 789) suggests, 'research indicates that untrained adult listeners cannot discern differences based on target languages in the babbling of 6-month-olds, while they can in 8-month-olds (de Boysson-Bardies et al. 1984)'. Crucially, then, the speech production and perception of babies seems to involve an interaction of innate and learned abilities.

However, we must also ask what happens when the normal exposure to language data does not happen. Skuse (1988) reports on a range of cases of neglected and deprived children, who were not given the expected level of linguistic or other social stimulation during the critical period. Isabelle, for example, was discovered in November 1938, in Ohio, at approximately 6½ years of age; she was an illegitimate child and had apparently been rejected by most of her family and kept in a dark room with her deaf and mute mother. Isabelle was hospitalised after her discovery and initially seemed to have no spoken language at all; but after a week she began to move away from the gestures she had used to communicate with her mother, and started to learn spoken words with even greater than usual speed, attaining a vocabulary of 2,000 words after 18 months. Adam was abandoned in 1972 in Colombia; he was kept in a girls' reformatory school until he was 16 months old, with no toys, very limited social contact, and a very restricted diet. He was, however, adopted at 34 months and at 14 years old was found to be developing normally in physical, social, linguistic and educational terms. The so-called Koluchova twins, identical boys born in 1960 in the then Czechoslovakia, were initially put into a children's home after their mother died in childbirth, but at around a year old went to live with their father and step-mother. Here, they were isolated, being kept in the house, mainly in an unheated cupboard or the cellar until the age of 7. After 18 months in a children's home, during which they began a swift process of cognitive development, they were successfully fostered; by adolescence, their language skills were above average.

Genie is perhaps the best known of these isolated children, and certainly one of the more distressing case histories (Curtiss 1977). Genie was isolated from the age of about 20 months to 13 years 7 months; she was kept restrained in a small bedroom, fed only baby food, and was essentially not spoken to. She was also beaten by her father when she made any noise, as was her nearly blind mother. After her mother escaped with her and Genie was discovered, in 1970, she was taken into care and underwent a long process of sometimes intensive tuition and rehabilitation; much of this is reported in Curtiss (1977). Reports during her initial hospitalisation indicate that Genie understood individual words, but did not seem to grasp that their different syntactic contexts might add to or change their meaning (Curtiss 1977: 11); she seemed to respond better to gestures than to speech. Standard tests showed that Genie had a level of cognitive development which would typically allow language to be acquired normally – at an earlier age. The question was what would happen when someone aged more than 13½ began learning a first language.

Genie quickly acquired vocabulary. In January 1971, she was spontaneously saying just *stopit* and *nomore*. By June 1971, when Susan Curtiss began working with her, Genie was demonstrably recognising hundreds of words, and 'by May 1971 she was spontaneously saying numbers from one through five, color words, the verbs *open*, *blow*, and others, and many, many nouns' (Curtiss 1977: 15). She progressed, after two years, from single-word to two-word utterances; but her progress beyond this grammatical level was uncertain, with longer strings typically involving set phrases (like 'may I have + Noun'), and more complex structures eluding her:

> Genie could not memorize a well-formed WH-question. She would respond to *What do you say?* demands with ungrammatical, bizarre phrases that included WH-question words, but she was unable to come up with a phrase she had been trained to say. For example, instead of saying the requested *Where are the graham crackers?* she would say *I where is graham cracker.* (Curtiss 1977: 31)

Curtiss (1977) reports that in the time she was working with Genie, Genie did not succeed in learning to ask grammatical questions; she did not use third-person pronouns, or auxiliaries. In short, while normal language acquisition proceeds to a command of the major syntactic structures found in adult speech within the first three years, Genie's 'speech remains largely telegraphic – 4 years after she began putting words together. The great explosion has simply not occurred' (Curtiss 1977: 193).

Genie's grammatical difficulties are echoed in Curtiss's later (1994) report of the case of Chelsea, whose social interactions and upbringing had been essentially normal, but who was thought to be unable to learn language, until it was discovered, when Chelsea was already in her thirties, that she was deaf. With the help of hearing-aids, Chelsea very quickly acquired an extensive vocabulary; but combining her new words into more complex utterances seems beyond her grasp: 'her multiword utterances are, almost without exception, unacceptable grammatically and quite often propositionally unclear or ill formed as well' (Curtiss 1994: 229).

All these cases of intentional or unintentional deprivation of early linguistic stimulus seem to point in one direction: where children were discovered in infancy or relatively early childhood, they were able to go on to acquire language following normal patterns and milestones, ultimately achieving normal proficiency by adulthood, where follow-up evidence is available. However, where there is not exposure to substantial amounts of linguistic data by adolescence, acquisition seems less complete, with vocabulary learning proceeding relatively quickly, but syntactic acquisition being limited at best, as with Genie and Chelsea. This seems persuasive evidence in favour of a critical or sensitive period for language acquisition.

It is clear, then, that there is an environmental factor at work in language acquisition: children who do not receive or are deprived of linguistic data during the period for normal acquisition will be restricted in the extent to which they can be expected to succeed in learning and using language thereafter, being limited in severe cases to learning vocabulary. The chances of learning a second language (again, vocabulary apart) seem to atrophy as we get older; and Genie and Chelsea show that this also holds for first-language acquisition in the absence of early exposure to linguistic data. We turn in 2.7 below to the rather different question of whether children with developmental disorders and genetic conditions of various kinds might be unable to learn language regardless of the input data they might receive: in other words, sometimes the problem is the lack of data, and sometimes it is our capacity to process, arrange or receive it. This point is emphasised by consideration of a further set of twins discussed by Skuse (1988): Mary and Louise were initially brought up by their microcephalic mother, who appears also to have suffered from a psychiatric disorder and kept them isolated and almost silent. The girls were taken into care, but their subsequent development was very different, for while Louise went on to acquire language normally, Mary required special educational help and showed incomplete language acquisition and a range of autistic behaviours. Exposure to data is important, but inherited conditions may determine what use we can make of it. In this sense, it is perhaps inevitably unclear where Genie should be discussed. Mogford and Bishop (1988a: 24) note that language acquisition in exceptional circumstances can provide us with potentially enlightening 'natural experiments', but equally that 'one must beware of an over-simplistic approach that ignores the interaction of factors that occurs when environmental or organic context of development is disturbed or atypical'. The environment does not necessarily stay constant; 'The child's handicap will affect the language environment' (Mogford and Bishop 1988a: 24). We turn in the next section to the nature of the linguistic data the child receives as part of that environment.

2.4   The argument from poverty of the stimulus

In the heyday of behaviourism, in the 1930s and 1940s, the agenda was to account for as many human behaviours as possible as sequences of environmental stimulus and response; and this extended to language. Simplifying for the sake of brevity, the behaviourist account of language learning involved repeating what is heard in the linguistic behaviour of others, and gradually forming habits of pronunciation and grammar which encourage production to follow particular learned patterns. Perhaps the most famous statement suggesting that there must be more to language acquisition than this is from Chomsky (1965: 25): 'It seems clear that many children acquire first or second languages quite successfully even though no special care is taken to teach them and no special attention is given to their progress. It also seems apparent that much of the actual speech observed consists of fragments and deviant expressions of a variety of sorts.'

We have established so far that exposure to what Chomsky calls primary linguistic data, in other words, signed or spoken utterances in the child's immediate environment, is essential for the full acquisition of a native language system. If the child receives no data, or too little data before the end of the critical period (which seems, on the basis of the thankfully rare evidence from abandoned and neglected children, to be around the age of puberty), then he or she may be able to learn vocabulary but will not acquire the nuances and complexities of grammatical structure. However, what Chomsky is suggesting here is that there is an even greater mystery going on in language acquisition: we need the primary linguistic data to allow the process to get going; but that data turns out on closer inspection, putting it crudely, to be full of holes of various kinds.

Children, as we know, acquire language quickly and efficiently, following a series of developmental milestones, though the timing of these does vary to a degree depending on the individual child, and on the target language(s) she is exposed to (Dąbrowska 2004). In the normal case, the quantity of data is not the main issue: indeed, Cameron-Faulkner *et al.* (2003) suggest that the average toddler will hear between 5,000 and 7,000 utterances every day. However, when we consider the nature of the input data on which this acquisition process is based, we find that it underdetermines the grammar which children are clearly building, albeit more seriously in some parts of language than others. If you think about, or read a transcription of a normal conversation, it is not typically at all like the series of complete, grammatical sentences we find in written language. In talk, one participant often starts before the last has finished, so there are overlaps and interruptions; utterances break off part-way through as a participant's attention is drawn to something else, or he loses the thread of what he is saying, or struggles to express it accurately; and we find many abbreviated or incomplete utterances. And yet adults, and very soon children too, can produce the full, grammatical versions of such utterances in the appropriate sociolinguistic contexts. Even more extraordinary is the fact that children do not make apparently natural mistakes, and that they come to share community norms for what is grammatical and what is ungrammatical without explicit correction or teaching. Taking a well-known example of Chomsky's, consider the data in (1).

(1) a.   John is easy to please.
    b.   John is eager to please.
    c.   It is easy to please John.
    d.*   It is eager to please John.

Although purely superficially the examples in (1a) and (1b) seem to have the same structure, this cannot be the case, since (1c) can be produced on the basis of (1a), but the equivalent (1d) formed from (1b) is ungrammatical (indicated by the asterisk). Children do not hear (1d) in normal speech; but they also are not explicitly taught *not* to produce such utterances; this is not necessary, because they do not produce such forms anyway. This is the problem of negative evidence.

To take another example, the utterances in (2) are all grammatical and acceptable.

(2) a.   The train went off.
    b.   The bomb went off.
    c.   The cheese went off.

However, in (3), only the first is unquestionably grammatical; there tend to be differences of opinion on (3b) (so, neither of the authors likes it, but roughly half the staff in Linguistics and English Language at the University of Edinburgh think it's fine); and (3c) is completely out for English speakers, provoking laughter when it is presented to students as a possible example.

(3) a. Off went the train. (E.g. 'The guard blew his whistle, the doors locked, and off went the train'.)

b. ? Off went the bomb. (E.g. 'We had just crossed the road and were passing the gardens when off went the bomb'.)

c. * Off went the cheese. (E.g. 'We went away for the holiday weekend, and off went the cheese'.)

This underdetermination of the grammar the child acquires by the data she receives seems very clear for syntax and to a slightly lesser extent for morphology. We obviously learn words by hearing and repeating them, but, as Jackendoff (2002: 88) observes, hearing and learning the form of a word is not all there is to learning lexical semantics: a parent might point and say 'Doggie!', but,

> does *doggie* refer to that particular dog (like *Rover* or *Snoopy*), to dogs in general, to (say) poodles in general, to pets, animals, furry things, animate things? Worse, does it refer to the dog's tail, its overall shape, its color, the substance of which it is made, what the dog is doing right now (sitting, panting, slobbering), or (less plausible but still logically possible), the collection of the dog's legs, or the combination of the dog and the carpet it is sitting on? And *dog* is just a concrete count noun.

When we turn to phonology, we find Carr (2000: 93) arguing strongly against the poverty of the stimulus: 'Phonological knowledge is *not* "knowledge without grounds", as Chomsky has described linguistic knowledge. If anything, the sensory data are *more than complete*: there are … more data available to the neonate than is strictly required for phonological acquisition'. There are two caveats to bear in mind here. First, this does not mean there is nothing innate lying behind our knowledge and use of phonology: indeed, we have already seen that babies seem able to perceive and produce the sounds found in any human language, suggesting some internal, prespecified capacities (for an overview and references, see Blevins (2004: 9.1.2). What Carr is suggesting is that the distinctive sounds of a child's own language, and the accent-specific variants of these, are evidently and relatively straightforwardly learned from the primary linguistic data: even data that are highly impoverished in syntactic terms are likely to quickly provide evidence for the main phonemes and allophones of the language in question. In any case, these capacities need not be specific to language: vocalising and recognising patterns are key to phonological development, but to development in other domains too. Secondly, the discussion here is about acquisition of segmental phonology, vowels and consonants; and it is by no means clear that the same assumptions can be made about suprasegmentals. McMahon (2005a) argues that suprasegmental phonology, such as stress and intonation, may be more like syntax than like segmental phonology in this respect, being clearly underdetermined by the primary linguistic data. This is backed up, for example, by Dresher and Kaye (1990), who argue that there are relatively few types of stress systems, but that children cannot learn generalisable stress rules by observation alone, since the position of stress often depends on foot structure and syllable and foot boundaries, which are not explicitly physically signalled in spoken language. We return to the distinction between prosodic or suprasegmental and segmental phonology in Chapter 9.

In short, arguments from the poverty of the stimulus accept that children develop a grammar on the basis of the primary linguistic data, but further argue that 'on the basis of' needs a certain amount of additional interpretation. Children cannot be constructing an internal grammar that allows them to produce all and only the utterance types they hear, first because some of these might actually be ungrammatical (because they are curtailed, for example, or just plain mixed up through everyday slips of the tongue or brain), and secondly because there will be grammatical ones children do not hear. Somehow, children are also succeeding in not making superficially obvious possible mistakes, despite the fact that they are not hearing the full spread of ungrammatical utterances, and are not typically being alerted to ungrammatical cases explicitly either. Such poverty of the stimulus arguments make a case for Universal Grammar, some kind of innate predisposition to language learning which assists children in analysing and generalising from and beyond the data they hear. It is worth contrasting this pattern of normal first-language acquisition in children with the outcomes of the many and varied experimental attempts to teach human or human-like language systems to other primates. These experiments involved symbols placed on boards (as in the case of Sarah), keys pressed on computers (as with Lana and Sherman and Austin), or American Sign Language (as with the chimpanzees Washoe and the fabulously named Nim Chimpsky, Kanzi the bonobo and Koko the gorilla; see, for instance, Terrace 1979, Gardner *et al.* 1989, Savage-Rumbaugh and Lewin 1994). Whatever the system, the results are broadly speaking the same: the primates succeeded in learning to use individual signs for communicative purposes, and in some cases they progressed to producing two-item sequences which might be said to exhibit some consistency of structure. There are certainly some regularities of precedence among signs, for instance among Kanzi's short utterances. However, neither Kanzi nor the others progressed beyond this stage; longer strings, where they occurred at all, tended to involve substantial repetition rather than additional complexity (see Terrace 1979 for Nim). As Curtiss (1977: 193) noted of Genie, 'the great explosion' towards greater structural complexity and demonstrable command of syntax simply did not happen for these primates; nor did they appear to use their elementary language-like systems in the social way children grasp very early, for instance by initiating conversations. If these other primates so uniformly reach a stage beyond which they seem unable to progress even with carefully designed input and considerable explicit training, it is even more inexplicable in the absence of evolved, genetically prespecified and species-specific assistance that children acquire language via such clear developmental pathways, but with so little in the way of teaching, and such evident limitations in the primary linguistic data. Moreover, as Crain and Pietroski (2002: 173) argue, 'it's not enough to mention ways in which children could learn some things without Universal Grammar. To rebut poverty-of-stimulus arguments, one has to show how children could learn what adults actually know; and as close investigation reveals, adults know a lot more than casual inspection suggests. *That* is the nativist's main point.'

This does not mean that we can or should simply take the poverty of the stimulus argument as read: Pullum and Scholz (2002), for instance, propose that each case must be considered carefully and empirically to show whether innate priming is ever really required. Not all linguists support the innatist, or nativist view (Sampson 1999, 2002), and alternative, usage-based accounts are under development (see Dąbrowska 2004, Tomasello 2003). More specifically, it has been suggested that the primary linguistic data children receive cannot be equated with normal spoken language; instead, so-called 'motherese', or more usually today, child-directed speech (CDS), is a simplified register used by care-givers in speech directed to children. Snow and Ferguson (1977) suggest that the pronunciation of CDS tends to be exaggerated and careful, while syntactic structures are simple and become more complex in step with the child's stage of linguistic development. If CDS minimises some potentially interfering factors like hesitations and pauses, and maximises exposure to basic and straightforward structures, some poverty of the stimulus arguments might fall. However, Newport, Gleitman and Gleitman (1977) found that much of CDS still does consist of sentence fragments, and that many of these are not simple declaratives but questions and commands. It seems therefore that CDS might be instrumental in assisting children to acquire speech sounds (see also Kuhl 2004), precisely the area where the stimulus seems most complete in any case, but does not bridge the gap between stimulus and grammar for syntax. In any case, the CDS suggestions do not resolve the issues of negative evidence. As Mogford and Bishop (1988b: 240) put it, the fact that normal spoken language is often degenerate, involving gaps and errors, is one problem, and CDS might conceivably address that; but the other and more substantial issue is that 'the deep structure which the child must learn to become a competent language-user is not transparently obvious from the surface structure of the sentences provided by parents, and further, the child must learn grammar from positive instances only'. Even if parents wanted to provide examples and explicit correction for children through CDS, they generally could not do so: it has probably never crossed most non-linguist parents' conscious minds that *went off* is structurally and semantically different in *the train went off* and *the cheese went off*, or that *eager to please* works differently from *easy to please*; and it is highly unlikely that they would be able to express these differences to their children even if it had.

The nativist argument, then, is that children cannot acquire language as they do on the basis of just what they hear. Many linguists would take the further step of claiming that children cannot acquire language as they do by just being good general learners, though indubitably this is what children are. Jackendoff (2002: 82–3) notes that Bates and Elman (1996), and others,

> argue that learning is much more powerful than previously believed, weakening the case for a highly prespecified Universal Grammar. I agree that learning which makes more effective use of the input certainly helps the child, and it certainly takes some of the load off Universal Grammar. But I do not think it takes *all* the load off. It may allow Universal Grammar to be less rich, but it does not allow Universal Grammar to be dispensed with altogether ... Vision seems intuitively simple too, yet no one in cognitive science believes any more that the visual system has a simple structure or that the brain just 'learns to see' without any specialized genetic support.

We can also use this opportunity to spell out what Universal Grammar (also known as the Language Acquisition Device, or the language faculty) is intended to be, and what it is not. Jackendoff (2002: 72) defines UG as 'the prespecification in the brain that permits the learning of language to take place. So the grammar-*acquiring* capacity is what Chomsky claims is innate'. There is a frequently encountered misunderstanding here that Universal Grammar is the set of surface structural characteristics languages all have in common; but as we argued in Chapter 1, we could not and should not be talking about evolution of such surface systems. Instead (Jackendoff 2002: 75),

> Universal Grammar is ... supposed to be the 'toolkit' that a human child brings to learning any of the languages of the world ... When you have a toolkit, you are not obliged to use every tool for every job. Thus we might expect that not every grammatical mechanism provided by Universal Grammar appears in every language. For instance, some languages make heavy use of case marking, and others don't; some languages make heavy use of fixed word order, and others don't. We would like to say that Universal Grammar makes both these possibilities available to the child; but only the possibilities actually present in the environment come to realization in the child's developing grammar.

It is also worth raising a final concern here which will be explored in more detail in Chapter 9. Some of the neurological and physical structures involved in language acquisition and use might turn out to be specialised for language; others might be more general characteristics (like, say, pattern recognition or certain aspects of general learning). Even those which are now dedicated to language might have started out as involved with something else entirely, and become specialised through evolutionary time and evolutionary processes. But recognising that there may be some machinery that is specialised for language – because that is what Universal Grammar is – does not give us carte blanche to assign any feature of Language or languages we discover or hypothesise to Universal Grammar; we must recognise that general principles and cognitive functions may also provide explanations for some recurrent structural properties of languages, and bear in mind that in evoking Universal Grammar we are committing ourselves to an evolutionary account of all its ingredients.

2.5   Creativity and creolisation

A further argument for Universal Grammar, and therefore for some evolutionary input to language and language acquisition, comes from children's creativity: in a sense, this is a rather specialised extension of the argument from the poverty of the stimulus. Several different kinds of creativity are manifested during children's acquisition of language. The first is more a case of productivity than creativity: children learn that certain patterns can be generalised, and use this technique to produce words they have never heard before, so that in English, for instance, they will generalise the regular plural *–s* affix from *cats*, *dogs*, *houses*, which they may have encountered, to *hats*, *pens* and *buses*, which they might only have met in the singular. We can tell that they are generalising creatively by the fact that they also attach this affix to make plurals they cannot have heard before, since in adult terms they are ungrammatical, so that we find **mouses*, **tooths*, **deers*. Similar conclusions can be drawn on the basis of occasional backformations: our then three-year-old, on being told that we would be the *audience* for his brother's school play, asked if their baby sister would be an **audien* too. Similarly, children produce novel sentences based on patterns they have encountered: a child playing with toys might entirely plausibly and grammatically utter *Now the unicorn is kissing the princess*, though the chances of having heard exactly that before must be fairly slim.

Creativity shows itself too in instances of spontaneous compounding or word-coining: indeed, it has been reported that Washoe, one of the chimps taught American Sign Language, signed *water bird* when she saw a swan, though there is some controversy over whether this was intended as a label for that specific kind of creature or indicated that she saw some water, and she saw a bird (see Wallman 1992: 96). However, there are cases where children's linguistic creativity seems to be much more strongly implicated in the introduction of structural complexity on a grander scale: this involves the shift from a pidgin to a creole language.

When we turn to the development of creoles, we face an intriguing question. Recall Jackendoff's (2002: 75) contention that Universal Grammar presents the child with a range of possibilities, although 'only the possibilities actually present in the environment come to realization in the child's developing grammar'. But what happens when a child's linguistic environment is necessarily impoverished and restricted? What might be the outcome if the primary linguistic data children are hearing does not contain any of the possible structural options permitted by Universal Grammar, because none of them are available to adult speakers?

At first glance, this might seem like a most unlikely thought experiment: after all, we have already claimed that no language is more 'primitive' or structurally simple than any other. Pidgins, however, are the exception that proves the rule: we can define them as restricted in their functions and correspondingly simple and restricted in their structures, and, crucially, a pidgin is nobody's native language (Singh 2000). Typically, pidgins will arise in social situations (including trade or slavery) where those interacting have no single common language, and will be used in particular domains only, with speakers reverting to their respective home languages in other contexts. As compared with full languages, pidgins typically have restricted lexicons, little or no inflectional morphology, CV syllables with limited segmental phonology, and simple syntax without more complex constructions. Pidgins are therefore considerably simplified both formally and functionally when compared to the various full languages from which they are historically derived; and they are simplified across the board (not just in some areas of the grammar, with compensatingly large inventories or complex structures elsewhere).

It is entirely reasonable, of course, that part-time languages used only for special purposes might differ structurally from those systems which are first languages and require functional and structural flexibility. The interesting point, though, is what happens when pidgin speakers with different first languages have children of their own, who require a fully functioning first language. As Bickerton (1981: 5) strikingly observes,

> children of pidgin-speaking parents have as input something which may be adequate for emergency use, but which is quite unfit to serve as anyone's primary tongue; which, by reason of its variability, does not present even the little it offers in a form which would permit anyone to learn it; and which the parent, with the best will in the world, cannot teach, since that parent knows no more of the language than the child (and will pretty soon know less).

Bickerton's proposal here is for a modified version of Universal Grammar, namely the language bioprogram: this will guide the child to make sense of the primary linguistic data and to organise the resulting knowledge into a grammar, but also contains default structural options which will surface if the primary linguistic data is persistently lacking in some domain. There is no need to consider here the precise details of what the bioprogram is said to provide: suffice it to say that it is general rather than specific, for example, setting down a requirement for one degree of embedding (to allow subordination, for example, or relative clauses), and a series of morpho-semantic distinctions to do with tense, mood and aspect. Of course, the actual lexical material to be used in carrying these distinctions is not innately specified in the bioprogram – this would again be evolutionary nonsense. In any case, it would be over-simplistic to argue that children whose primary linguistic input is from a pidgin will nonetheless be shielded completely from hearing and understanding any other language; a child's mother or father will probably speak to her, when they are alone or with other members of their respective families or language communities, in their own first language, and elements of this may easily serve as the lexical model which can 'carry' innovatory creole structures. Nonetheless, creolising children do converge on agreed structures regardless of their parents' language backgrounds; and they do adopt certain options from the surrounding languages into the creole while ignoring other available complexities. Moreover, although clearly there is not identity across creoles, these adopted structural options do appear to fall into certain general and repeated categories for creoles around the world.

Bickerton's Language Bioprogram Hypothesis is controversial, and has been criticised extensively by scholars opposed to the innatist viewpoint (see Aitchison 1983, 1989a, and the points for discussion at the end of this chapter). However, the speed and extent of innovations during creolisation remain forces to be reckoned with. An explanation which, whatever the detail might be, invokes some kind of evolved language faculty is also encouraged by recent work on sign language in Nicaragua by Judy Kegl and her co-workers (see Kegl, Senghas and Coppola 1999).

Nicaraguan Sign Language (hereafter ISN, or Idioma de Señas Nicaragüense) is a new language, which has developed in schools for deaf people, as Kegl, Senghas and Coppola (1999: 180) argue, 'from the jumble of idiosyncratic homesign/gestural systems in use by students who entered the schools in the late 1970s and early 1980s'. Until 1979, access to education in Nicaragua was generally restricted, but for deaf people it was not available at all: being deaf was stigmatised, and deaf people tended to be kept closely within their families, with their means of communication often restricted to *mimicas* 'mime', 'gesture', the label for 'homesigns, the gestural communication that typically develops within a family or limited social sphere where one member is deaf and no preexisting signed language is available' (Kegl, Senghas and Coppola 1999: 183). Mimicas is highly individual and variable, so that even if deaf people met, they would have found signed communication difficult; there was no deaf community to teach and foster a shared sign language.

However, after the revolution of 1979, the new government began a literacy campaign which included provision for special education: deaf children began to be admitted to such schools in Managua, the capital, and elsewhere. Each child brought his or her own idiosyncratic system of homesigns, which they quickly began to share and build on to produce the Lenguaje de Señas Nicaragüense, a peer-group pidgin. Since children were not encouraged to sign in class (since their teachers were working on the basis that they should be assisted to learn spoken Spanish), their interactions were spontaneous and typically took place in unsupervised contexts, and the new system was not taught. Although the pidgin Lenguaje de Señas Nicaragüense allowed much more efficient communication than the homesigns, it was still typically quite slow and restricted, especially grammatically. However, children younger than 7 who joined the schools very soon began to develop a co-existing but much more complex and flexible system in the form of Idioma de Señas Nicaragüense, which Kegl, Senghas and Coppola (1999: 181) describe as a 'full-blown signed language'. Given the involvement of children, the speed of the developments, the evident impoverishment of the 'pidgin' model, and the increase in both formal and functional resources in ISN, it is perhaps unsurprising that this is seen as equivalent to creolisation:

> did this language evolve gradually over time as homesigns became more conventionalised with use by a wide range of Deaf people? Or, did ISN arise

abruptly when very young children radically restructured a highly variable, less than optimal signed input by bringing their innate language capacities to bear in acquiring it? The results of our narrative studies support the latter possibility. (Kegl, Senghas and Coppola 1999: 201)

It is worth noting, with one eye still on the question of the critical period for language acquisition, that 'very young' here means under 10: 'only children who were exposed to the community of signers before age 10 were able to benefit from the richer signed input … Signers over the age of 10 gain gradual fluency in LSN but never show native-like mastery of ISN as reflected by fluency and grammatical complexity in their signing' (Kegl, Senghas and Coppola 1999: 201–2).

Of course, the emergence of ISN is not a straightforward case of creolisation, but where it diverges from most such cases, it is in a direction which makes it even more interesting from the point of view of Bickerton's bioprogram and evolutionary linguistics. Most importantly, pidgins typically arise when native speakers of different languages, lacking any common language, need to communicate about some specific situation; but many of the deaf Nicaraguan children joining the first special schools in a sense had no language at all to build on beyond their homesigns, which were too variable and degenerate to count as full language systems. These children were starting school with no first language; the school system was designed to teach them to speak, lip-read and write Spanish, but in this it was notably unsuccessful (Kegl, Senghas and Coppola 1999: 204). On the other hand, the younger children did achieve full command of ISN, their own new language, which emerged from interactions among themselves and with the older pidgin signers of LSN. It appears that, given sufficient primary linguistic data at a sufficiently young age, the degeneracy of that data can go considerably further than the usual *er*s and *um*s, underdeterminations and fragmentary utterances a hearing child acquiring a full spoken language has to deal with; and fast, fluent, flexible, rule-governed grammatical systems can still emerge. Kegl, Senghas and Coppola (1999) would contend, and we would agree, that this emergence requires the interaction of the primary linguistic data with the so far unspecified something we might call Universal Grammar. Whatever it is, and wherever exactly it came from, there is enough evidence for its general presence in our species to assume as a starting point that it has developed through the mechanisms of evolution.

2.6   Language and the brain

In Chapter 1, we drew a distinction between genes, structures and systems. Evolution works at the level of genes, since mutations affect the genetic level and are the units of heredity. Selection then operates within populations, since those organisms with certain physical or behavioural characteristics arising from a given genetic variant may be more successful than others, leaving more offspring and hence conserving and propagating that variant. The physical or neurological structures resulting from such variants shape how we look and what we can do.

If we apply this model to language, we have to ask what the structures are that shape our language systems. One of these is clearly the set of physical structures which make up our vocal tract, lungs, and the other organs implicated in speech. These may well have evolved for other purposes, and still in some cases have other uses: the lungs are of course fundamental to respiration. They have certainly been further modified over evolutionary time in ways which maximise their use-fulness for spoken language, as we shall see in Chapter 5. However, rather than focusing in the current chapter on the readily observable vocal organs, we shall continue our theme of uncovering evidence from less immediately accessible structures by considering language specialisation in the brain.

We shall provide a more detailed overview of the structure of the brain in Chapter 6; for the moment, we focus on the cerebrum, the largest part of the human brain, which is composed of a left and a right hemisphere joined by bun-dles of nerve fibres. The most important areas of the brain for language are typ-ically in the left hemisphere. It is worth noting at this point that we should not expect to find self-contained little areas of the brain, each responsible for one aspect of language. This kind of pigeonholing might initially seem attractive, but it radically oversimplifies the situation. When we think of physical activities, there might be a single part of the body primarily responsible, but it will not be working alone: you might put in your contact lenses with the tip of your mid-dle finger, or kick a rugby ball with your right foot (and mostly the big toe, at that); but the fingers and toes are attached to the hands and arms or feet and legs, and a whole network of supporting bones, ligaments and muscles actually effect and control the movement. If this is true at the physical level, we should also expect it to be true of the neurological level, where the movements are ultimately activated.

Studies of the brain's involvement in language, as for other functions, have typically progressed by concentrating on cases where things go wrong. The first serious studies in this area date from the mid nineteenth century, when Paul Broca (1865), for instance, observed that language loss happened far more com-monly in cases of left-side brain injury than when the damage was on the right. We now know that right-brain-damaged people do have language deficits, though these are of a different kind from those, discussed below, which arise when the left hemisphere is damaged. Notably (Obler and Gjerlow 1999: Ch. 7), damage to the right hemisphere seems to lead to difficulties with intonation; trouble in understanding multiple meanings of words, or appreciating that they can have more than one meaning; and a range of pragmatic deficits, including problems in understanding the emotional state of a speaker from his or her language. Studies suggest that children with early and catastrophic damage to their left hemisphere, or who have had their left hemisphere removed, appear to go on to develop lan-guage near-normally, with subtle differences in performance only revealed in grammatical testing (Dennis and Kohn 1975). It seems that in young children,

most language functions can be taken over by undamaged parts of the left hemisphere or by the right hemisphere (Satz, Strauss and Whitaker 1990); but the evidence reviewed below on adult brain damage and its consequences for language
show this so-called plasticity is not available for adults, a further point bearing
on the question of the critical period for language acquisition.

Turning to more specific types of brain damage, Broca (1861) and Wernicke
(1874) identified regions within the left hemisphere where injury or stroke
seemed to lead to specific deleterious effects on language; these conditions of
partial language loss are known as aphasias. In the case of Broca's area, interference is mainly with language production, while comprehension is largely spared;
speech is slow, and seems to require a lot of effort; there are some difficulties in
finding the right word, and most of the words used tend to be full words, like
nouns or verbs, with function words and affixes quite rare. Wernicke's aphasics,
on the other hand, tend to experience greater difficulty with perception and comprehension; and although their speech is often apparently fluent, with both lexical and grammatical items and normal speed and intonation, the content is often
perplexing and may make little sense.

However, as noted earlier, we must take care not to simply equate damage in
Broca's area or in Wernicke's area with a predictable list of symptoms. Clearly,
the extent of disruption will depend on the extent of damage; but in addition, if
the sites of damage or lesions in the brain are found in the part of Broca's area
nearer to Wernicke's area or the part of Wernicke's area nearer to Broca's, the
types of disruption found can start to overlap. These are also by no means the
only parts of the brain where damage can lead to particular clusters of disruptions: for instance, individuals with damage to the arcuate fasciculus, which links
Broca's and Wernicke's areas, often suffer from a condition known as conduction aphasia (though some conduction aphasics have damage elsewhere instead).
Conduction aphasia involves difficulties in repeating words, along with a high
frequency of phonological errors, notably substitution of sounds within words. It
has been suggested (Obler and Gjerlow 1999: 62–3) that this implicates the arcuate fasciculus in phonological planning, where phonemes are put together in the
right sequence to make words.

In addition, many aphasics have deficits of varying degrees and types in finding words, especially nouns: this is known as anomia, and provides a useful clue
to the way our vocabulary is stored and structured in the brain. Damasio *et al.*
(1996) carried out two parallel studies, one of 127 subjects with lesions in different parts of the brain, and the other involving 9 normal subjects. Tests were carried out to work out whether individuals in both groups could retrieve the names
of individual people, words for types of animals, and words for tools. In the case
of the brain-damaged subjects, the location of their lesions was then correlated
with failure on these tests; 'failure' here means they provided some description
or other indication that they were familiar with the person, animal or tool concerned, but could not find the name. For the normal subjects, positron emission
tomography (PET) was used to assess which areas of the brain showed most

activity when words from these different categories were successfully retrieved. PET involves injecting subjects with a radioactive tracer, and then measuring the X-rays given off; these will be most concentrated in areas where tissue is most active (see further Chapter 6).

Damasio *et al.* (1996) found that thirty of their subjects with brain lesions experienced serious difficulties in retrieving words in at least one of these categories. All but one of these 30 subjects had damage to the left hemisphere. Moreover, the location of their lesions correlated strongly with the types of word-retrieval difficulties they experienced: those who failed to retrieve people's names had damage to the left temporal pole; impairment in animal words mapped onto damage in the left inferior temporal lobe; and impairment in words for tools involved damage to the posterior inferior temporal lobe and the temporo-occipito-parietal junction (Damasio *et al.* 1996, and see also Caramazza 1996 for a summary). None of the normal subjects had damage in these areas; though three of the subjects with brain damage had lesions elsewhere. Some subjects had trouble with more than one category, in which case the combination was always people plus animals, or animals plus tools, but never people plus tools unless the problem also extended to animal words. The geography of the left hemisphere accounts for this observation: the sites involved with deficits in retrieving words for people and tools 'are not even contiguous, do not overlap cortically or subcortically, and are so distant as to make it virtually impossible for a single lesion to compromise them without also compromising the intervening region' (Damasio *et al.* 1996: 502). The corresponding PET study of normal subjects supported the hypothesis that lexical items from single conceptual categories are stored together in specific areas of the brain, since the sites activated most strongly when subjects retrieved names of people, words for animals, or words for tools matched the sites where lesions were present in patients who failed to retrieve words from these same categories. This parallel study of two subject groups is particularly important, because not all the evidence considered by Damasio *et al.* comes from brain-damaged individuals with consequent disruptions to normal language; such evidence will always be slightly clouded with doubt since where there is one area of damage we can easily perceive, there might be other things going on that we can't (yet) observe. Increasingly, novel methods of scanning like PET give us further vital insights into the workings of a normal brain using language normally.

This study provides strong support for the insight that anomias may involve difficulties in retrieving words from particular categories, and that this maps onto damage at specific sites. Such category-specific deficits have been known since at least Nielsen (1946; cited in Caramazza 1996), who describes the case of an individual with far more serious difficulties retrieving the names of living things than of non-living things. Ashcraft (1993) further confirms that a highly specific language function may sometimes be interrupted when there is a highly specific piece of damage. Ashcraft, an experimental psychologist, describes his own 45-minute transient anomia which left him able to think of concepts like 'printout', 'data' and 'experiment', but unable to find the words; he notes that the word-retrieval problem, as far as he is aware, was limited to the professional vocabulary he would normally use in his office, which is where the episode took place. He explains the experience as 'one of being fully aware of the target idea yet totally unable to accomplish what normally feels like the single act of finding-and-saying-the-word' (1993: 49). The anomia resolved itself spontaneously, but medical tests after the episode showed that an arterio-venous malformation had taken place in the left temporal lobe; this means additional blood was flowing through one of the arteries, causing a bulge, which also involved blood being 'stolen' from brain tissue in the area. Ashcraft reports that his research assistant, who had come in to ask what he wanted to do with a particular printout during the episode, had not initially been particularly perturbed by his failure to comment coherently, supposing that he was engrossed in another piece of work and unwilling to turn his attention to something else!

The brain is clearly a highly complex organ, and we have scarcely touched on its structure, function and evolution here: we return to these issues, along with more detail on methods for investigating the brain, in Chapter 6. However, the key argument here is that the brains of individuals, though showing some limited variability, nonetheless appear to share a very general map, where geographical areas correlate with particular functions. This is shown by investigations of language deficits with their roots in brain lesions, and increasingly supported by imaging of normal brains processing language. Where strong inter-individual correlations of this kind appear, we have a further basis for arguing in favour of an evolutionary account.

2.7   'Language genes'

In a sense, the division between this section and the last is an artificial one, since the structure of the brain and the division of functional labour within it are clearly strongly responsive to instructions provided by our genes – as indeed is the case for all our physical and neurological structures. We recognise this connection between brain and genes explicitly by covering both in somewhat greater detail in Chapter 6 below. However, separating brains and genes into two sections here does point up an important distinction. While the cases in the previous section involve brain injuries which lead to a disruption of language, invoking evolution in the construction of neurological structures in our species implies that some such disruptions should also be hereditary. While the transient or permanent damage to language reviewed earlier reflects difficulties with the linguistic environment or with the brain, some language deficits should also have the potential for inheritance. And they do.

Of the inherited conditions which affect language, many also affect other neurological functions. Fragile X-Linked Mental Retardation Syndrome, for example, involves physical characteristics such as a long narrow face with large or prominent ears; and behavioural difficulties including attention problems, hyperactivity and strong negative reactions to change in their environment. There is also general mild to moderate mental retardation, with IQ scores typically around 70, though in some cases IQ scores may be in the normal range, and in others they may be in the range corresponding to severe mental retardation (IQ score of less than 40). In linguistic terms, many males with Fragile X Syndrome repeat words, especially at the end of a phrase, which can make it virtually impossible for them to finish a sentence, and such individuals frequently fail to meet developmental milestones in language as elsewhere. Both boys and girls with Fragile X Syndrome tend to experience pragmatic difficulties and find conversational situations difficult (www.fragilex.org, accessed 13 May 2011).

On the other hand, there are also dissociations of language and other cognitive processes, and these operate in both directions. In some cases, language seems relatively spared though many other systems show severe deficits. Individuals with Williams Syndrome, for example, tend to have IQs around 50–60 ('normal' is around 100), and would be unable to look after themselves independently; they have considerable learning difficulties, and physical defects in systems including the heart and circulation. However, their language skills seem considerably better developed than might be anticipated if cognitive function were equally affected across the board (Bellugi *et al.* 1988, 1994, 2000). Of course, as Dąbrowska (2004: 55) notes, these are partial rather than complete dissociations. People with Williams Syndrome tend to have developmental delay in acquiring language, and their language in adulthood is not completely normal. For instance, Bellugi *et al.* (1988) report on three adolescents with Williams Syndrome, including Crystal, aged 15, with an IQ of 49. Crystal, like many individuals with Williams Syndrome, showed a penchant for unusual words and phrases, which it was clear she understood very well from the context or from definitions she provided. For example, when one of the researchers asked, 'May I borrow your watch?', Crystal responded, 'My watch is always available for service' (Bellugi *et al.* 1988: 183). Likewise, when asked as part of a standard test to give as many animal names as she could within 60 seconds, Crystal scored at the level of her chronological age – but the names she provided, rather than being more frequent, basic or generic, included *koala bear*, *moose* and *anteater* (while 11-year-old Van, also with Williams Syndrome, suggested *sea lion*, *hippopotamus*, *chihuahua*, *owl* and *reptile*). The language skills of people with Williams Syndrome seem exceptional, but this must be seen in comparison with their other cognitive capacities.

On the other hand, it is also possible to identify cases where language seems to be the major or only cognitive function affected by some inherited condition. Our understanding of such instances is for the most part in its infancy, and the genetic, environmental and behavioural aspects of such conditions are yet to be teased apart. Perhaps most notably, there is a good deal of discussion in the literature of Specific Language Impairment (SLI), though this may ultimately turn out to be more than a single condition. Here again, the dissociation is typically partial, so that individuals with SLI are of normal or above-normal intelligence and show some deficits in their language abilities, but may also have some additional non-language deficits, including problems with short-term memory or auditory processing (Dąbrowska 2004: 53 for references).

It is worth saying a little more about one particular family with SLI, the so-called KE family (Hurst *et al.* 1990). About half the members of this family, over three generations, have been diagnosed as having a range of speech and language problems, including severe difficulties with articulation, and with the use of affixes to signal grammatical information including tense and agreement (Gopnik and Crago 1991, Gopnik 1999). These deficits are inherited in a dominant fashion, so that every affected individual will also have an affected parent; this makes the genetics of the condition relatively easy to trace. The gene involved is on the long arm of chromosome 7 (Fisher *et al.* 1998), and affected family members have been found to have mutations in this gene, now named *FOXP2* (Lai *et al.* 2001).

Finding the gene involved is obviously a crucial part of understanding inherited diseases and deficits. In the case of Fragile X Syndrome, a particular gene is always switched off or absent; this has enabled genetic tests to be developed so that affected families can assess the risk of their children or future children carrying or inheriting the condition. Williams Syndrome results from the loss of part of chromosome 7, which contains many genes; this means that although the location of the genetic defect responsible is known, the exact gene responsible has not yet been identified unambiguously. In the case of SLI, the stakes are in a sense lower for individual families, since individuals with SLI are often relatively unaffected by other deficits and are typically able to lead normal lives, though they may benefit from some speech and language therapy. However, the stakes for evolutionary linguistics are very high indeed, since SLI provides the first evidence that a particular genetic mutation may underlie interruptions to normal language function, and consequently that a particular gene may therefore be linked with normal language development.

We return to *FOXP2* and its evolutionary history in Chapter 7, where we will also be considering the issue of dissociation further, along with its implications for whether language is an independent cognitive 'module' or intrinsically linked with other functions; but for the moment, it is essential to dispel a few myths before they get going. The fact that *FOXP2* has been implicated in SLI in a particular family does not show that it is damaged in all families with SLI (though this has been observed in some other cases; Lai *et al.* 2000), or that it is unscathed in all families without SLI. It has been argued that SLI is just one of the difficulties faced by members of the KE family, and therefore perhaps by extension affecting other SLI families; Vargha-Khadem *et al.* (1995) have argued, for example, that the KE family manifest a more general developmental verbal dyspraxia which extends beyond language. Perhaps most importantly, even if it turned out that we had a perfect dissociation of language from other cognitive capacities for the KE family (which it appears we do not); and even if this was found to hold for every family with any indication of SLI (which it does not), then we would still not have found in *FOXP2* 'the language gene' or 'the grammar gene'. As we shall show in Chapter 7, even the apparently simplest physical traits will usually involve many genes, which, as we have seen, will also interact with environmental effects; any assumption that one gene ever corresponds to one trait at the phenotypic level is erroneous. For example, more than twenty genes are involved in blood clotting, and many more interact to determine their level of expression (McMahon and McMahon 2005). Any more complex physical, cognitive or behavioural system is almost by definition going to involve a serious number of genes: more than 900 are involved in smelling (Strachan and Read 2004: 261), and many of these will also be implicated in other physical, cognitive and behavioural systems.

The KE family and *FOXP2* are vitally important, then, because they show that a language deficit (whether or not it involves other kinds of deficit too) can be inherited in families, and that we are able to begin to locate some of the genes implicated in that inheritance. But this is only the start. *FOXP2* will not be found to contain a tiny blueprint for Chomsky's Universal Grammar or Bickerton's Language Bioprogram. Genetic life, like linguistic life, is vastly messier and more complex than that. However, without evidence that some aspects of language deficit can be inherited, we have fewer grounds for hypothesising that some aspects of our normal language capacities are inherited, as must be the case if we are to invoke evolution in a meaningful way in our explanation of how and why we have become linguistic animals. Progress in the genetics of language and of language difficulties is less advanced than in some of the areas we have reviewed in this chapter, but we can add these first tentative results to our mosaic of evidence which suggests that the capacity for acquiring and using language, alone or in association with other capacities, has indeed evolved during the history of our species.

2.8   Summary

In this chapter, we have surveyed a range of evidence types which suggest that some capacity for language has evolved in our species. Much of this evidence is indirect, and since a good deal of it involves observing situations where something has gone wrong neurologically, socially, environmentally or genetically, there is very little prospect of controlling our results or of being sure whether language problems arise on their own account or as a result of some other deficit. Taken together, however, the results from these very different domains are persuasive. It appears that there is a critical period (or perhaps more accurately, a sensitive period, since we are dealing with gradual slowing down rather than abrupt switching off of the capacity to learn language) during which exposure to language data in the child's environment is essential for normal acquisition to proceed (2.3). The linguistic stimulus itself, in many areas of the grammar, does not seem to provide enough explicit clues to stop children making apparently natural errors which they do not in fact make, or to allow them to make judgements of grammaticality which they do make (2.4). Children's creativity is extraordinary under these normal circumstances, but is all the more remarkable when they are acquiring language on the basis of a pidgin; children's input into the more complex creole that results was set out for Nicaraguan Sign Language in 2.5. Finally, we know from a range of studies of brain-damaged subjects, and a small but increasing number of normal ones, that certain areas of the brain, rather loosely defined, are associated recurrently with certain types of linguistic deficit, or correspondingly with brain activity in the retrieval or processing of particular aspects of language (2.6). Although many such deficits result from a brain injury or from illness, others are inherited, and are beginning to be tracked through families (2.7). Some of these familial patterns of language deficit appear primarily to affect language, while others are intrinsically connected with difficulties in other areas.

What is not clear from this discussion, then, is whether the capacity for language has evolved independently, as a specific module or device which humans develop in response to particular instructions in our genes, or whether it has developed as a result of, or on the basis of, other capacities. We return to this question in subsequent chapters. For the moment, we would suggest that a biologist, hearing that this book is about a capacity which seems to be acquired by members of a particular species, along approximately parallel developmental lines and with a critical period beyond which environmental exposure will not trigger normal acquisition; where input data are essential to the process but underdetermine the results; and where some deficits run in families, and some properties can be localised to particular regions of the brain, would have no hesitation in considering such a capacity to have evolved. We cannot rest on the Argument from Design alone; but it may have been leading us in the right direction.

Further reading

A brief but very readable overview of some of the same areas covered in this chapter is in Jackendoff (2002: Ch. 4). For the argument from design, see Dawkins (1986) and Dennett (1995). Gerken (1994) is an overview of normal language acquisition; Vihman (1996) provides detailed coverage of acquisition of phonology in particular. Hauser (1996) includes details of critical periods for other communication systems. Bishop and Mogford (1988) is an excellent survey of acquisition in a range of unusual circumstances. The poverty of the stimulus argument is raised first in Chomsky (1965); Ritter (2002) is a special journal issue containing an excellent set of papers discussing the history, evidence for, validity and future of the poverty of the stimulus argument; and Dąbrowska (2004) provides a usage-based account of some of the relevant data. References are given in the text for many of the ape language experiments, but see Savage-Rumbaugh, Shanker and Taylor (1998) for a view which seeks to minimise the distance between what the apes have learned to do and human language, and Terrace (1979) and Wallman (1992) for the opposite position. Readers interested in the differences between spoken and written language (in this case specifically English) might like to look at Miller (2006).

Overviews of pidgins and creoles are provided in Singh (2000), Romaine (1988) and McWhorter (2005), and their relevance from the perspective of historical linguistics is summarised in Chapter 10 of McMahon (1994). The language bioprogram hypothesis makes its appearance in Bickerton (1981, 1984), and there is a particularly interesting negative review (see also points for discussion below) in Aitchison (1983). Kegl, Senghas and Coppola (1999) provide an excellent summary of the Nicaraguan Sign Language case, and you might also like to consult other work by these authors, and to look on the web, where there is plentiful discussion of this case. Obler and Gjerlow (1999) provide an admirably accessible, but by no means over-simplistic, overview of what we know about the localisation of linguistic functions in the brain. Readers interested in finding out more about inherited diseases affecting language can consult OMIM (Online Mendelian Inheritance in Man, hosted at www.ncbi.nlm.nih.gov/omim, accessed 13 May 2011). This front page allows searches to be carried out; readers should note that the entries in this database are arranged chronologically, so that references to the same condition may be discontinuous (and sometimes apparently contradictory). Nonetheless, this is a constantly updated resource.

Points for discussion

1. Do we really need a bioprogram, or the equivalent, to account for creolisation? Doubts have certainly been raised over this. For instance, Elman *et al.* (1996) (quoted in Jackendoff 2002: 101), on the subject of Nicaraguan Sign Language, comment:

    > We would agree that these phenomena are extremely interesting, and that they attest to a robust drive among human beings to communicate their thoughts as rapidly and efficiently as possible. However, these phenomena do not require a preformationist scenario … If children develop a robust drive to solve this problem, and are born with processing tools to solve it, then the rest may simply follow because it is the natural solution.

    What does 'a preformationist scenario' mean here? How far is this a neutral way of describing what e.g. Bickerton is suggesting? And what might 'a robust drive' be? Aitchison (1983, 1987, 1989a) proposes a similar 'common-sense approach' to creolisation, suggesting that 'an obsession with innateness can lead to a neglect of other possible reasons behind language universals' (1987: 14), and that 'one fairly obvious observation about human life is that, for most problems, there are usually a finite number of possible solutions' (1987: 17). (There is some discussion in McMahon 1994: Ch. 10.) Compare Aitchison's suggestions about creolisation with those of Bickerton or Kegl, Senghas and Coppola; you might also want to compare the case of

Al-Sayyid Bedouin Sign Language, as discussed, for example, in Sandler *et al.* (2005). Which do you find more persuasive? Is there some way of reconciling these innatist and 'common-sense' views? Are they as far apart as they might initially appear?

2.   Find out whatever you can about at least three so-called 'wild children' or 'wolf children'. How much can these cases tell us about the critical period hypothesis? In what ways is the evidence limited or inconclusive? How far are the insights and limitations shared across the cases you have considered? (If using this suggestion in class, you might like to divide groups up to consider one or two case studies each.)

3.   How far would you support the idea, putting it in artificially basic and oppositional terms, that while in SLI only language is affected, in Williams Syndrome virtually everything except language is? We recommend that you read Karmiloff-Smith *et al.* (1997), Stojanovik, Perkins and Howard (2004) and Vargha-Khadem *et al.* (1995) before embarking on an answer.

4.   In section 2.7, the notion of a dissociation between language and other cognitive functions was raised. Making sure that you understand what is intended by this term, find out about at least one linguistic savant (you might like to read about Christopher, in Smith and Tsimpli (1995), for example). What is a savant, and what can such cases tell us? How does this relate to the question of whether language is or can be dissociated from other cognitive functions, and why might such dissociations be important?

3 The comparative methods

3.1 Overview

If you were told to find out first-hand about the roots of a plant, you would be likely to go out into the garden, stand at the level of today's topsoil, and dig downwards. In this chapter, we tackle the apparently common-sense idea that to find out about the roots, or ultimate origins of languages, we should equally start from languages we know today and work backwards to their roots; and we show that this approach cannot work. First, the good methods we have for linguistic reconstruction (notably the comparative method) appear to be time-limited and cannot take us back far enough to be of evolutionary relevance. There are less-constrained methods, but we cannot rely on their results. Second, and perhaps more important, any such method will necessarily be working at the level of behavioural, output language structures, and we have seen that if we really want to talk about evolution, we need to get behind such superficial behavioural characteristics to the underlying physical and neurological structures, and ultimately to the genetic instructions which have configured them. We turn, therefore, from linguistic reconstruction to the comparative method in biology, and consider some similarities and differences between human language and aspects of other primate communication systems.

3.2 Going backwards to move forwards

When Jespersen (1922; discussed in 1.3.2 above) has finished noting his disappointment with early 'speculative' theories of language origins, of the bow-wow and yo-he-ho variety, he suggests that we should instead turn to evidence from the historics of languages. On the basis of earlier stages of present-day languages, and reconstructions of languages which no longer exist, we may 'attempt … step by step to trace the backward path. Perhaps in this way we may reach the very first beginnings of speech' (1922: 418). Jespersen suggests, as we saw earlier, that change typically involves reductions in the phonological shapes of words, and in word length, along with a trend towards greater analysis, and division into small, individual forms, each with its own meaning. He therefore proposes that 'Primitive linguistic units must have been much more complicated in point of meaning, as well as much longer in point of sound, than those with which we are most familiar'

(1922: 425); moreover, these long words were rather unstructured internally, with each expressing the meaning of a whole sentence today.

We are unlikely to agree in detail with all of Jespersen's conclusions here: for one thing, our knowledge of change suggests that it tends to be more of a spiral than a linear progression. For example, when sounds are lost, others frequently develop: in the history of English, many varieties (including the more prestigious accents of England, though not those of Scotland or the USA) have become non-rhotic, losing realisations of /r/ where they do not directly precede a vowel, so that /r/ will be pronounced in *starry* and *star is*, but not *star*, *start*, or *star light*. However, a speaker of a non-rhotic English variety will not pronounce *star*, or *here*, or *bird* just as a Scots speaker would, but without the /r/: instead, the vowel system has changed in step with the consonantal change, so that English speakers typically have /ɑː/ at the end of *star*, a new centring diphthong/ɪə/in *here*, and the 'long schwa' /ɜː/ in *bird*. Scots speakers have none of these vowels, their equivalents being /star/, /hir/ and /bɪrd/ (or /bʌrd/), with just the same vowels as would be found in words with a consonant other than /r/ (like *sat*, *heat* and *bid* or *bud*). In other words, the loss of /r/ in non-rhotic varieties is signalled by developments in the vowel system; we might almost think of this as some kind of compensation.

Stepping away from the phonology, one of the types of grammatical change which has received most attention in the last decade or so has been grammatical-isation (Hopper and Traugott 1993, Heine and Kuteva 2005), which involves a lexical word like a noun or verb becoming more grammatical, turning into a clitic, or a function word, or an affix. Again, however, this tends to be a cyclic process: we do not see a shift, or progression in one direction only. To take a well-known example of grammaticalisation, in Classical Latin the future tense of verbs was formed by adding a series of suffixes, for instance *–bo* for the first-person singu-lar, so that *cantabo* meant 'I will sing'. However, this fell into disuse, and was replaced by a range of different constructions including the verb infinitive and a reinforcing extra verb, as in *cantare volo* 'I want to sing', or *cantare habeo* 'I have to sing'; and in due course, the latter became the most frequent and in fact the expected way of saying 'I will sing'. By this point, we can say that the *habeo* part has grammaticalised; it does not really have its own meaning, but contributes the idea of futurity to the whole construction, though futurity would not be its meaning if we encountered it alone.

In the daughter Romance languages, the *cantare* part and the *habeo* part continued to coalesce, giving forms like French *chanterai*, also meaning 'I will sing', where the *–ai* suffix is all that remains of the originally independent *habeo*. However, this is not the end of the story: the cycle is continuing, so that in French it is increasingly common to hear *je vais chanter*, a literal match for English 'I – am going – to sing'. Grammaticalisation may be unidirectional, if we define it as the part of the cycle which involves independent items becoming less and less independent; but it works in a productive tension with other kinds of change which operate in the opposite direction. Jespersen seems to be assuming, counterfactually, that unidirectionality is something we can assume for all the changes from the earliest human language until the languages we see today (see also Newmeyer 2003, Hurford 2003).

3.3   An outline of the comparative method in linguistics

Jespersen, then, seems to make some simplistic assumptions about how straightforward it should be to reconstruct the features of the earliest language on the basis of recurrent and unidirectional historical linguistic trends. However, we could still give reconstruction itself the benefit of the doubt, and investigate whether there are methods which might, if they cannot take us directly to the first human language, at least get us part-way there. The best-known and most thoroughly tested method of this kind is the comparative method (see Campbell 2004: Ch. 5, McMahon and McMahon 2005: Ch. 1, Durie and Ross 1996).

The comparative method is intended as a means of reversing linguistic history, taking variation in daughter languages and retracing it through a series of changes to arrive at the putative original form in the mother, or protolanguage. It goes without saying that these changes must be regular if we are to stand any chance at all of reversing them: sporadic changes which affected every word differently would be indistinguishable from chance. However, many changes in morphology and phonology, the classical stamping-ground of the comparative method, are indeed regular, and leave regular patterns in the daughters of the sort we see in (1).

(1)   Some illustrative Indo-European correspondences (partly after Durie and Ross 1996: 6)

Ancient Greek	Sanskrit	Gothic	Modern German	English
pod-	*pad-*	*fōt-*	*Fuss*	*foot*
pénte	*pañča*	*fimf*	*fünf*	*five*
polús	*pulu-*	*filu*	*viel*	*(full)*
ed-	*ad-*	*it-*	*ess-*	*eat*

The comparative method is applied to languages we already know or suspect to belong to a single family (that is, to come from a single common ancestor): in the case of the languages in (1), the family is Indo-European, and the protolanguage, of which we have no direct record, is Proto-Indo-European. We know these languages are genealogically related because they share a very large number of recurrent correspondences of sound and meaning, of the sort in (1): that is, wherever we find a /p/ at the beginning of a word in Ancient Greek, we are highly likely to find a /p/ in Sanskrit too, and an /f/ in the various Germanic languages. Of course, there are specific exceptions: in the third set of examples, all the forms mean 'much, many' except the English one, which is in brackets because it has shifted in meaning. But there are many, many sets of complete cognates of this kind, on a range of different patterns – in the final row, for instance, you can see that stem-final /d/ in Greek and Sanskrit tends to correspond to /t/ in Gothic and English and to /s/ in German; and you can test these correspondences by thinking of other English words like *fish* with initial /f/, or *forget, hot* with final /t/, to check if they correspond to forms with /p/ in Latin and /f/ in German (*piscis, Fisch*) or to German stem-final /s/ (*vergess-, heiss*). The comparative method does not require that these correspondences will always hold in an exceptionless way; but it does say that where there are apparent exceptions, we should be able to find an explanation for them.

Evidence like (1) bears on the relatedness of languages and tells us something about the subgroups they fall into; but in order to reconstruct the most likely ancestral form for the whole group we also need to use our knowledge of linguistic typology and of typical patterns of language change. In the /p/ versus /f/ cases, evidence from a whole range of language families tells us that it is much more common for /p/ to become /f/ than the other way around (and indeed, more generally, sounds more often weaken or lenite from stops to fricatives than strengthen from fricatives to stops). We therefore reconstruct *p here for Proto-Indo-European. As for the /d/~/t/~/s/ cases, we can use aspects of the same argumentation to suggest that /s/ is unlikely to be the protoform; and as for /d/ and /t/, we know that devoicing sounds at the end of words is a very common change cross-linguistically, so our reconstruction will be *d, the voiced stop, rather than the voiceless *t. Following the same steps for the other sounds in the daughter languages will ultimately allow us to reconstruct whole proto-morphemes or proto-words.

There are strong indications that the comparative method works. McMahon and McMahon (2005: Ch. 1) give an overview of evidence to this effect, but, in brief, the method has been confirmed after the fact by, for instance, the decipherment of Linear B (which turned out to be a writing system used for a very early stage of Greek, and to provide orthographic evidence for a series of labio-velar sounds reconstructed for Proto-Indo-European but not until then attested in the history of Greek). Furthermore, attempts to reconstruct Proto-Romance have produced a system strikingly like Latin – though interestingly in some respects more like Vulgar than Classical Latin, as we would hope since Vulgar Latin represents a rather later and more colloquial set of variants, which are more likely to have been the basis for the formation of the modern Romance languages.

However, there are also questions remaining. For one thing, when we reconstruct, we idealise; and there is a considerable controversy over whether Proto-Indo-European as reconstructed is close enough to the real language for us to hold a conversation with a Proto-Indo-European if we found one, or whether reconstructions are just convenient formulae for historical linguists, bearing little or no relation to any actual language. We also idealise by factoring out variation – indeed, this is the whole point of the exercise, since doing the comparative method means taking different forms now and figuring out their most likely shape when they were the same. This, of course, entails assuming that they really were the same, and therefore idealising out any dialect variation at the protolanguage level. Reconstruction of phonology and morphology has also been notably more successful than reconstruction of syntax, where patterns and directions of change are not so clear, so that even the best reconstructions may have gaps in the grammar.

Finally, when we say languages are related, we typically express this by constructing family trees, with subgroups and higher-level branchings, and the protolanguage at the very top; but this kind of diagram can easily be taken too literally and obscure or deny the fact that similarities between languages can be caused by another, very pervasive factor apart from genealogical relatedness. When speakers borrow words and structures from one another, their languages can be affected; and although some borrowings involve only individual lexical items or clusters of particularly marked and easily spotted technological or cultural words, intensive contact can have much more profound effects, and can easily create apparent regularities which can be mistaken for evidence of relatedness. The comparative method is based on the family-tree model, but there has been a great deal of very active research in historical linguistics recently, indicating that contact must also be recognised as an important factor in language histories (see Thomason 2001, Heine and Kuteva 2005, Aikhenvald and Dixon 2001, and Matras, McMahon and Vincent 2005). Recent work in quantitative historical linguistics (see McMahon and McMahon 2005, the papers in McMahon 2005b, and Heggarty *et al.* 2010, for example) has shifted noticeably towards network rather than tree representations, where networks have the great advantage of accommodating both tree-like patterns resulting from genealogical relationship, and more complex connections resulting from contact.

3.4   Reconstruction and comparison are time-limited

Even if we could resolve all these outstanding problems, though; even if the comparative method had been and could be equally successfully applied to all language families, not just those with plentiful and well-analysed evidence like Indo-European; and even if we knew which families all the world's languages belong to in the first place, which is by no means clear, there is still an intrinsic limitation which disqualifies this method, or indeed any means of reconstructing or extrapolating backwards from today's languages, from contributing to our investigation of evolutionary linguistics. This is the fact that methods of reconstruction are time-limited.

By 'time-limited', we mean that the comparative method and other means of comparison and reconstruction cannot operate completely freely, being applied to a subgroup, then a larger family, then to the protolanguages of those larger families and so on without limit and back to the beginnings of human language. This cannot be done for the simple reason that after a certain period of time, there will not be enough relevant evidence left for us to base any serious and robust conclusions on. To see why this should be, consider that each language will only have a relatively small inventory of phonemes, for instance – on the average, let's say between 10 and 50 segments (though there are outliers in both directions; see Ladefoged and Maddieson 1996). If we could guarantee that each sound would be affected by one and only one sound change in the history of a language, then we would be secure in our ability to reconstruct that change and the previous form; but changes overlap and interfere with one another and, given the smallish number of phonemes they have as their possible domain, are very likely to operate on the same original sound sequentially. To take an example, we looked earlier at the loss of /r/ in non-rhotic varieties of English, and noted that the development of the centring diphthong/ɪə/in *here* acts in a sense as a signal that /r/ was originally pronounced in this word. But centring diphthongs also developed in *door* and *sure*; yet these /ɔə ʊə/ diphthongs are now rather archaic and are giving way in the speech of younger people in particular to [ɔ:] a vowel which is found in many items that have never had a historical /r/, like *saw, law*. If English was an unwritten language, and we had no clear evidence from relatives like German (where we can find cognate *die Tür, sicher* with /r/ in each case) or from rhotic varieties like Scots, we would not be able to reconstruct the presence of historical /r/ in these lexical items for earlier stages of the language. Even leaving aside all these ifs, we could still reconstruct historical /r/ in the appropriate cases if change was not cumulative; but it is, and that fact means that over time we will inevitably lose access to information about ancestral states.

Accepting the logic of this argument does not, of course, mean that we are able to pinpoint the temporal threshold for the operation of the comparative method to the year, or even to the millennium: Renfrew (2000: xii) notes that estimates from 6,000 to 12,000 or even 14,000 years have been proposed. Trask (1996: 207) spells this out further, ending on a note which nicely encapsulates the unclarity around the answer:

> for how long can two or more languages diverge from their common ancestor until we can no longer see the slightest evidence of their common origin? There can be no hard and fast answer to this question, since it's always possible that some languages will change more slowly than average and thus preserve remnants of their common ancestry longer than average. If you put this question to a group of historical linguists, though, most of them will probably give you an estimate of about 6000–8000 years, even in the most favourable cases, the families for which we have lots of languages and some substantial early texts. A few will go as high as 10 000 years ago, which is a nice round number.

Fortunately, we do not have to resolve this question here, since a difference of several thousand years still gets us absolutely nowhere near the shift from historical into evolutionary time on a species-level timescale. We return in later chapters to exactly what the various strands of evidence suggest about the time when something approximating modern human language developed, but we can be pretty confident that it was not later than 40 kyBP (that is, thousand years Before Present), the point when we know initial human migrations to Australia, for instance, were taking place, and when we start, in some areas, to see evidence of accelerated cultural development like cave-painting. Nobody is suggesting that we can use the comparative method to get as far as 20,000 years back into the past, let alone twice that far or more.

Interestingly, these arguments for time-limited applicability of the comparative method converge to an extent with similar discussions in lexicostatistics. While the comparative method uses mainly phonological and morphological evidence, lexicostatistics involves the collection of lists (usually 100 or 200 items in length) of the words which express basic meanings like small numbers, parts of the body, natural phenomena like 'sun', 'water', 'earth', and verbs for universal human experiences like 'drink', 'eat', 'say', 'walk'. These are often known as 'Swadesh lists', after Morris Swadesh, who pioneered the idea of basic vocabulary (see Slaska 2006 for an overview and suggestions on future meaning-list methodology). The words expressing these basic meanings in a range of languages are collected and compared to assess how many are cognate between pairs of lists; the percentage cognacy is calculated and used to express the degree of similarity between the languages involved. In a further and much more controversial step, it has been suggested by proponents of glottochronology that these percentages can be translated by an equation into the time separating these two languages from their common ancestor (see Embleton 1986 and the papers in Renfrew, McMahon and Trask 2000). This is contentious because vocabularies demonstrably do change at different rates, depending primarily on social factors, such as the degree of contact with other languages or the prevalence of taboo (Bergsland and Vogt 1962), and it has to an extent been abandoned, though there are current attempts to use other techniques to date particular linguistic developments, notably the break-up of Indo-European into its daughter subgroups (see the papers in Clackson, Forster and Renfrew 2005). Again, however, there is some consensus that even under optimal conditions we must regard the time over which lexicostatistics or glottochronology can operate as essentially limited. For example, Nichols (1998: 128) calculates that, by the usual glottochronological rates of lexical replacement, 'after 6,000 years of separation, two languages are expected to exhibit only 7% shared cognates; and 7% represents the lowest number of resemblant items that can safely be considered distinct from chance'. Kroeber (1955) proposes an outer time limit on glottochronology of 11,720 years, while Campbell (forthcoming: 8) suggests that virtually all basic vocabulary will have been replaced within 14,000 years, so that no recognisable cognates will remain in languages which separated before that time. Again, there is a certain margin of error here, but the crucial point is that evidence will become unavailable after a certain amount of divergent change, which may or may not correlate well with a precise period of years; and this seems robust.

Our discussion above of the limited scope of the comparative method rested in part on the relatively small number of phonemes in each language, and therefore on the propensity of change to be cumulative; but even in the case of lexical semantics and vocabulary, where we are dealing with a very much larger population of ostensibly independent items, it appears that methods are time-limited and cannot succeed in tracing relatedness far into prehistory. It is true that lexicostatistics is crucially restricted to basic vocabulary, to facilitate cross-linguistic comparison; but this is likely to increase the reach of the method rather than decreasing it, since non-basic words may well be even more changeable and ephemeral. In recent years a number of different research groups from a range of different disciplines and perspectives (Lohr 1999, Starostin 2000, Pagel and Meade 2005, McMahon and McMahon 2005) have sought to identify meanings that seem to be particularly conservative and resistant to replacement of the lexical item 'carrying' that meaning. However, although the resulting lists do overlap to a considerable extent, they are not identical; and they will also inevitably form a very small and restricted subset of the language's lexicon, so that any conclusions based on these super-conservative meanings alone will be limited.

Evidence from the comparative method and lexicostatistics seems to converge, then, on a time limit for the successful application of methods of comparison and reconstruction of somewhere between 5,000 and 15,000 years. This might seem like a pretty substantial margin of error, and it is true that this suggestion of disagreement reflects the somewhat volatile nature of the field at present. However, wherever historical linguists eventually agree within this range of likely dates, we will be nowhere within reach of the beginnings of human language. Although there is also currently lively debate over methods for language comparison and classification, these do not seem likely to stretch the period over which such methods can productively work by more than a few millennia – certainly not enough to propel us across the great leap into the period of evolutionary time when predispositions for language as we know it were evolving in our species or our relatives.

3.5   Reconstruction and comparison of morphosyntax

Much of the discussion in the sections above has involved phonology and the lexicon; yet in many ways the greatest mysteries of evolutionary linguistics concern morphosyntax, or grammar. In fact, in many ways the lexicon in particular is less interesting than other areas of the grammar in evolutionary terms, since it is on the whole more culturally specific, is less susceptible to poverty of the stimulus arguments, and is essentially learned rather than acquired. As we shall see in section 3.8 below, the lexicon is also the part of modern human language which has closest analogues in aspects of vocal communication in non-human primates, so that the challenge in accounting for the evolution of vocabulary is arguably less than for, say, phonology and syntax. As for phonology (or at least segmental phonology), it is likewise at least substantially learned from environmental input, which will under normal circumstances be plentiful in this domain; and although clearly humans make a range of sounds far more extensive and differentiated than those produced by any of our primate relatives, there is a fairly straightforward physical explanation for this: as we shall see in Chapter 5, humans have a very different shape of vocal tract. Many of the arguments in Chapter 2 above on the critical period, the poverty of the stimulus, and deleterious effects of brain damage or inherited disease have to do with morphosyntax.

The problem is that methods for language comparison and linguistic reconstruction focus for the most part precisely on phonology and lexis. True, the comparative method is morphophonological: but the emphasis is on the phonological expression of grammatical morphemes, not on the inventory or typological possibilities for those grammatical morphemes themselves (perhaps because the method has been developed for use in individual and well-characterised families, the members of which tend to be broadly typologically similar). There are many reasons for this restriction, not least the fact that we understand far more about the directionality and recurrent patterns of change in phonology than in syntax in particular. Furthermore, the options in syntax tend to be fewer, so that accidental resemblances are potentially more common and more of a problem: it is hard to imagine that the recurrent correspondences of /p/~/f/ or /d/~/t/~/s/ could have arisen by accident, but, as Campbell (forthcoming: 13) observes, either Objects will precede Verbs, or Verbs will precede Objects, so there is not a lot of conceptual space to play with (see also Newmeyer 2003). There are also many possible reasons for a shift from one to the other – and more worryingly for methods of reconstruction, for multiple such shifts, given that changes are frequently cyclic.

Just as there have been attempts to identify meanings which seem to change more slowly than the average, there has also been some suggestion that we might identify a set of particularly stable or conservative morphosyntactic features (Nichols 1992, 1995, 1998). Nichols has suggested that features of this kind might include morphological complexity; the existence of an inclusive/exclusive distinction; head or dependent marking; whether a language has nominative/accusative or ergative/absolutive alignment; and the existence of noun classes. If such features are relatively unlikely to be borrowed or to develop in parallel by chance, they will be good indicators of deep genealogical relatedness; and we might then also expect them to be strong candidates for features of the earliest human language systems. However, there are growing indications that these features may not be so stable at all; Mithun (2004) argues that morphological complexity and head or dependent marking, for example, characterise linguistic areas in the Central Northwest Coast of the USA, and have therefore spread by borrowing, while Campbell argues that 'The claim of stability for a number of other traits is also unsupported' (forthcoming: 14, and see also Campbell 2003).

To digress slightly, there may nonetheless be cases, though they are few and far between, when we can observe the introduction of constructions into a language or language family apparently from scratch. This requires, of course, that we have a long historical record for the language concerned, and historical records do not come much longer than in the case of the Semitic language Akkadian.

Deutscher (2000) argues, on the basis of a substantial corpus of letters, that there is clear evidence showing the development of sentential complementation during the history of Akkadian. The earliest letters date from the Old Akkadian period, around 2000 BC; they had not been studied extensively before Deutscher's work because they were seen as having little or no literary merit, but are particularly important for linguists because they were dictated by the 'author' to a scribe, and are therefore just about as close to the spoken language as we can expect written language to get.

In Old Babylonian, from around 2000–1500 BC, the form *kīma* appears in the role of a causal adverbial conjunction; in the earliest letters, it means 'as', 'like', 'when', and 'because', as shown in (2).

(2)   *kīma udammiqak-kunūši   dummiqā-nim*
         as I did favours to you (pl.)   do favours (imp. pl.) to me
         'as I have done you favours,   do me favours' (Deutscher 2000: 40)

Some cases of *kīma* are found in ambiguous contexts, where they could be interpreted as either 'because' or 'that', as shown in the example in (3).

(3)   *kīma   šuddun-am   lā   nileû ana bēli-ja   aqbi*
         that/because   collect. (inf.-acc) not   we can to lord-my   I-said
         'I spoke to my Lord because we were unable to collect (them)'
         OR 'I said to my Lord that we were unable to collect them' (Deutscher 2000: 44)

Semantic bleaching is clearly going on even at this early stage, as *kīma* is used in some cases (like (4)) where only the complementiser meaning of 'that' makes any sense.

(4)   *kīma eql-am...ana rēdî   iddin-u   anna ītapal*
         that field (acc.) to soldiers (gen.)   he-gave (sub.)   yes he-answered
         'he acknowledged that he gave a field to the soldiers' (Deutscher 2000: 45)

By the Middle Babylonian period, around 1500–1000 BC, *kīma*, sometimes now reduced to *kī*, is clearly marked as a complementiser by word order, since by this stage complement clauses are postverbal, while adverbial clauses are preverbal.

Deutscher's evidence suggests that there was an earlier, attested period of Akkadian without finite complementation, and a later period where such structures do appear. This raises the question of whether, in such admittedly very unusual cases, we are seeing language change, or the end of linguistic evolution. This would suppose, of course, that the initial stages of human or hominid language were simpler, and that they proceeded to become gradually more complex; though Campbell (forthcoming), for example, points out that this is not a necessary assumption; and recall that Jespersen (1922) suggested language may have its roots in greater complexity and that it then progressively became simpler. Furthermore, it assumes that complementation is an essential part of language, something that all human languages must be able to do however it is expressed in each specific case; and this is by no means clear either.

It appears from the Akkadian case that, where we are particularly privileged in the amount and antiquity of the linguistic data we have, we can sometimes observe a new construction being introduced into a language. However, frustratingly, we cannot make full or properly controlled comparisons here because data for different languages at the same or a comparable period are lacking. Our conclusions are therefore necessarily limited. Akkadian might have had some other strategy for complementation which had already vanished without trace by the first written records; or complementation might not be essential to language in any kind of evolutionary sense. If we are not suitably circumspect here, we risk drawing conclusions about what structures and possibilities must be absolutely core to language and which could have been introduced later on the basis of a relatively small quantity of accidentally preserved written data.

Although the Akkadian case is a single example, the questions raised here are more generally relevant. Linguists have a long-established tendency to look for linguistic universals: but what if certain constructions we think are essential actually turn out to appear only in given cultural circumstances? What consequences will it have for evolutionary linguistics if apparently necessary elements of language are really optional? If we base our conception of what language is, and therefore of what is required in order for it to be learned and used, on assumptions that turn out to be erroneous, we may be trying to build an evolutionary model of something which need not, in its entirety or in certain details, have evolved at all. We risk, in other words, perpetrating exactly the confusion between evolution and history that we have been arguing against. We return to these questions in 3.7 below, but first consider one final, and highly controversial, method of reconstruction.

3.6   Global etymologies

If it seemed that reconstructing back from languages we know now to find out about language origins was in principle a good idea, but not possible because the methods we have are time-limited, the solution would seem obvious, if challenging: we should find or invent another method which is not time-limited, or at least which would work over a far more extensive period of time. In the next section, we shall be arguing that in fact reconstructing is not a valid part of evolutionary linguistics, whatever particular method we use; but for the moment, if we give it the benefit of the doubt, there is one other method we can consider, for which quite spectacular claims have been made.

Merritt Ruhlen has been foremost in a small group of linguists arguing that all known languages are descended from a single common ancestor, and indeed that we can reconstruct actual traces of the forms of that ultimate protolanguage, which has variously been called 'Proto-World' or 'Proto-Sapiens'. Ruhlen's work (see in particular Bengtson and Ruhlen 1994, Ruhlen 1994a, b) draws on Greenberg's proposals of superfamilies, or megafamilies like Amerind (Greenberg 1987), which includes the great majority of indigenous languages of the New World, and Eurasiatic (Greenberg 2000), which includes Indo-European, Uralic-Yukaghir, Altaic and Eskimo-Aleut, among others. The subtitle of the latter book, 'Indo-European and its closest relatives', suggests strongly that Greenberg himself did not see the hypothesis of Eurasiatic as the limits of the enterprise.

It goes without saying that megafamily proposals of this kind, given the timescales involved in their subsequent differentiation into the families and languages found today, could not have been developed using the classical comparative method. Greenberg's hypotheses rest on another, and much more controversial, method, which he himself called multilateral comparison, though many historical linguists know it as mass comparison. Rather than proceeding by gradual comparison of pairs of languages, via lists of cognates and, crucially, the common correspondences they exhibit, Greenberg's mass comparison involves looking in a much more superficial way at a very much larger number of languages, over potentially many more words, to ascertain whether there are immediately apparent resemblances. These resemblances do not need to be recurrent; in fact, they need not extend beyond the individual 'etymology' in which they are observed. What matters is only that there is observable similarity – not that these similarities are regular, or recurrent, or found in any particular part of the vocabulary, or based on any particular requirements about the degree of closeness in form or meaning among the daughter-language elements compared. Just the existence of similarity is enough to justify a proposal of relatedness; and the prospect of missing a true relationship is seen as far worse than the possibility of suggesting an invalid one.

Needless to say, this method of mass comparison has been extremely controversial. It is not that historical linguists object to the idea of superficial inspection as a starting point: this is often the way any valid comparison must begin. But the difference is that applications of the comparative method then go on in a systematic, bottom-up way, noting regular correspondences and requiring that these be validated by reconstructing common ancestral forms and checking in each case that a series of plausible changes can convert that protoform into the attested reflexes found now. Mass comparison explicitly denies the need for this kind of cross-checking via reconstruction of the protoform and the requisite changes; rather than suggesting that superficial inspection is a starting point, on the basis of which we decide whether to move forward to a full examination of the evidence, Greenberg and Ruhlen argue that only a single step is necessary.

Although reconstruction is not a necessary part of the process in mass comparison, it is used selectively in the proposal of so-called 'global cognates' (Ruhlen 1994a) or 'global etymologies' (Bengtson and Ruhlen 1994). Bengtson and Ruhlen (1994: 277) quote with approval Greenberg's dictum that synchronic breadth can lead to diachronic depth: in other words, the more languages we compare, the further we can proceed into the past. Their proposal is to compare not specific languages, but language families; and their hypothesis is that 'Current contrary opinion notwithstanding, it is really fairly simple to show that all the world's language families are related' (Bengtson and Ruhlen 1994: 280). The language families providing their evidence for this assertion are shown in (5).

(5)
Khoisan	Niger-Congo	Kordofanian	Nilo-Saharan
Afro-Asiatic	Kartvelian	Indo-European	Uralic
Dravidian	Turkic	Mongolian	Tungus
Korean	Japanese-Ryukyuan	Ainu	Gilyak
Chukchi-Kamchatkan	Eskimo-Aleut	Caucasian	Basque
Burushaski	Yeniseian	Sino-Tibetan	Na-Dene
Indo-Pacific	Australian	Nahali	Austroasiatic
Miao-Yao	Daic	Austronesian	Amerind

On the basis of evidence from these thirty-two families, Bengtson and Ruhlen (1994) propose twenty-seven global etymologies, complete with a label in the form of 'a phonetic and semantic gloss'; as Bengtson and Ruhlen (1994: 291) admit in a footnote, 'We do not deal here in reconstruction, and these glosses are intended merely to characterize the most general meaning and phonological shape of each root.' However, it is clear that these 'glosses' are intended as guides to the ultimate, original protoform for these roots: for instance, Ruhlen (1994b) includes material on global cognates in his chapter 'The origin of language', and for the gloss 'AQ'WA "water"', for instance, he states that (1994b: 115):

> An original form very similar to that reconstructed for Proto-Afro-Asiatic, *aq'ʷa, could, through the simple phonetic processes mentioned above, have given rise to the dozens of slightly different shapes we have just examined. I believe it did.

Likewise, for Bengtson and Ruhlen's (1994: 322) global etymology 23, 'TIK "finger; one"', Ruhlen (1994b: 115) suggests that:

> Another striking resemblance among the world's language families is a word whose original meaning was probably 'finger' (though it has evolved to 'one' and 'hand' [= 'fingers'] in many languages), and whose original form was something like *tik*.

It seems, therefore, that reconstruction is the intention here. True, the 'glosses' are selected to provide a general idea of the original form and meaning of the global root in each case, rather than an absolute commitment to the initial shape in every phonetic and semantic detail; but then this is exactly what the asterisked forms in reconstructions are intended to convey in any case, so it is odd that Bengtson and Ruhlen should be quite so coy about this.

As further illustration, and to provide some data for the discussion below, (6) gives the full entry for Bengtson and Ruhlen's (1994: 326) global etymology 26, 'TSUMA "hair"' (omitting only the lists of abbreviations for the sources, such as grammars, from which each form is taken).

(6)   TSUMA: 'hair' (Bengtson and Ruhlen 1994: 326).

KHOISAN: !Kung *čum* 'shell', *šum* 'skin', Eastern ≠ Hua *č'ū* ~ *tˢ'ū* ~ *dtˢ'ū* 'skin'; G//abake *čā* ~ *čo* 'skin'; /Xam *tũ* 'shell'.

NILO-SAHARAN: Nyangiya *sim-at* 'hair'. Nandi *sum*.

AFRO-ASIATIC: Omotic: Proto-Omotic **somm-* 'pubic hair'; Cushitic: Sidamo *šomb-*, Proto-Southern Cushitic **se?em-* 'hair'; Old Egyptian *zm3*; Semitic: Proto-Semitic **šmġ* 'fine hair shed by a camel'; Chadic: Hausa *suma* 'growth of hair'.

CAUCASIAN: Proto-Caucasian **tˢ'fiwĕme* 'eyebrow', Proto-Lezghian **tˢ'ʷem*, Proto-Nax **tˢ'a-tˢ'?Vm*.

BASQUE *zam-ar(r)* 'lock of wool, shock of hair'.

YENISEIAN: Proto-Yeniseian **tˢəŋe* 'hair'.

SINO-TIBETAN: Proto-Sino-Tibetan **tˢʰām* 'hair'; Archaic Chinese **sam-* ~ **ṣam* 'hair, feather'; Tibeto-Burman: Proto-Tibeto-Burman **tsam* 'hair', Lepcha *ătsom*, Tibetan *(?ag-)tshom* 'beard of the chin' (= [mouth]-hair), Kanauri *tsam* 'wool, fleece', *(mik-)tsam* 'eyebrow' (= [eye]-hair), Magari *tśham* 'hair, wool', Burmese *tsham*, Lushei *sam* 'hair (of the head)', Dhimal *tśam* 'hide, bark', Garo *mik sam* 'eyebrow', Nung *əŋsam* 'hide'.

MIAO-YAO: Proto-Miao-Yao **śjām* ~ **sjām* 'beard, moustache'.

AMERIND: Almosan-Keresiouan: Pawnee *ošu* 'hair', Dakota *šũ* 'feather', Woccon *summe* 'hair'; Penutian: North Sahaptin *šəmtai* 'pubic hair', Nez Perce *simtey*, Kekchi *tˢutˢum* 'feather', *ismal* 'hair', Mam *tsamal*, Quiche *isumal*; Hokan: Proto-Hokan **čʰemi* 'fur', North Pomo *tˢime* 'hair', Kashaya *sime* 'body hair, fur', Northeast Pomo *čʰeme* 'body hair', Mohave *sama* 'root', Cocopa *išma* 'hair', Tlappanec *tsũŋ* 'hair, root'; Central Amerind: Tubatulabal *tˢomol* 'hair, head'; Chibchan-Paezan: Matagalpa *susum* 'beard', Xinca *susi* 'beard'; Andean: Tsoneka *čomki* 'pubic hair', Quechua *sunk'a* 'beard'; Equatorial: Caranga *čuma* 'hair', Quitemo *čumi-či*, Aguaruna *susu* 'beard', Candoshi *sosi*.

Greenberg's methods and findings, in the Americas and elsewhere, have been challenged in the historical linguistics literature (see, for instance, Campbell 1988, 2003, 2004, McMahon and McMahon 1995, Matisoff 1990, Trask 1996). Many of these criticisms are directed at the general practice of mass comparison; but they hold equally (or even more validly) for the particular application to global etymologies, and since this is the aspect of the programme which is allegedly relevant to evolutionary linguistics, it is most appropriate for us to illustrate them in this context.

First, Bengtson and Ruhlen (1994) argue that they are comparing language families, and not individual languages; but what does this mean? One reasonable assumption would be that only the reconstructed protoforms for each family in the comparison should be used; then we would be comparing like with like. A slightly laxer criterion would be that reconstructed forms should be used whenever they are available, and only substituted by actual, attested forms when a family has not been fully reconstructed yet (or where, as in the case of Basque or Burushaski, we are dealing with language isolates, or in other words families with only a single member). However, neither of these limitations seems to be in place, and instead we find sometimes reconstructed forms, sometimes attested ones, and in other cases a mixture of the two, both within an etymology and even within a family.

For instance, in the global etymology in (6) above, we find within Afro-Asiatic, in Omotic, only a reconstructed form from Proto-Omotic; but in global etymology 22, there is no Proto-Omotic form, though there are attested forms from the Omotic languages Male, Koyra, Kachama, Bambeshi, Nao and Dime. Does this mean the reconstructed form is irrelevant or unhelpful in the global etymology 'leg, foot', though it is in 'hair'? Or, perplexingly, that the protoform is a good match in 'hair', but none of the daughter-language reflexes are? And how can we weigh the relative significance of one protoform for a subfamily, as against six attested examples in another etymology? Returning to 'hair', we find that of the thirty-two families in (5) from which Bengtson and Ruhlen have taken data, only nine are represented; there are forms from thirteen in the next global etymology, 'water', but five of these only involve a single item. Etymology 4, 'nose; to smell' is supported by data from eighteen groups; etymology 10, 'who?', has material from twenty-four. How do we evaluate the relative security of these different etymologies? Again, if there is only a single item for a family or subgroup, should our view depend on whether that item is reconstructed or not? Or on how far afield the languages retaining a transparently similar reflex might be found, either within families or across them? If evidence can be found only in a small number of daughters in a large family, how confident can we be that something similar can be reconstructed for the protolanguage of that family anyway, rather than invoking contact or independent development, in which case no amount of daughter-language evidence can be carried back to Proto-World? This is the case in the *TIK 'one; finger' etymology, for example, where, as Salmons (1992: 210) notes, the Proto-Austro-Tai form cited is actually one variant of three possible reconstructions in the source, all of which 'appear to be based on only 18 languages out of the more than 1,000 in that proposed grouping'. Allowing complete freedom in whether attested or reconstructed forms are used means multiplying the number of available languages; furthermore, data from vastly different time periods will be included, meaning we are not comparing like with like.

All these strategies, however, do ensure that Bengtson and Ruhlen are considerably increasing their chances of finding some kind of match in each case. Ringe, in particular, has proposed in a series of publications (1992, 1996, 1999) that the results of mass comparison are indistinguishable from chance, partly because the forms cited are often rather short, increasing the possibility of a random resemblance, and partly because the more languages we allow as candidates, the more likely such chance matches will be. Salmons (1992: 207) likewise argues that 'the method of mass comparison excludes the traditional means of eliminating chance correspondences'. Leaving aside the phonologically unmarked shape of *tik, which would again allow a greater likelihood of chance matches, 'by invoking at least six general meanings and using at least four different time depths, the method of mass comparison multiplies the number of possible forms, and the chances of accidental similarities, at least two dozen times' (Salmons 1992: 220).

The only way to restrict the likelihood of chance resemblances being confused with real and meaningful ones is to ensure that our criteria for what counts as a meaningful resemblance are spelled out and are reasonably difficult to attain.

If virtually anything can count as a match with anything, then we are unable to evaluate the claims being made. Yet it appears that Bengtson and Ruhlen's criteria for matching are lax phonetically and semantically; though this must be deduced from examples since there is very little explicit comment to be found. For instance, Bengtson and Ruhlen (1994: 288) claim that 'we have constrained the semantic variation of each etymology very tightly, and few of the semantic connections we propose would raise an eyebrow if encountered in any of the standard etymological dictionaries'. True, this is not exactly a quantifiable criterion; but it might seem a priori reasonable, until we recall that the standard etymological dictionaries are usually based on demonstrable and proven connections between languages, with strong support from the comparative method, regular and recurrent correspondences, and attested semantic changes. None of these are available for the global etymologies. Within the 'hair' etymology, for example, we find 'hair', 'pubic hair', 'beard' and 'fur', straightforward enough semantic specialisations or transfers; but also 'shell', 'skin', 'root' and 'bark'. In a detailed critique of the TIK 'finger, one' etymology 23, Salmons (1992) notes that even within Indo-European, the reflexes mean 'say, tell', 'proclaim', 'teach, instruct', 'show', 'point', 'finger', 'sign, mark', 'throw', and, as Trask (1996: 394) suggests, possibly 'toe'. Yet because we have 'finger' (admittedly only in Latin), and 'point' (in Sanskrit), Ruhlen can select these minority senses and reconstruct 'finger, one' as the original meaning.

The evidence on what counts as a phonological match is even more problematic. In the 'hair' etymology in (6) above, what seems to count is a template of something alveolar/palatal/otherwise front and voiceless at the start, some vowel, and then something nasal. This provides a lot of scope for detecting matches. For KUNA 'woman', Campbell (forthcoming: 2) suggests that 'the target is approximately $KVN(V)$, where "K" is any velar-like sound, "N" some n-like sound. However, matches are not tight, since for the "K" any of the following fits: k, k', g, q, x, h, w, b, $ž$, $?$, $č$. For the final "N", any of the following count: n, r, m, $ã$, $w?$, $??$, and $Ø'$. Trask (1996) goes further, suggesting, for instance, that any vowel matches any other, and that consonant place is really all that matters, though even here adjacent or near-adjacent places can be lumped together. Salmons (1992) estimates that about 6 per cent of all CVC forms in any language could therefore count as a match for TIK 'finger, one'; and, as Trask (1996: 395) points out, Ruhlen also allows CV or VC matches, as well as substantially longer forms. Worst of all, none of this can actually be tested. If we complain that the evidence is not strong enough, a proponent of Proto-World can simply enthuse about how amazing it is that we should find so much relevant data after such a long period of time. We cannot falsify a global etymology, because we cannot know why particular items have been included, while other potential cases have not. Trask (1996: 395) sums up:

> Since Ruhlen recognizes no systematic correspondences, there is no way
> of checking the validity of proposed instances of *tik*, and there can be no

such thing as counter-evidence. Instead, there are only confirming instances; finding one language that presents a suitable word counts as evidence, while finding 100 that fail to do so counts for nothing.

3.7   Limitations on reconstruction

It seems, then, that we cannot hope to find out about the first human language, or languages, by starting from surface features of languages we know about today and working backwards. This is clearly a superficially attractive way of finding out about the linguistic past, and something we have had to examine quite carefully, especially because it seems, at least on the face of it, to mirror practice in other evolutionary disciplines. The biologist Vincent Sarich is widely quoted as observing that 'I know my molecules had ancestors: the palaeontologist can only hope that his fossils had descendants.' Yet we must be careful how we interpret this quotation. Starting from the present, and potentially at the genetic level, does not mean we have to gain information about the past by reconstruction. What Sarich seems to be advocating here is comparing current organisms genetically, rather than focusing on superficial, phenotypic traits which may have evolved in a common ancestor, or in a parallel manner in different species, or which may be controlled in different species now by completely different genetic mechanisms. When we consider the surface traits, we can certainly speculate about their origins and significance; and we can make these hypotheses stronger by making cross-species comparisons. On the other hand, when we examine the genetic level directly, we can see just how similar the current systems are, and trace the mutation(s) involved back to a common ancestor.

Let us then summarise the situation for language. Even if working backwards was the obvious and perhaps the only way of proceeding, the sections above show that this could not be achieved. Either the methods we have do not reach far enough back, or they have relaxed relevant criteria so far that we cannot rely on their results. In addition, even the best methods are limited to the reconstruction of individual words, morphemes and phonemes, with perhaps some extrapolation to the likely shape of phonological systems. But these are, as we have seen, the elements of language which are perhaps least interesting in evolutionary terms.

Campbell (forthcoming: 17) suggests that we might contemplate one further way to extrapolate backwards towards the origin of language, this time starting from the design features of human language. These are the definitional properties of human language, features every language seems necessarily to incorporate: the classical list of design features is usually attributed to Hockett (1960), and includes those in (7).

(7)   **Design features of human language**
       Vocal–auditory channel
       Rapid fading (once a signal is produced, it vanishes quickly)

Interchangeability (participants can play both roles, as speaker and hearer)

Arbitrariness (the connection between the signal and its meaning is
   conventional)

Displacement (it is possible to talk about actions and entities which are not
   part of the immediately present situation)

Productivity (new signs and utterances can be created)

Semanticity (association between a signal and a particular object, action or
   concept)

Duality of patterning (sounds can be recombined into new meaningful
   units)

Recursion (one element can be embedded in another of the same category:
   this allows for subordination).

However, there are two problems with taking this set (or any set) of design
features as a point of departure, whereby 'it is argued that the earliest human
language will have had the design features of human language and this gives us
some clues to its nature' (Campbell forthcoming: 17). This kind of uniformi-
tarian approach, which involves a claim that what we call language now must
have the same kinds of features as anything in the past we could call language,
is intrinsically limited in not being able to deal with the really interesting question, which is emergence. As Campbell continues (forthcoming: 17), 'We can
assume that the earliest language(s) did meet the design feature requirements
of human language, but this is in a sense a definitional demarcation which says
anything else is not human language, which cuts off access before emergence,
leaving unaddressed the question most fascinating to many, of how human
language originated and evolved from something that was not (yet) human
language.'

Worse, the set of design features is not fixed. We might be projecting onto
our putative Proto-World a set of properties which are not criterial for human
language at all. Some of the features in (7) are problematic because they do not
apply to human language in all modalities: sign language does not use a vocal–
auditory channel (and yet in Chapter 2 we used the development of Nicaraguan
Sign Language as evidence for the partially innately driven acquisition of language, so it would be completely incoherent to argue now that sign language
somehow doesn't count). Written language, or recorded spoken language, does
not fade rapidly. Of course, we could argue that the availability of written or
otherwise recorded evidence reflects history or cultural development rather than
evolution; but then using the design features, if Campbell is right, restricts us
precisely to history rather than evolution anyway.

There are also recent claims that even the most central design features are
not, in fact, universal: Everett (2005) seems initially to be arguing that Pirahã
lacks displacement, productivity and recursion. He attributes these, and a range
of other linguistic limitations (such as the absence of numerals and counting,
lack of quantifiers, and the highly restricted kinship system) to a general 'Pirahã
cultural constraint on grammar and living' (Everett 2005: 622), whereby 'Pirahã

culture avoids talking about knowledge which ranges beyond personal, usually immediate, experience or transmitted via such experience' (2005: 623). On
closer inspection, these design features are not absolutely absent from Pirahã.
Thus, 'Pirahã of course exhibits displacement in that people talk regularly about
things that are absent from the context at the time of talking about them, but
this is only one degree of displacement' (Everett 2005: 633). Consequently, displacement 'is severely restricted in Pirahã' (Everett 2005: 633), but is available.
Likewise, productivity is said to be 'severely restricted by Pirahã culture, since
there are things that simply cannot be talked about, for reasons of form and content, in Pirahã in the current state of its grammar' (Everett 2005: 633); but again,
it is not unavailable in principle. As for recursion, Everett shows that Pirahã
syntax typically consists of sequences of main clauses, with strategies like nom-
inalisation used instead of embedding – so, 'he knows how to make arrows well'
would in Pirahã translate literally as 'he sees attractively arrow-making' (Everett
2005: 629). Even here, however, Everett (2005: 629) concedes that there are two
alternative analyses of this construction: 'The first is that there is embedding,
with the clause/verb phrase "arrow make" nominalized and inserted in direct
object position of the "matrix" verb, "to see/know well". The second is that this
construction is the paratactic conjoining of the noun phrase "arrow-making" and
the clause "he sees well".' Everett supports the latter analysis because it 'seems
to fit the general grammar of Pirahã better'; but this is potentially self-fulfilling,
insofar as analyses lacking embedding have been preferred elsewhere.

What Everett has certainly demonstrated is that culture can affect the way
languages are structured; this has been argued elsewhere, as Newmeyer (2003:
73) summarises:

> Trudgill (1992) … has argued that the languages of small, isolated commu-
> nities tend to be overrepresented by grammatical constructs that pose diffi-
> culty for non-native speakers, such as complex inflectional systems that are
> characterized by multiple paradigms for each grammatical class. They also
> tend to manifest the sorts of feature that one might expect to be present in a
> tight social network, such as weird rules with enough exceptions to mystify
> non-native learners. Nettle (1999), by means of computer simulations, has
> come to very much the same conclusion.

It is notable in this context that Pirahã has particularly complex systems of verbal
morphology and prosody.

The question is how we interpret this impact of culture on language. Everett
(2005: 634) claims that 'For advocates of Universal Grammar, the arguments
here present a challenge – defending an autonomous linguistic module that can
be affected in many of its core components by the culture in which it "grows".'
However, for one thing, we have not yet established whether, in evolutionary
terms, the innate component underlying language is autonomous: this is still
an open question. And for another, the availability of genetic underpinnings
for a system does not require that all the potential should be employed in

every case. Phenotypic traits are precisely the result of an interaction between genes and environment, and we should expect both parts to matter. Far from undermining a scenario where language relies in part on an evolved genetic component (which may, remember, use physical or neurological elements originally dedicated to some other purpose), Everett provides excellent evidence for precisely that.

If design features can be partly or wholly suspended under particular cultural circumstances, we have another reason for not relying on in-depth studies of specific present-day languages in attempting to infer the structure of the original human language(s). We must bear in mind the opposition we introduced in the first two chapters between genes, the physical and neurological structures those construct, and the various behavioural systems these in turn enable. The linguistic comparative method is inevitably restricted to the third and most superficial level, namely the characteristics of languages themselves. But when we are talking literally about evolution, we mean something that has happened at the first, deepest, genetic level; which then constructs or recycles available material at the structural (neurological, physical) levels; which then constrain or enable what is produced at the behavioural level. Looking at properties of languages might tell us something about what the underlying structures and genetic instructions allow (though, as we have seen, some of these may be minimised or suspended depending on the cultural context), but it is not so likely to tell us very much about the properties or histories of those evolved structures themselves. To make progress here, we still need a comparative approach, and indeed a comparative method – but the biological rather than the linguistic version.

3.8   The comparative method in biology

As Harvey and Pagel (1991: 1) suggest,

> Evolutionary biology shares with astronomy and geology the task of interpreting phenomena that cannot be understood today without understanding their past. Stars in the Milky Way, mountains in the Swiss Alps, and finches in the Galápagos Islands each have their own common histories which give them characteristics that set them apart from the stars of other galaxies, mountains in other regions, and the finches of other archipelagos.

We can add to this list of disciplines both historical linguistics and evolutionary linguistics: the former is concerned with accounting for why particular language systems are different from other equally possible systems; and the latter seeks to account for the development of the capacity for such language systems in our species through the processes of evolution. We have already established that it is not possible to apply the techniques of comparison and reconstruction from historical linguistics in the pursuit of these evolutionary goals; so it seems reasonable to see whether there is a set of techniques in evolutionary biology which might be better designed for that purpose.

The biological comparative method is certainly not new: Darwin's theory of natural selection was firmly rooted in a comparative framework, which underlies his conclusion in *The Descent of Man* (1871) that humans are anatomically, physiologically and behaviourally just another species of primate. More specifically, the aim of comparative biological studies is to identify the characteristics of organisms which make them particularly suited to the kinds of lives they lead, in the kinds of environments where they lead them. To explain these characteristics, they must then be integrated into a historical framework, asking how they have been arrived at, and which evolutionary forces have shaped them. Traditionally, this approach has focused on observable structures or behaviours; but more recently comparative work has shifted from the superficial, phenotypic level to direct examination of the plethora of genetic data now available following genomic sequencing projects.

The biological comparative method in fact covers two rather different approaches. The first starts by identifying characters or traits which are unique to a particular species, or shared only by species doing the same kind of job (like disproportionately large front feet for creatures that dig), or operating in the same kind of environment (like arctic foxes, arctic hares and snowy owls, which have white coats or plumage as cryptic camouflage for snowy conditions). These characters are assessed from a strongly functional perspective; the underlying assumption is therefore that they are adaptations which have been developed through natural selection precisely because they improve success or efficiency in some particular task, lifestyle or environment. This kind of explanation can be strengthened by comparison with related species whose lifestyle or environment is different, to assess whether they share the adaptation or not; and by further comparison with the fossil record and our knowledge about environmental change, which can tell us when and where novel structures emerged.

Such arguments are widespread in biology, and the best examples have led to outstandingly strong and convincing cases for the adaptive status and evolutionary history of otherwise obscure features of organisms. For example, von Frisch (1967) assumed that the fact that flowers are coloured and patterned must have some adaptive explanation; he then set out to demonstrate what that adaptive rationale might be. His conclusion was that colour and pattern are evolutionary strategies developed by flowers to attract bees, and therefore to improve the chances of pollination. In the context of evolutionary linguistics, von Frisch is remembered primarily as the discoverer of the bee 'waggle dance', which is used to communicate to conspecifics at the hive about the distance and direction to flowers; but he made this discovery while seeking evidence for his adaptive account of flower colour.

On the other hand, the drawback in adaptationist thinking is the ever-present danger of descent into 'just-so stories' (Gould and Lewontin 1979), like the idea that elephants have long trunks because an ancestral elephant, being prone to curiosity, put its nose into the river and had it pulled to its current extent by a crocodile. These folktales are charming but not remotely scientific; indeed, they violate one of the most important tenets of evolutionary thinking, since they involve an acquired character (the elongated nose) being inherited by subsequent generations. In proposing natural selection and adaptation, we must constantly be on our guard against telling such engaging and potentially plausible but unprovable stories. As Dawkins (1983: 32) puts it, 'Adaptationist conviction cannot tell us about physiological mechanisms. Only physiological experiment can do that. But cautious adaptationist reasoning can suggest which of many possible physiological hypotheses are most promising and should be tested first.'

Furthermore, there has been considerable criticism of the adaptationist paradigm on the grounds that specific characters may appear less than optimally adapted for their apparent niche, or sometimes even maladaptive (Gould and Lewontin 1979, Gould and Vrba 1982, Lewontin 1978). Moths, for example, fly into flames; arguing that this is adaptive does not seem an immediately obvious strategy. These critiques are absolutely right in noting that non-adaptationist evolutionary processes have also contributed to the state of present-day organisms, but they are incorrect in claiming that an adaptationist approach requires every character to be optimally adapted if it is adapted at all. Returning to our moths, it appears that moths flying at night use bright objects (that is, the moon or stars) to navigate, since maintaining a constant angle relative to a light in the sky ensures the moth travels in a straight line. This was a good adaptive strategy before people invented candles and electric light bulbs; but when moths apply it to such terrestrial lights, which emit non-parallel beams of light, they will inevitably spiral inwards until they hit the light source.

The problem is that, as we have seen, history works far faster than evolution, and once an adaptive behaviour is in place, it will take a long time to get rid of it: it follows that some apparent maladaptations will result from a mismatch between the environment where a character evolved and the environment now. This is one of many constraints on evolution that can limit the current optimality of a character; these are perhaps most eloquently expressed in Dawkins (1983: Ch. 2), where we find the list shown in (8).

(8)   Constraints on evolution
   a.   Time lag (character has not caught up with change in environment; moths and lights).
   b.   Historical constraints (accidents of past history: in some ancestral vertebrate, the shortest route from the brain to the larynx was via the aorta. Consequently, all mammals now route the recurrent laryngeal nerve via the aorta, which means the nerve in, say, giraffes, is a great deal longer than it might optimally be).
   c.   Available genetic variation (evolution can only operate on mutations which have already occurred spontaneously; mutation and variation cannot be preordered).

d.   Constraints and costs (size matters, but it's easy to overshoot: a giraffe's long neck is an adaptive plus for reaching high foliage, but if it grows too long the animal will be unable to hold up its head).

e.   Variable selection at one level affecting others (producing too many offspring too early means earlier ageing; but waiting around too long before reproducing is risky in other ways).

f.   Unpredictable or variable environment.

It follows that changes in environment or other characters may lead to an originally near-optimal character or behaviour becoming less so with time, and that constraints on one system may lead to necessary modifications in another. Although these may mean the latter character is less than optimal, they may allow the whole system to work; and evolution is full of compromises and 'deals' of this kind, whereby organisms can go about their business in a reasonably ordered and internally well-resourced way, though the history of the organism and the systemic interactions within it will inevitably lead to elements functioning below their ideal and isolated best.

In addition, we must bear in mind that the fossil record, by its nature, is patchy: we cannot reliably tell accidental gaps from systematic, meaningful ones. We cannot therefore absolutely disprove a particular adaptationist argument because we lack fossil evidence to substantiate it: soft tissues, brain contents and behaviours clearly do not fossilise at all; bones and shells, which do, are preserved by circumstances beyond our control and may be partial or in a poor state; and there may be whole species, important events, or long periods of time which have (as yet) no fossil representation at all.

This historical comparative approach, which, as we have seen, is biased towards adaptationist accounts, has more recently been supplemented by a quantitative comparative method which begins by looking for statistical trends between traits and environments across species. On the one hand, we may find that organisms which are only very distantly related live in similar environments, and share traits not present in their closer relatives which occupy different niches. Conversely, if large numbers of species belonging to many different families share similar characters, then any that deviate from this common pattern may well be doing so for interesting biological reasons. Of course, there may be a subsequent step in the argument here to the effect that we have adaptations in each case; in other words, the outcome of this more quantitative comparative method and the historical comparative method may be exactly the same, with both converging on an adaptationist explanation. What is different is the starting point: the historical version began with a particular salient feature and attempted to explain it by invoking its adaptive function and evolutionary development; but the quantitative approach begins by taking more global measurements and considering a whole batch of traits, some of which will turn out to be totally uninteresting. When trends, or obvious exceptions to trends, are unearthed on the basis of these correlations, researchers can then return to the historical comparative method and to the fossil record to establish why and when the relevant trait evolved. However, the basis

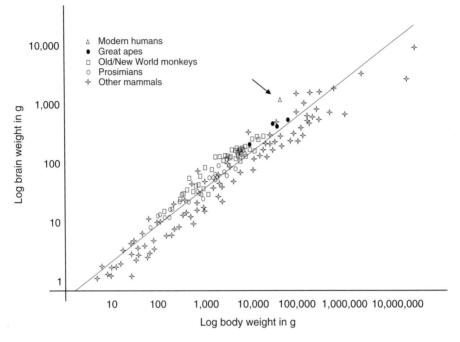

Figure 3.1 *The encephalisation quotient, data from several sources*

for this reasoning (which may, indeed, be adaptationist) is stronger since we begin by demonstrating that a trait is common or unusual rather than isolating a particular trait without knowing how common or unusual it might be. Foley (1987: 74) characterises this as 'a higher order of model … where the rules governing behaviour, not the behaviour itself, are extrapolated back through time'.

As an example, take the so-called 'encephalisation quotient' (Clutton-Brock and Harvey 1977), which relates body weight to brain weight: more specifically, it therefore indicates the proportion of overall body weight taken up by brain weight. The graph in Figure 3.1 shows this relationship for a group of mammals, including primates.

Figure 3.1, as expected, shows a general trend whereby animals with bigger bodies also have bigger brains. However, primates in general have bigger brains relative to body size than similarly-sized non-primates. When we turn to humans, we can make two observations based on the data in the graph. First, and most obviously, humans are big primates; but secondly, and more strikingly, humans in fact fall even further off the scale than the first observation would lead us to expect; so while primates are particularly big-brained mammals, humans are even more particularly big-brained primates. Biologists would interpret this as indicating that brain size in humans requires further investigation, and may be worth exploring as potentially adaptive.

Both versions of the biological comparative method require phylogenetic information for proper evaluation. We must in particular be careful in establishing the appropriate basis for comparison, and in general it is important to begin with closer relatives in assessing how special a particular trait might be. For example, if we compared humans with cats, dogs, hedgehogs and pangolins (which are effectively armour-plated, tree-climbing armadillos with the engaging property of rolling up in a ball at the slightest provocation), we would find that humans alone have the ability to see accurately in three dimensions with good colour discrimination. We might then jump to the conclusion that humans are unique among mammals in having this skill. However, if we instead compare humans with our much closer relatives, the Old World monkeys and apes, we find that three-dimensional colour vision is a widespread character. If this is to be explained as an adaptation, it would be to our ancestral, arboreal lifestyle, rather than to any subsequent, specifically human, context. On the other hand, if we look even further away in the biological family tree, we find that colour vision is also widespread in fishes and reptiles, so that the issue at hand is not why primates might have developed this character, but why they have retained it while other mammals have lost it. This may be because the ancestors of primates benefited from colour vision to tell whether fruit, one of the main components of their diet, was ripe or not; but this is admittedly hypothetical. However, if we are limited by the vagaries of our fossil evidence to hypothetical scenarios, we can at least ensure, by comparing phylogenetically as widely as we can, and beginning with immediate relatives, that we are trying to explain the right history (as retention rather than innovation, for example) and not the wrong one.

How, then, can we use the biological comparative method to cast light on the evolution of the capacity for language in humans? One possibility is to return to the design features, properties like displacement, productivity, recursion and so on, and to ask which communication systems in other species have which of these properties. This has the potential to turn into a rather sterile debate about which species 'have language' and which do not; and because historically there has been a trend to assume rather than to argue that only humans do, this has led to a progressive moving of goalposts as design features are demoted when the communication system of another species gets too dangerously near to satisfying the criteria for human language we first thought of. On the other hand, this should not stop us from assessing the shared properties of human language with other systems, and the properties which differentiate them, so long as we make this assessment in an objective way, and remember that human language is not necessarily a single, autonomous and monolithic system, but results from interactions at a number of levels.

What we are trying to do in this case is to go back beyond human language, without any assumptions about what a pre-language would need to be. We need to consider what communication systems some of our primate relatives might have, and what physical or neurological structures distinguish us. Sometimes, the similarities are more marked than we might initially expect. To take one famous example, Seyfarth, Cheney and Marler (1980) have demonstrated that vervet monkeys give different alarm calls depending on the particular predator sighted:

playing back these alarm calls elicits appropriate escape behaviour from other vervets, so that the 'leopard' call leads vervets to run into trees, while following the 'eagle' call, they all look up. These calls are not structurally complex, so there is no argument that recursion or duality of patterning is involved, but they may well be arbitrary, and certainly seem to show semanticity, or referentiality. Seyfarth and Cheney (1986) further show that these vervet signals are learned, with immature vervets developing both their own increasingly adult-like repertoire of calls and appropriate responses to the calls of others over the first four years. Infant vervets tended to overgeneralise the species to which alarm calls should apply; a 5.5-month-old infant, for instance, produced a recognisable variant of the adult 'eagle' alarm call in response to a marabou stork, which is not a predator, while another immature vervet gave the 'leopard' call when confronted with a warthog. Adult vervets do not appear to overgeneralise in this way; but Seyfarth and Cheney established that they do produce their own alarm calls in response to appropriate calls made by immature vervets. When adults hear an immature vervet making an alarm call, they do not seem to respond automatically by calling themselves. For example (Seyfarth and Cheney 1986: 1648):

> on 12 out of 17 (70.6%) of the occasions when an immature was the first member of its group to give an alarm to a martial or crowned eagle, adults followed the immature's alarm with alarm calls of their own. By contrast, second alarms occurred in only three out of 60 cases (5.0%) when an immature was the first member of its group to give an alarm call to a non-predator … Such second alarms by adults could potentially have served as reinforcers, guiding the infant's developing recognition of the relation between alarm-call type and predator species.

Our next step, then, is to examine the properties human language may have in common with communication systems in other species; and to consider, via the various manifestations of the biological comparative method, the ways in which we and our immediate and more distant relatives may differ from each other, or share features not found in other lineages. Determining the phylogenetic relationship of a group of organisms is, of course, not a trivial exercise in itself; nor does it depend on behaviour and characters alone. We return to these issues for humans in Chapters 4 and 6 below, with a more detailed account of where humans fit into the phylogenetic species tree, and further information on the genetic data which are increasingly becoming available to fine-tune our hypotheses and to provide evidence where the fossil record is impoverished or absent.

For the moment, however, we have established an initial modus operandi. In following the biological comparative method for language, we must initially stop seeing language as a single phenomenon, and break it down into the different capacities that make it possible. Remember that we cannot be interested at this evolutionary stage in specifics of linguistic systems or behaviours, like glottal stops, or subordinate clauses, or affixes, or first person: any of these might have arisen in historical rather than evolutionary time, and as we have seen from Everett's

(2005) recent work on Pirahã, the interplay between culture and language may be complex. We need first to establish how we as a species are able to produce the carriers for these specific features, whether in the form of vocal communication or the neurological signals which allow us to store and retrieve individual linguistic elements and organise them into highly structured utterances. In each case, following the biological comparative method, we will be asking which of these physical and behavioural traits are shared with other species in closer or further-flung branches of the species family tree, and how they are used in these different organisms.

These questions will occupy the next three chapters. In Chapter 4, we investigate the evolutionary history of the human species: who are, or were, our relatives; what does the fossil record tell us about them; and where and when are humans likely to have evolved in something like our current form? In Chapter 5, we turn to the prerequisites for vocal communication of the type underlying all spoken human languages: it is a truism that humans have a specially configured vocal tract, but nonetheless 'many of the subsystems that mediate speech production and perception are present either in our closest living relatives or in other, more distantly related species' (Hauser and Fitch 2003: 178). Chapters 6 and 7 deal with interacting questions involving genes and brains: what do we know about the localisation of linguistic functions in the brain and the genetic instructions which are implicated in language production, perception and acquisition? What questions remain, and how can we hope to know more in future? In Chapters 8 and 9, we will return to the later stages of the biological comparative method, asking whether, in view of this phylogenetic, physical and genetic evidence, human language or the capacity for language can be seen as adaptive: are we still simply telling just-so stories, and what non-adaptive alternatives might there be?

3.9   Summary

Our main question in this chapter has been whether there is a place in evolutionary linguistics for comparative research. We have established that, tempting though it is to try to reconstruct the earliest stage(s) of human language by starting with language systems we know today and working backwards, this is not feasible. If we apply rigorous and well-supported methods like the comparative method, the cumulative nature of change will mean we relatively rapidly reach a time limit beyond which we cannot see. But relaxing the criteria that make the comparative method reliable and trustworthy is not the solution either: what is the point in developing a method that can operate over longer periods, if it lacks these essential properties? However, we have shown that there exists a biological comparative method which is even more appropriate, insofar as we are concerned with the evolution of the human capacity for language, and not in changes within behavioural linguistic systems. In biology, phenotypic traits, ideally supported by genetic evidence, can be compared to show when and where particular structures and capacities developed, and to identify species as outliers worthy of further investigation in cases where they have a certain characteristic not present in their immediate or more distant relatives. In the next chapter we therefore turn to the biological family tree of modern humans.

Further reading

Good overviews of grammaticalisation can be found in Hopper and Traugott (1993) and Heine and Kuteva (2005). For more detail on English /r/, and its patterns of distribution in varieties today, see Wells (1983) and Giegerich (1992). The comparative method in linguistics is set out in operational detail in Hoenigswald (1960). A textbook concentrating on reconstruction is Fox (1995); and a very clear and accessible account of the comparative method in particular is in Campbell (2004: Ch. 5 and especially 5.2 'The comparative method up close and personal'). McMahon and McMahon (2005) provide an overview, in Chapter 1, of existing methods of language classification, including cases of validation of the comparative method; arguments against mass comparison are also included here, as well as in McMahon and McMahon (1995), Matisoff (1990) and Campbell (1988). The first four chapters of Joseph and Janda (2003) provide helpful overviews of reconstruction and classification: these are Rankin (2003), Harrison (2003), Ringe (2003) and Campbell (2003).

On contact, and its relevance for the comparative method and the family-tree model, see Thomason and Kaufman 1988, Thomason 2001, Heine and Kuteva 2005, Aikhenvald and Dixon 2001, and Matras, McMahon and Vincent 2005. For recent work in quantitative historical linguistics introducing network rather than tree models and methods, see McMahon and McMahon (2005) and the papers in McMahon (2005b), Clackson, Forster and Renfrew (2005), and Heggarty, Maguire and McMahon (2010); you might also try searching for work by Søren Wichmann, Tandy Warnow, Don Ringe, Luay Nakhleh, Quentin Atkinson, Russell Gray or Simon Greenhill, to name but a few of those currently active in this very lively field. Arguments that the comparative method can be applied sequentially, up to the level of comparison of protolanguages, is a keystone of the Nostratic hypothesis, which is debated in Renfrew and Nettle (1999). For discussion of whether the comparative method is time-limited, see Renfrew, McMahon and Trask (2000). Global etymologies are proposed by Ruhlen (1994a, b) and Bengtson and Ruhlen (1994); critiques of this approach can be found in Trask (1996), Campbell (2003, forthcoming) and Salmons (1992).

The design features of human language are put forward by Hockett (1960), and discussed in Aitchison (1989b), where there is also an introductory outline of the communication systems of a range of other species; a more detailed account is in Hauser (1996). Everett (2005) confronts the design features with evidence from Pirahã, and there is further debate in Nevins, Pesetsky and Rodrigues (2009) and

Everett (2009). Harvey and Pagel (1991) outline the comparative method in biology, and Foley (1987) is a superb account of its application to humans, by a biological anthropologist; Foley's Chapters 1 and 4 would be excellent preparatory reading for our Chapter 4. Overviews of the ideas behind adaptationist thinking can be found in Pinker (1994); perhaps the outstanding defence of this line of argument is in Dawkins (1983, especially Chs. 2 and 3), though this can be hard going if you do not know very much biology. Dawkins (1986) is less detailed but more approachable (though somewhat more polemical), and includes a particularly good account in Chapter 4 of how evolution tends to work with what it has, even if that might not be exactly right for the job. The seminal papers on communication in vervets are Seyfarth, Cheney and Marler (1980), Seyfarth and Cheney (1986), Cheney and Seyfarth (1988), and Seyfarth and Cheney (1990), and see further the points for discussion below.

Points for discussion

1. Consider the following set of data, from five Tupi-Guarani languages.

Asurini	Kamayura	Siriono	Guarani	Guarayo	gloss
[poko]	[huku]	[hook]	[puku]	[puku]	'long'
[iapoʔa]	[iʔahuʔa]	[hua]	[ijapuʔa]	[apua]	'round'
[poka]	[huka]	[ika]	[puka]	[puka]	'laugh'
[tʃin]	[cin]	[ʃin]	[tʃĩ]	[tʃĩ]	'white'
[pin]	[pin]	[ĩ]	[pĩ]	[pĩ]	'rub'
[ʔaw]	[ʔap]	[a]	[ʔa]	[a]	'hair'
[iwaŋ]	[iwak]	[iba]	[iba]	–	'sky'
[awatʃi]	[awaci]	[abatʃi]	[abatʃi]	[abatʃi]	'corn'

   Using the linguistic comparative method, reconstruct plausible protoforms for these items, and state the sound changes which each language is likely to have undergone. In which cases is there a clear answer, and where are you left in more doubt? What kind of evidence and arguments do you find yourself using to decide on the likely ancestral forms?

2. Consider the list of design features presented in this chapter. Compare it with the lists given in other sources – these might include Campbell (forthcoming), Everett (2005), or Aitchison (1989b), but it should not be difficult to find others, either in print or on the web. Which features appear more consistently? Do these also seem to you to be the best-supported design features? Can you find out about other communication systems which have each of the design features? Are there any which seem genuinely to be unique to human language?

3. Read Cheney and Seyfarth (1988) and Seyfarth and Cheney (1990). Summarise the argument of each of these papers. In particular, what do these papers tell us about the ability of vervets to detect unreliable signals? What is meant by 'unreliable' here, and why might detection of unreliability be relevant for evolutionary linguistics?

4. Make a list of six features or behaviours you might intuitively see as particularly characteristic of humans. In each case, research whether these are also

found in other primates. This would also be a good opportunity, in preparation for the next few chapters, to make a list of other primates and find out something about their particular distinguishing features and interrelations, and how they fit into groups: you could consider chimpanzees, langurs, gorillas, bonobos, baboons, vervets, lemurs, orang-utans, bush-babies and howler monkeys, for example. One place to start might be Jones, Martin and Pilbeam (1992: Ch. 1.3) (which also has some great pictures).

5.   There is currently a great deal of interest in constructing typological databases (see http://wals.info/, for example). When might it be appropriate to use typological data in historical linguistic investigation, and how reliable might such data be in drawing historical inferences? Have a look at Wichmann and Saunders (2007), and the debate in Dunn *et al.* (2005), Donohue and Musgrave (2007), and Dunn *et al.* (2007) for one particular case-study.

4 Who, where and when?

4.1 Overview

This chapter is about 'the most compelling detective story in science, the story of where we came from' (Gribbin and Cherfas 2001: 3). We begin by setting out the usual basis for biological classification into kingdoms, classes, orders, families, genera and species. This gives us a straightforward picture of where humans fit into the greater family tree of life; but a quick look at the problematic and controversial aspects of the tree model of language classification will suggest by analogy that such straightforwardness is truly deceptive. We return to construct a much more complex (and more realistic) picture of human ancestry, adding a number of less readily defined stages and sub-branches to our immediate family tree, and building up a picture of the debates and controversies that remain in the field.

4.2 Biological family trees

4.2.1 The big picture

Most readers of this book will feel entirely comfortable looking at pictures of family trees, because they will have encountered such structures before, either as genealogical pedigrees of their own or other immediate human families, or in linguistics, where they appear in family classifications, and also as organisational devices for complex structures (for example, as immediate constituent diagrams in syntax or syllabic phonology). The point of trees is that they provide a classificatory framework, which might variously tell us that elements at a lower level belong to the same unit at the next level up; or that those lower-level units share particular characteristics which identify them as belonging together; or both.

We can safely have exactly the same expectations of biological family trees. They show us whether the groups under the same node and at the same level share particular characteristics, or belong to the same group in the sense that they have the same ancestry, or both. Strictly speaking, that potential dichotomy sums up the two possible ways of drawing biological trees in the first place. The earliest generally accepted classification of the biological world was put forward by Carl von Linné (1707–78), more famous under his Latinised name of Carolus Linnaeus, who developed his tree of species within an essentially static and creationist paradigm (Linnaeus 1758). His taxonomy (which included stones and minerals as well as animals and plants) was completely phenetic; in other words, it was based entirely on whether the creatures or objects he was classifying shared particular superficial characteristics in their phenotypes (their physical appearance or behaviour). He was not operating on the assumption, which emerged only with Darwin a century later, that some of the nodes in his tree were descended from others, but rather assumed that all had been created with just the features they currently have. The biologist's job, in Linnaean terms, was to classify according to present features, with no reference to shared ancestry. Linnaean trees are hierarchical, but this reflects the similarities between species, not how recently those species shared a common ancestor.

The trees we shall be drawing here, on the other hand, are typically based on genetic rather than phenetic information, and the main emphasis is very firmly on which groups share more recent common ancestry. This post-Darwinian position holds that all biological species are ultimately related and have developed from a common initial form of life via millions of years of descent with differentiation through the forces of evolution, mutation and selection. Species may look alike, or share common patterns of behaviour, without being particularly closely related at a genetic level, since these superficial, phenetic characteristics might have arisen through common reactions to similar environments; on the other hand, they may look and behave very differently, but share enough tell-tale genetic signals for us to trace them back to a common ancestor in the (relatively) close evolutionary past.

With this difference in classificatory practice in mind, we can turn to the rudimentary tree of life shown in Figure 4.1. This is, if you like, the highest-level branching structure within the tree of life, showing the most widely accepted division into the three major kingdoms of Eukaryotes, Eubacteria and Archaebacteria. This deepest, tripartite division is based crucially on molecular genetic comparisons, which indicate a highly tentative date for the break-up of the three kingdoms of around 1,000 myBP. More specifically, the three-way split reflects differences in one small segment of RNA (small subunit ribosomal RNA), which is involved in turning DNA into proteins (Gribaldo and Philippe 2002).

At first glance, it is quite difficult to see where we and our human ancestors fit into this supertree, since we (along with not only chimpanzees, cats, pangolins and duck-billed platypuses, but also birds, insects, reptiles and amphibians) fall into the small and (from the branch-lengths) rather recent group of animals. It is important to recognise, however, that the tree has not had this overall shape for very long. As Dawkins (2004: 460) notes, 'Traditionally, and understandably, the tale was told from the point of view of big animals – us. Life was divided into the animal kingdom and the vegetable kingdom, and the difference seemed pretty clear. The fungi counted as plants because the more familiar of them are rooted to the spot and don't walk away while you try to study them.' Major reanalyses throughout the twentieth century introduced a kingdom of Fungi alongside the Animalia and Plantae, and

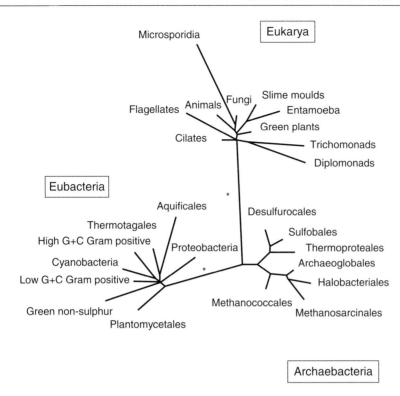

Figure 4.1 *Outline Tree of Life showing the three majors domains. This tree is shown unrooted, with two positions for the last universal cellar ancestor indicated by * (redrawn from Gribaldo and Philippe 2002)*

subsequently struggled with single-celled organisms larger than bacteria, such as 'Amoeba, once thought to be close to the grand ancestor of all life – how wrong we were, for an *Amoeba* is scarcely distinguishable from a human when viewed through the "eyes" of bacteria' (Dawkins 2004: 460–1).

We, along with the *Amoeba*, have most recently been relegated to supporting roles within the Eukaryotes by the relentless advance in identification of new species of bacteria. It is tempting to see the tree of life anthropomorphically (after all, humans drew the thing, so surely we should get star billing); but from a genetic point of view, the antiquity of life on earth, and the length of time such life was limited to single-celled organisms, mean that two bacteria living in the average municipal waste site (say, *Escherichia coli* from the gram negative Proteobacteria branch of the Eubacteria, and *Methanobacterium thermoautotrophicum*, one of the Archaebacteria) are more different from one another than humans are from slime moulds or parsnips, let alone mongooses or zebras. In fact, Marks (2002: 28) estimates that at the DNA level, humans and daffodils are about 35 per cent identical. Revising our viewpoint radically, we should probably work up a degree of gratitude that the most recent trees of life allow animals still to constitute a different branch from plants at all.

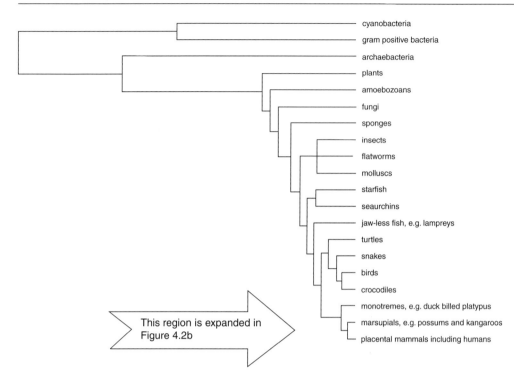

Figure 4.2a *Animals (rooted using bacteria as outgroups) (drawn from data in Dawkins 2004)*

4.2.2   The Animal family tree

Having been properly put in our place by this genetic picture ('Notice, with proper humility', says Dawkins (2004: 436), 'where you and I belong'), we can proceed to focus a little more closely on that node labelled 'Animals' in Figure 4.1. If we zoom in on that part of the Eukaryote subtree, we find the more differentiated and perhaps more familiar story of Figure 4.2.

Figure 4.2 develops the tree from the highest level of the kingdom down to the lower level of the order, with a selection of species, the lowest level of all, sketched in for each order in brackets. Using these trees together, we can establish that humans belong to the kingdom or even superkingdom of Eukaryotes; the kingdom of Animals (note the ambiguity of 'kingdom' here, generated as a by-product of the recent addition of the higher-level structure in Figure 4.1); the phylum of Chordata, or animals with backbones; the class of Mammalia, which are warm-blooded, have hair, and feed their young with milk; and the order of Primates, which share a large brain, flexible fingers and toes, and good vision. We can, however, make further distinctions even lower in the tree, as shown in Figure 4.3.

Of all these groupings, the species is the level with the most straightforward biological definition, since it denotes a group which can interbreed but cannot successfully breed with members of any other species. However, there are agreed

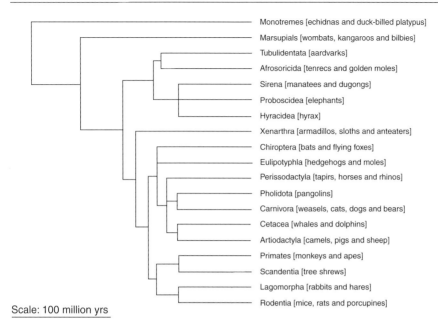

Scale: 100 million yrs

Figure 4.2b *Expanded subtree for mammals (redrawn from data in Springer et al. 2004 and references therein)*

rules for naming at all levels, controlled by the International Commission on Zoological Nomenclature (www.iczn.org/; accessed 13 May 2011), which has developed the ZooBank website, launched in 2008 (the 250th anniversary of Linnaeus's death) as a one-stop shop for those seeking to name and register new species (Polaszek *et al.* 2005). Certain aspects of the naming system are essentially formal, so that labels for classes are Latinised forms ending in *–a*, while species names still follow the binomial system developed by Linnaeus, with a genus and so-called 'trivial' name, both italicised, and with the former capitalised. Our own trivial name, *sapiens*, means 'wise'; as Luria, Gould and Singer (1981: 666) appositely suggest, 'No rule dictates that names be appropriate, only that they be distinct.' As North (2005: 20) notes, there are few restrictions currently operative on the content of the names actually chosen, so that 'there are a remarkable number of improbably named creatures in the canon of animal taxonomy'. Take, for instance, *Erechthias beeblebroxi*, a moth with a false head named after the two-headed Zaphod Beeblebrox in Douglas Adams' *Hitchhiker's Guide to the Galaxy*; *Pulchrapollia*, an extinct parrot whose name translates as 'pretty Polly'; or *Rostropria garbo*, a solitary female wasp named for the actress Greta Garbo, who famously wanted to be alone. Although the International Commission has a mission to sort out some confusions, and carefully vet proposed new names, it is comforting that North (2005: 21) cites its spokesman Andrew Polaszek as not seeking to stamp out silliness altogether.

Returning to the order of Primates in Figure 4.3, we find a two-way split into prosimians and simians or anthropoids. The prosimians, such as the lemurs and

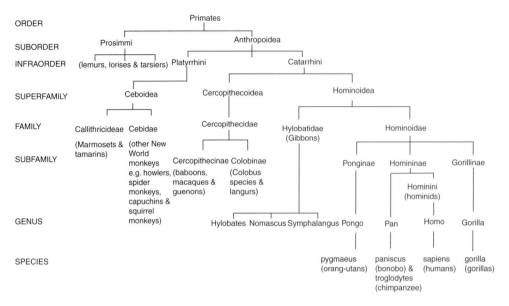

Figure 4.3 *A primate taxonomic classification (adapted from Figure 3 of Shoshani et al. 1996, with additional information from Stringer and Andrews 2005: 16)*

lorises, indris, aye-ayes and bush-babies, appear to be closest to the ancestral state for primates in general and are typically nocturnal, very reliant on their sense of smell, and with smaller brains than other primates of similar size. The simians, in turn, divide into two major groups, differentiated mainly by features of their teeth and ears, namely the New World platyrrhines or broad-nosed monkeys (including the spider monkeys, marmosets, tamarins, howler monkeys and squirrel monkeys), and the Old World monkeys and apes. Again, the Old World monkeys or catarrhines ('narrow noses') fall into two groups: the monkeys have tails and are quadripedal (these include macaques, baboons and vervets), while the apes are tail-less and have long arms and fingers. Within this group of apes, we find a branch for gibbons, then a further group of great apes, comprising chimpanzees (both common and bonobo), gorillas, orang-utans and humans. Humans, then, belong to the (sub)family Homininae, which have an even larger brain than the average for primates and walk upright; the genus *Homo*, with a brain size exceeding 800 cc, small teeth, and a larger body; and finally, the species *Homo sapiens*, with the same characteristics taken a degree further, so that the brain size exceeds 1000 cc, and the teeth and bone structure are smaller and finer again.

There is some debate here on the finer details of the classification at the genus level, since older classifications tend to oppose the Hominidae (humans) to the Pongidae (all other great apes). However, genetic data show that, despite phenotypic differences involving characteristics like bipedality and hairlessness (and language, come to that), humans are clearly most similar genetically to chimpanzees. Our tree in Figure 4.3 therefore shows humans and both sub-species of chimpanzees as clustered most closely together, indicating that these shared a common ancestor more recently than either did with gorillas. The most recently agreed, approximate date for the chimpanzee–human split, based on molecular genetic information, is around 5–7 myBP (Hedges and Kumar 2003); we return to this below.

4.3　Beware: falling trees!

So far, we have presented these biological family trees as essentially agreed and relatively uncontroversial; true, there have been some changes and expansions, but this is an entirely natural response to our developing knowledge of the natural world. However, a closer look at the trees reveals unclarity and division: we have already seen that some of the terminology we use is uncertain (since what counts as a kingdom, for instance, might fluctuate as we start to accommodate higher levels of classification; and then what happens to groups like the Animalia, which used to be recognised as a kingdom in earlier trees?); and we have encountered a difference of opinion over whether chimpanzees should be classified with gorillas as against humans, or humans with chimpanzees as against gorillas.

We can start to put these questions in perspective by returning briefly to linguistics, since here we encounter parallel difficulties with trees, and can make these problems more accessible to readers by introducing them first in association with linguistic issues. When historical linguists draw family trees to display genealogical affiliations among languages, they recognise that they are using an excellent visual tool; but they also recognise, for the most part, that trees have a range of deficiencies which make them less than literal representations of the truth. Trees, in short, have to be understood as heavily idealised approximations to reality, for a number of good and unavoidable reasons.

First of all, when we draw family trees for languages, the nodes at the apexes of branches are often not 'known' but hypothesised. The grammar we propose for, say, Proto-Indo-European is based on the structures we observe in its daughter languages, plus our knowledge of language change, which we use to extrapolate those attested features into the past. There are recognised methods (notably the comparative method) which allow us to undertake that process of extrapolation with a degree of confidence; but at the same time, we must recognise that there will inevitably be elements of the protolanguage grammar that we will miss, because relevant evidence for them is absent from the languages we know today. So, for instance, the oldest, written Indo-European daughters, like Sanskrit, Latin and Greek, retain evidence of certain cases in the noun system (like the locative and instrumental cases), allowing us to reconstruct those for Proto-Indo-European; if we had evidence from only the daughters spoken today, which preserve a smaller range of cases, the case system we could reconstruct would be less full.

It is understandable, then, that we may have a general tendency to trust evidence from the earlier attested languages over what we see in languages spoken today; but there are cases where that can lead us into error. For example, historical linguists are now confident that in the history of Sanskrit a series of changes in the vowel system caused several ancestral vowels to collapse into /a/; there is little or no trace of /o/ and /e/ in Sanskrit. In the nineteenth century, when Sanskrit had just been 'discovered' by western scholars and was hugely influential and popular, there was a period when evidence from Sanskrit was prioritised over data from other branches of Indo-European, skewing the reconstruction so that it was assumed /a/ was ancestral wherever it appeared in Sanskrit, and had split to /a e o/ in the other daughters (see Hock 1991). The balance has now shifted back so that */a e o/ are typically reconstructed for Proto-Indo-European, with a series of mergers in the vowel system specific to Sanskrit. The point is, though, that the priority we accord to certain kinds of evidence, from certain kinds of sources, can determine the structure we hypothesise for the nodes in family trees.

A further and more general problem lies in the structure of trees themselves. A tree, by its nature, can display only one cause of similarity between languages, namely the features they retain (albeit sometimes in a changed form) from their common ancestors. There are, however, at least two other major reasons for resemblances between languages, and these are homoplasy, or independent, common, parallel development, and contact. In the first case, two languages (regardless of whether they are related or not) may undergo the same sound change, simply because it is a relatively common and natural thing for sounds and human vocal tracts to do – take devoicing of sounds word-finally, for example, or the change of /p/ into a fricative like /f/ and subsequently into /h/ (Foulkes 1993), processes which can be observed the world over but which give no information about likely common ancestry precisely because they are so common. To propose that two languages belong to the same family, we have seen that we need a whole range of parallel patterns, ideally involving a certain degree of idiosyncrasy or quirkiness. Likewise, contact between speakers can have serious repercussions for languages, and this is certainly not just confined to the vocabulary, but works at the level of structure, too (see Thomason 2001). In both these cases, we face two difficulties. First, if we misinterpret a pattern that really results from contact or parallel innovation as signalling common ancestry, then we are at risk of drawing the wrong tree. And second, even if we rule out these signals of borrowing or homoplasy rigorously, isolate only those features resulting from common ancestry and subsequent descent with differentiation, and consequently draw the tree absolutely right, that tree by its nature is only telling part of the historical story of those languages. We have filtered the rest out in order to get a neat tree.

Put as starkly as that, we can see why this kind of observation makes historical linguists nervous. There have been a range of reactions, ranging from the equivalent of government health warnings (trees are idealisations; do not take this tree literally), to the more recent and arguably more productive shift towards adopting quantitative methods for generating trees and alternative visual representations.

If we are to have any faith in trees, we need to be clear that they are being drawn on the basis of objective and repeatable observations, not just to match a linguist's strong instinct that those languages belong together; McMahon and McMahon (2005a) argue that this can be achieved by adopting programs from population biology which generate all possible trees and select the best one. But even more to the point, as many of the contributors to McMahon (2005b) argue, we can adopt programs which draw trees only when the data provide a clear and tree-like signal, but networks where the signals are more complicated and potentially conflicting; and we can increasingly use this technology to distinguish signals reflecting common ancestry from those resulting from contact or parallel development. All of this depends, of course, on the availability of sufficient good data to allow us to discern signals of anything in the first place, let alone to then categorise those signals according to their likely causation. We cannot develop these ideas further here, but should not be surprised if we find a trend in biology, too, towards increasing the repertoire of visual and diagrammatic representations, with a shift away from the conventional tree in certain circumstances.

Finally, a difficulty inherent in trees is their branching structure. If we took this literally, it would suggest that languages split absolutely, immediately and irrevocably, or put in more social terms, that one day the speakers of a certain language split into two groups and never spoke to one another again. There are certain areas of the world where, under very particular circumstances, something approaching this scenario might just have happened – the Pacific, with its clear history of island-hopping expansions, is a good candidate. But even here, we cannot rule out the odd visit back whence a population came, and there is evidence of ongoing trade between even very distant islands (Bellwood 1987).

In fact, the idealised representation of absolute branching covers two problems for tree representations. In the first place, if branching is actually gradual and incremental, we are missing out a whole range of potentially interesting stages between the node at the top of the tree, the Proto-Language, and whatever appears at the next level down. If Proto-Germanic split into Proto-West-Germanic, Proto-North-Germanic and Proto-East-Germanic, and Proto-West-Germanic then split ultimately into Old English, Old Saxon, Old High German and so on, what about the pre-Old English or pre-Proto-West-Germanic periods, which might have lasted some considerable time and been full of interesting and significant changes, but which appear as undifferentiated straight lines on the tree? This problem is mirrored in our second difficulty further down the tree. Because typically language family trees include only languages in their higher reaches, they tend to branch at the lowest level only into languages, so that the ultimate twigs, if you like, will be English, German, Frisian, Dutch and the like. Yet as any sociohistorical linguist knows, the real linguistic action culminating in split and new language formation would happen at an even lower level, among the accents and dialects, so that there is a powerful motivation for including Scots, and High German versus Low German, and varieties of Dutch from the Netherlands as opposed to Belgium.

Family trees for languages, then, are idealised. This does not mean we cannot or should not use them, but does mean we have to interpret them carefully and accept that they cannot show the whole picture, which will have to be expressed in other ways. The critique above has left us with the following problems:

- we cannot always be sure about the characteristics of the apex nodes we draw for protolanguages. Sometimes we will have insufficient evidence to allow us to do more than sketch an outline of the pro-tolinguistic system; and inequalities of preservation may lead us to overprioritise certain types of evidence, which may colour that picture further.
- contact and parallel innovation cannot be included in trees. If we include them by mistake, we may draw the wrong trees. If we wish to include them in their own right, we will have to draw something other than trees.
- trees often include only major nodes, which at higher levels means we may miss out on important historical stages and intermediate representations; lower down, it may exclude accents and dialects, where much of the variation and change important for descent with differentiation is actually happening.

In the next section, we return to human evolution from proto-primates through Miocene apes to modern *Homo sapiens*. Although initially we shall be setting the scene using the familiar, simple tree architecture of stepwise splits and absolute divisions, we should keep in the back of our minds the three problems raised above for language family trees. In section 4.4.2, we shall revisit these limitations from a biological perspective, showing that all are equally relevant here; accepting and exploring them brings new dimensions of detail and complexity to the sometimes controversial question of how and when we became human.

4.4   Hominid histories

4.4.1   Proto-apes to humans: one simple story

The order of modern primates has its origins towards the end of the Cretaceous period (Clemens 1974), at around the time (65 myBP or so) when dinosaurs ceased to roam the earth. All current primates share a range of characteristics, which by the biological comparative method we can assume to be ancestral, as they are highly unlikely all to be parallel, independent developments: these include the presence of a grasping foot with a big toe that can be moved independently; relatively forward-facing eyes with stereoscopic vision; large brains relative to body size; relatively long gestation with slow foetal growth; and a distinctive tooth pattern lacking one incisor and one premolar when compared to the mammalian norm (Stringer and Andrews 2005). Many of these traits can be seen as adaptations to an arboreal environment, allowing us to hypothesise that the common primate ancestor lived in the trees.

There is very little fossil evidence from the period before 50 myBP, but much more from the Eocene (Martin 1993), which extends from approximately 55 myBP to 35 myBP; this period begins with a general warming and consequent expansion of the equatorial forests. Even so, it is worth noting at this point that 'fossil' does not mean 'complete skeleton in approximately lifelike posture and configuration', but rather 'isolated bone or two, often a jaw'. Jaws fossilise so well and so completely because teeth are particularly resilient; hence also the classificatory importance of the distinctive primate tooth pattern discussed above. Eocene primate fossils typically strongly resemble modern primates, and are found in the Americas, Eurasia and Africa. They form two primary groups, one (the Adapidae, also known as Lemuroids) which may include the ancestors of lemurs and lorises, and the second (the Omomyidae or Tarsoids) much more likely to be related to present-day anthropoids and Tarsiers (Fleagle 1999). Global cooling at the end of the Eocene seems likely to have virtually wiped out the primates (among other groups), and fossil evidence again becomes scanty, except for the anthropoid apes in Africa, which appear to have thrived and expanded as the climate cooled. By 20 myBP, the fossil evidence shows considerable variability in these anthropoid apes; it seems likely that the monkeys had split from the apes by this time, though it is difficult to prove this conclusively from the fossil record, since the main distinguishing feature is the presence of tails in monkeys, and tails include a number of small bones which are quite likely to be lost prior to fossilisation. Ape fossils from the Miocene have been assigned to a range of species, some of which are shown at the extreme left of Figure 4.4. One early genus, consisting of several putative species, is Proconsul, the earliest proposed ancestor of the Hominoidea (Walker and Teaford 1988), which, as shown in Figure 4.3 above, includes the modern gibbons, orang-utans, gorillas and chimpanzees, as well as the hominin group of upright-walking apes to which modern humans belong.

Readers might expect the next step to be a report of increasing diversification among these early apes, with an unbroken line of fossils documenting the transitions resulting in the modern species. Nothing, unfortunately, could be further from the truth. Whereas working on the basis of the odd jaw and a few scattered teeth might have seemed a hardship for palaeontologists investigating Proconsul, such fossils would be a positive embarrassment of riches to those seeking to fill in the African fossil gap which extends from approximately 14–7 myBP. In this period, there are fossils of a small number of primate species from elsewhere, including Dryopithecus and Sivapithecus from Northern India and Ankarapithecus from Turkey; these are the first examples of ape-like fossils to be found outside Africa, reflecting the continental drift which brought Africa into contact with Eurasia in the late Miocene. However, there is no fossil ape evidence at all from Africa until *Sahelanthropus tchadensis* from around 7–6 myBP; and this species consists only of a single well-preserved skull (TM 266–01–060–1)

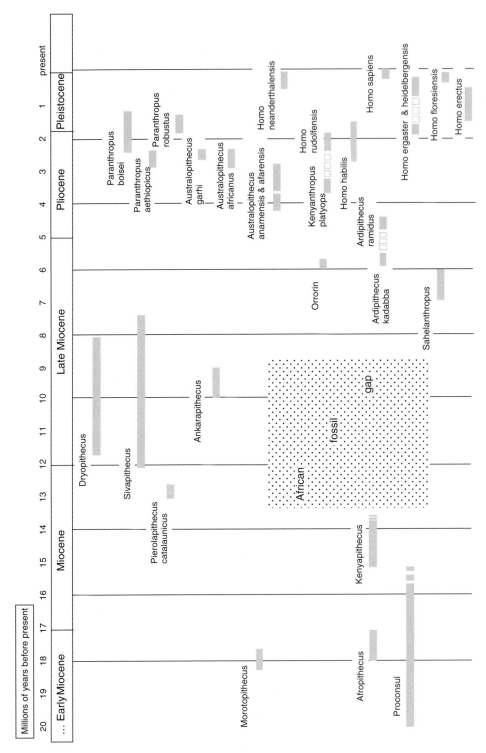

Figure 4.4 *Evolutionary time line and aspects of the ape fossil record. Grey boxes indicate known fossils and their associated ages; for discussion see text.*

and a few teeth reported by Michel Brunet in 2002 (Brunet *et al.* 2002, Zollikofer *et al.* 2005).

The start of the Pliocene (which extends from 5–2 myBP) sees a much wider range of better-evidenced early ape species in Africa and elsewhere. From here on, the human line is the best researched and most plentifully attested among the Hominoidea (though of course these two factors may not be totally unrelated). In contrast, the chimpanzee line is represented in the fossil record by precisely two teeth from Kenya, dating from around 500 kyBP, which were reported only in 2005 (McBrearty and Jablonski 2005). Genetic evidence dates the chimp–human split to approximately 5–7 myBP, and the first significant group of hominins, ancestral to humans but not to chimpanzees, are the Australopithecines, attested from around 4 myBP. It is possible that the earlier Ardipithecus (with specimens bridging the Late Miocene and Early Pliocene, 6–4.5 myBP) (White *et al.* 1994, Haile-Selassie 2001) may be ancestral to Australopithecus, but there is too little evidence at present to be sure. Ardipithecus has been assigned tentatively to the human line by virtue of its relatively small teeth, but it is not at all clear whether it walked upright (we can say rather little about the possibility of bipedality on the basis of molar size).

If Ardipithecus may or may not have been bipedal, the Australopithecines certainly were, though not fully so – they are intermediate, and could have stood and walked upright but may have had some difficulties in balancing, and spent some time moving in trees (Ward 2002). We are fortunate in that the first example of *Australopithecus afarensis*, for instance, consisted of a knee joint, the architecture of which must be very different for animals which walk upright or on all fours; and likewise, a set of footprints of three *afarensis* individuals, two adults and a juvenile, in volcanic ash from Laetoli in Tanzania, can only indicate bipedality (Leakey and Hay 1979). Lucy, one of the most complete fossil apes (a whole 40%), discovered by Don Johanson in Ethiopia (Johanson and Edey 1981) and dating from around 3 myBP, is also an example of *Australopithecus afarensis*. Although she was clearly bipedal, the form of her hip joints and proportions of her legs show that her stride length would have been restricted relative to that of modern humans. The first evidence of *anamensis*, in turn, is from approximately 4.2–4.1 myBP, and consists of an elbow and lower leg (Leakey *et al.* 1995); both of these are vital in demonstrating bipedality at this relatively early period, the elbow because it lacks the specialised ridges associated with knuckle-walking and found still in non-human apes.

Australopithecines seem subsequently to have diversified into two groups, essentially depending on how the different species adapted to changing climatic conditions in the Pliocene, as cold, dry periods led to the retreat of the African forests. Paranthropus (a group of species previously known as robust Australopithecus) adapted to the dry climate by growing larger and developing much more substantial jaws and teeth, allowing them to exploit a challenging diet of seeds and roots rather than the fruits which had been available in the disappearing forests. Australopithecus, on the other hand, remained more gracile or slender, and it is typically argued that they dealt with climate change more by behavioural than physical change, resulting in more extensive tool use. Either in *Australopithecus anamensis* or *A. afarensis*, or in their likely descendant species *Homo habilis*, meat-eating also developed.

Homo habilis, attested from around 2 myBP, marks the definite beginning of the hominins, the genus Homo – though there are no absolute and unarguable markers for membership of this group, which is characterised by even bigger brains and even smaller teeth (Leakey, Tobias and Napier 1964), clearly characteristics on a continuum. In fact, several *Homo habilis* fossils have been interpreted as variant Australopithecines (see discussion in Wood and Collard 1999). Over more than half a million years, we witness in the fossil record a gradual transition from *Homo habilis* to *Homo erectus*, the first hominin species to be attested far beyond Africa, with finds also from China, Java and the Caucasus. It is sometimes argued (see Stringer and Andrews 2005: 139) that the fossils formerly known as *Homo erectus* should in fact be split into three groups. The original label is retained only for examples found in Asia. Of the two further subsets, *Homo ergaster* describes the early examples found only in Africa, and *Homo heidelbergensis* is used for a group intermediate between *Homo erectus* and *Homo sapiens* – once called archaic *Homo sapiens*. The assumption underlying this relabelling is the out-of-Africa view that *Homo heidelbergensis*, the variant found both in Africa and elsewhere at around 500 kyBP, developed in Africa but then spread beyond. Although *heidelbergensis* shares a common ancestor (in the shape of *ergaster*) with the *erectus* populations found in Asia, *Homo sapiens* descends uniquely from one *Homo heidelbergensis* population. Retaining a single label, such as *Homo erectus*, for all these groups, would therefore lead us to anticipate a closer relationship between modern humans and the *Homo erectus* populations found in Asia in particular than is in fact the case. Elsewhere in the world at around the same time, different species of Homo were also evolving in tune with prevailing local conditions; these include *Homo neanderthalensis* in Europe (Stringer and Gamble 1993), and *Homo floresiensis* in Indonesia (Brown *et al.* 2004, Morwood *et al.* 2004), though both have now died out.

4.4.2   Introducing complexities

Given the criticisms of language family trees raised in section 4.3 above, we should perhaps not be overly surprised to find that exactly the same limitations of evidence, problems of interpretation and exclusion of non-tree-like processes are relevant for biological family trees. Some assertions in the previous section have already been modified, and some clues have been provided that the picture of human evolution is not such a linear and uncontroversial one as a tree representation might suggest. In this section, then, we introduce a range of more complex and tentative, though very likely more realistic, modifications to the outline above, using the limitations on language family trees as a framework.

In the previous section, it was asserted that the human–chimp split, and therefore the last immediate common ancestor for modern humans and modern chimpanzees, can be dated to around 5–7 myBP. However, we also observed that all the fossil evidence we have for chimpanzees, namely two teeth, could fit in a matchbox, and that these teeth date from much more recently, around 500 kyBP. The existence of the general African fossil gap, which must overlap with the beginnings of human–chimp diversification, also means we have no palaeontological evidence for the early stages of this process. It follows that the 5–7 myBP dating cannot rely on fossil evidence; on the contrary, it rests completely on molecular genetics; but the introduction of that new type of evidence did not entirely support earlier assumptions.

Up until the early 1970s, the physical differences between modern humans and chimpanzees were taken to indicate a long historical divergence (Lewin 1997: Chs. 4–6). This argument was largely based on the discovery in the 1930s of a range of Himalayan fossils from around 12 myBP; these had flatter faces and smaller teeth than contemporaneous species like *Sivapithecus*, and were therefore taken to be examples of a separate species, *Ramapithecus*, which was claimed to be the first distinctively human ancestor. However, in the 1960s, advances in molecular biology led to an understanding that proteins (present in, for example, the blood) differ between individuals. Such differences involve alternative patterns in the genetic conformation of the same protein; these are known as polymorphisms, and many have no effect on the function of the proteins whatsoever, so that their propagation through subsequent generations has no positive or negative consequence for the viability of the individuals concerned. Allan Wilson and Vincent Sarich then established that, when extended to the species level, the degree of difference between proteins could be mapped directly onto the time since these species shared a common ancestor (Sarich and Wilson 1967). This method, applied to modern primates, provided a date of 4–5 myBP (Wilson and Sarich 1969) for the human–chimpanzee split; and this has been both confirmed, and slightly extended in terms of its error range, by subsequent detailed work directly on DNA affinities and mismatches between species. In consequence, the putative *Ramapithecus* specimens were rejected as human ancestors; subsequent, additional fossil evidence led to their reclassification as a subtype of *Sivapithecus*, and to the recognition that these are most closely related to modern orang-utans (Pilbeam 1982, Andrews and Cronin 1982). It was work of this kind, and the controversies it initially generated, that motivated Sarich's famous (and much paraphrased) quote to the effect that 'I know my molecules had ancestors; the palaeontologist can only hope his fossils had descendants.'

That question of whether particular fossils had descendants arises at various points in the family tree. There is a first, general question of whether any of these fossils is the direct ancestor of any modern species at all (and this again has its linguistic analogue: we say blithely that modern English is descended from Old English; but which Old English? Can we be as confident that any variety of modern English is descended from, say, ninth-century West Saxon?).

As Stringer and Andrews (2005: 82–3) note, there are currently 194 recognised species of primates (going up all the time as new species are identified in parts of the world less extensively investigated up to now); but during the Eocene, even in the period with most fossil finds, there is evidence of fewer than half this number. Directly related to this issue is at least one aspect of the 'out-of-Africa' controversy.

In the previous section, we mentioned various hypotheses surrounding *Homo erectus*. In the earliest debates on human origins, this was seen as a single species which gave rise directly to modern populations of *Homo sapiens* in different parts of the world – so, Asian populations of *Homo sapiens* continue from Asian *Homo erectus*, and likewise for Africa. The assumption here is that *Homo erectus* originated in Africa, then expanded outwards, and differentiated locally *in situ*. This idea is now known as the multi-regional hypothesis, and is associated particularly with the work of Milford Wolpoff (Wolpoff *et al.* 1984, Wolpoff and Caspari 1997). The out-of-Africa hypothesis, in contrast, suggests that all modern human populations derive from a single, late *erectus* population, assumed to be African. On this view (championed by Chris Stringer and a host of molecular biologists; see for example Stringer and McKie 1996), *Homo erectus* is used specifically for fossils found in Asia, since this is where the first examples were discovered; but these are seen as later, derived variants of an earlier species now relabelled as *Homo ergaster*. *Ergaster*, then, has a two-part history. It spread quickly into Asia (becoming *Homo erectus*); and in Africa, this early species then developed into *Homo heidelbergensis*, and subsequently spread into other regions of the world. Here it developed into locally distinct populations adapted to local conditions; and these in turn are labelled *Homo sapiens*, *Homo neanderthalensis*, and *Homo floresiensis* – and we might hypothesise others, though they are not currently attested in the fossil record.

The critical difference between these hypotheses is not whether the ultimate ancestor of modern human populations evolved in Africa; this, despite the names of the competing theories, is accepted by all. What differs is the age proposed for that most recent common ancestor. Under the multi-regional hypothesis, the critical expansion from Africa took place much earlier, around 2 myBP, and there is scope for a good deal of differential evolution *in situ*; but in the out-of-Africa model, all modern populations of *Homo sapiens* outside Africa share a common ancestor in Africa less than 100,000 years ago. This has obvious consequences for the evolution of language. On the multi-regional hypothesis, either language must have evolved separately and independently in different areas of the world, or the genetic and anatomical prerequisites for language at least must have been in place 2 million years ago. The out-of-Africa alternative allows for language having developed much, much later: if the radiation of modern *Homo sapiens* (leaving aside *Homo neanderthalensis* and *Homo floresiensis*, for instance) happened 200–100 kyBP or later, then language or its underpinnings need have been in place only then to allow for its invariant occurrence in all modern populations. We return to this issue in later chapters.

A recurring issue in this section, evidenced for *Sivapithecus* versus *Ramapithecus*, and for the shifting reference of *Homo erectus*, is just when we recognise an independent species, and when we see variants as qualifying for co-existence within a single species. Of course, we do not need to look far for a linguistic analogue here either, since linguists are constantly and somewhat embarrassingly confronted with their inability to assess when we have two languages as opposed to two dialects of the same language. We have already seen that gaps in the fossil record mean some intermediate stages, and almost certainly a number of entire species, are unrepresented; but even when we do have some palaeontological evidence, the vagaries of fossilisation mean a species is quite likely to be represented by (a few parts of) a single individual. Just as linguists agonise over whether they can take the output of a single remaining speaker of a dying language as characteristic of the earlier full language system as a whole, or whether some aspects might be purely idiolectal, so palaeontologists struggle with the representativeness of isolated fossil remains. And just as historical linguists interested in classification can be divided into lumpers, who tend to pull all the data together into a single family based on sometimes tenuous similarities, and splitters, who work more incrementally and recognise additional groupings in the family tree by prioritising differences, so we find lumping and splitting tendencies in the recognition of biological species.

One recent example involves *Homo floresiensis*, the so-called 'hobbit' discovered in Indonesia in 2004 (Brown *et al.* 2004). Initially, part of one tiny skull and a few parts of the post cranial anatomy from one individual were found in a cave, at a level dating from around 18 kyBP. The date is surprising, because it is concurrent with finds of *Homo sapiens* in the same area; yet the physical features of the skull (and in particular the absence of a chin) are arguably more consistent with *Homo erectus*. Arguments (summarised from both sides in Morwood *et al.* 2005, and see also Argue *et al.* 2006, 2009) rage over whether this is really a new species of *Homo*, or whether the individual concerned might be a member of *Homo sapiens* suffering from the genetic disease microcephaly (which can result in abnormalities of the brain and small stature). Further finds more recently reported from the same cave suggest that several individuals over a relatively lengthy time period, from before 74 kyBP to 12 kyBP, had the same physical attributes. Either we are witnessing the result of a particularly high incidence of microcephaly, or these are not abnormal physical features at all; they are the perfectly normal defining characteristics of a new species, *Homo floresiensis*, an adaptation of *Homo erectus* (possibly via *Homo heidelbergensis*) to island life.

The problems raised by this case interact with the general difficulties of drawing species-level consequences from individual data: if we do not have data from enough individuals, we cannot determine the acceptable range of variation within a particular species. This opens the door to arguments over species boundaries and species numbers. We can determine which individuals belong to the same species today by checking whether they can interbreed; but this does not help at all when we are dealing with fossils. Furthermore, we can observe considerable variation within an individual present-day species. Sometimes this reflects geographical breadth of distribution, with adaptations to local conditions (such as body shape in the Inuit, for instance, as opposed to the Samburu of northern Kenya, who traditionally live in cold versus hot climates respectively; see Jones, Martin and Pilbeam 1992: 285). In other cases the variation depends on sex, with males and females having typical but different physical patterns, as in the sagittal crest on the heads of male gorillas, which is absent in the (generally much smaller) females. Consequently, if we find fossil remains of a male and a female, where these finds are not closely associated in space and time, we might analyse these as belonging to two distinct species, or as sexually dimorphic individuals from one species. The unclarities in such situations are reflected in the use of the term *chronospecies*, meaning a possible species defined on the basis of fossils from different time periods alone.

If we cannot establish where species boundaries lay in the prehistoric past, and consequently cannot be sure whether interbreeding between populations took place, we face the possibility that genetic interchange has happened between groups that appear as totally separate in the family tree, a difficulty analogous to the factoring out of contact and borrowing from linguistic trees. In Figure 4.4, we listed *Homo sapiens*, *Homo floresiensis* and *Homo neanderthalensis* as separate species; but we have already seen that the dates from *Homo floresiensis* overlap with those for *Homo sapiens* in the same area. If these were indeed different species, this is irrelevant, since they could not have interbred; but if in fact they were regional variants of the same species, there could have been genetic interchange. In the absence of genetic evidence from both groups, and ideally from other control groups outside the area, their ancestors and their descendants, we cannot be sure.

Just this kind of genetic approach has been taken to the possibility of interbreeding between *Homo sapiens* and *Homo neanderthalensis*. The latter are evidenced in Europe in relatively large numbers, indicating a lengthy period of occupation; the earliest fossils identified as Neanderthal date from around 200 kyBP, though the later, more extremely cold-adapted variants date from 120–30 kyBP. Neanderthals therefore clearly overlapped with *Homo sapiens* in geographical range and time. Neanderthals have hopped in and out of the human family tree: Stringer and Gamble (1993: Plates 2 and 3) show two reconstructions made from the same skeleton, one depicting a hairy gorilla-like ape, and the other an only slightly over-hirsute, thickset person who might not draw too much adverse attention in a suit on the No. 27 bus. It is likely that we can exclude significant levels of interbreeding between Neanderthals and *Homo sapiens*, on the basis of genetic studies: DNA may be preserved in tiny quantities in relatively recent Neanderthal skeletons and particularly the roots of teeth. From these samples, mitochondrial DNA can be retrieved and analysed. The mitochondria are the power sources of the cell, and have their own DNA, inherited only through the maternal line; mitochondrial DNA mutates extremely fast, and appears to provide reliable dates for species splits. Stringer and Andrews (2005: 181) note that mitochondrial DNA from five Neanderthals shows the range of variation within the group does not overlap with variants found in over 1,000 modern humans (see also Krings *et al.* 1999, 2000, Ovchinnikov *et al.* 2000, Höss 2000, and now Green *et al.* 2010 for a draft sequence of the entire Neanderthal genome).

It appears that while *Homo sapiens* and *Homo neanderthalensis* do share a common ancestor some 600 kyBP, this is much earlier than the most recent common ancestor for all modern human populations, around 200 kyBP. This assumes, of course, that the molecular tree constructed on the basis of these individuals matches the total population tree for Neanderthals versus modern *sapiens* of the time, and we shall revisit these issues in Chapter 6. But where one question might be answered, a host more remain: why, for instance, did the apparently numerous and successful Neanderthals dwindle away by 30 kyBP? Neanderthals were clearly cold-adapted, and this period does see climatic change and gradual warming in Europe; but is this sufficient to account for the increased dominance of *Homo sapiens*? And turning this question on its head, why did the non-cold-adapted *Homo sapiens* survive and prosper in the preceding Ice Age? What did they have going for them that the Neanderthals lacked? It is here that we start to add questions of cultural development and adaptations that may not have been purely physical to our lines of enquiry; and we shall be exploring these further towards the end of the book.

First, however, we turn in Chapters 5 and 6 to particular trends in the fossil record. From the highly complex and sometimes unclear picture painted above, with all its attendant confusion over boundaries and unclarities of nomenclature, we can nonetheless discern a number of clear trends from the earliest hominins to modern humans. In particular, the brain becomes progressively larger, and we see increasingly clear signs of habitual bipedality, tool use and cultural diversity. In the next two chapters we will begin with modern humans and with physical structures and genetic underpinnings which seem clearly associated with language, and aim to trace these back through the human family tree.

4.5   Summary

This chapter has had two interacting aims. First, we have sought to provide an outline of the human family tree, plotting our species' place in the tree of life generally, and in the tree of primates more specifically. We have noted the evolutionary changes, evidenced in the fossil record, from the earliest proto-apes through *Australopithecus* and early *Homo* to *Homo sapiens*. However, we have also used limitations in the construction, architecture and interpretation of linguistic family trees, with which many of our readers will be more familiar, to highlight parallel problems for biological family trees. In particular, we have noted the limitations of fossil evidence; the permeability of species boundaries; and the fact that speciation and nomenclature are typically determined by argument rather than proof. Modern humans belong to a very young species, with a very old pedigree. But our relative youth does not mean that all the questions about our origins, development and characteristics at each period have been, or at this stage can be, answered. Recognising the complexities in the family tree is important, as we shall return periodically to some of these controversies as we focus increasingly closely on possible adaptations for language.

Further reading

You can read more about language family trees and their limitations in McMahon and McMahon (2005a), and in Fox (1995); and because trees are closely linked with the comparative method in historical linguistics, you may also find the reading recommended in Chapter 3 relevant here.

A good deal of research into evolution goes on in the major museums, and a number of these maintain excellent websites with good graphics which should help you understand the major steps in human evolution and the controversies that remain. A particularly good example is at the Smithsonian Institution (www.mnh.si.edu/anthro/humanorigins; accessed 14 May 2011), which includes a terrific tree, the hall of human ancestors with some beautifully photographed skulls, and a chance to ask a palaeontologist whatever your heart desires. You might also want to look at the Natural History Museum (www.nhm.ac.uk/nature-online/life/human-origins; accessed 14 May 2011), which has links to evolution and palaeontology, and a range of online resources including videos and virtual reality presentations on Lucy and Neanderthals. The only problem with these sites is that they do not always include up-to-date lists of paper references to follow up the points raised, so although they will provide an excellent introduction, it is highly likely that you will also want to consult some of the following books and articles.

There are a number of books on human evolution aimed primarily at general readers or students, which deal with the controversies raised above: these include Lewin (1997, 2005), Stringer and Andrews (2005), which is an excellent resource with superb pictures and diagrams, and Lewin and Foley (2004). Oppenheimer (2003) provides a readable and sometimes polemical interpretation of the most recent stages of human evolution; this is sometimes at odds with accepted wisdom, but will make you think about the issues. Two books which seek to combine research overviews with more imaginative retellings of the human story are Sykes (2001) and Dunbar (2004). Sykes is particularly strong on the development and use of molecular evidence, and gives an interesting behind-the-scenes account of the human story of the researchers who developed molecular methods. But bear in mind that the second half of the book is fiction, and quite how close the fiction comes to any (pre-)historical fact is open to debate. Dunbar is more concerned with cultural aspects of human evolution, but the same caveat applies.

The wonderfully titled *Another Unique Species* (Foley 1987) is a highly readable account from the comparative perspective of a biological anthropologist.

Foley (2001) is a helpful overview article on potentially confusing issues of nomenclature, including an attempt to rationalise the entire question of the human family tree; this is part of a useful thematic issue of *Evolutionary Anthropology*. Dawkins (2004) is a compendious overview of the current state of knowledge on the history of life; that makes it sound like an impenetrable tome, but in fact it is highly engaging and full of quirky examples and marginal fulminations. Finally, a detailed, specific and technical treatment can be found in Relethford (2001: Ch. 3).

Points for discussion

1. The 'out-of-Africa' and 'multi-regional' hypotheses are often presented as diametrically opposed; yet both argue for an African homeland for the common ancestor of all modern human populations. On the basis of your own reading (and you might like to contrast proponents of each view, say Stringer versus Wolpoff), give a concise definition of each hypothesis, and say what might constitute convincing evidence for or against each.

2. As we have seen, Neanderthals are closely related to modern *Homo sapiens*, and there are various enigmas about their characteristics and demise. Find out when and where the first Neanderthal fossil was found; where the name for the group comes from; and what characteristics Neanderthals are generally thought to have had. You could begin with Stringer and Gamble (1993) or Shreeve (1995), and follow up some of the more recent genetic references included in the text above.

3. Find out about one important fossil in the human family tree. You might like to consider Lucy, or the 'hobbit' *Homo floresiensis*, or Turkana boy/ Nariokotome boy (note again the alternative nomenclature for the last specimen).

4. Fossils can rarely be dated directly and usually dates are estimated from the surrounding geological materials. Find out about methods used by archaeologists and palaeontologists for dating fossils. These methods might include carbon 14 dating; tree-rings; potassium-argon; indicator fossil species; and thermoluminescence. You could consult Renfrew and Bahn (2004), and Chapter 7 of Lewin (2005).

5 The vocal tract

5.1 Overview

This chapter begins to engage with a distinction that will be increasingly important in subsequent chapters, between language and speech. Language is something that happens at least in part in our brains, and might in some respects be controlled by our genes; we return to this in the next chapter. However, the primary modality of most language users is speech; it is the main way we externalise our knowledge of language and make use of it in the world. In this chapter, we outline the anatomy and physiology of human speech production, considering evidence from the biological comparative method to assess which aspects might be specific to humans and when these might have developed in our evolutionary history. We also, however, raise the question of modalities other than speech, and consider the possibility that the origin of language may lie in gesture as well as sound.

5.2 Producing speech sounds

Students learning acoustic phonetics used to be told to imagine the human vocal tract as a tube of uniform cross-section, approximately 15 centimetres long. However, this highly idealised description factors out most of what really matters for speech production. If our vocal tracts were really just single tubes, we would not only look rather different, but would not be able to produce some of the most important sounds of human languages.

To produce sound, we need three things: a moving body of air; vibration of some kind to create sound waves; and ways of modifying those sound waves in order to increase the variety of available sounds. None of these can work alone. Breathing produces moving air, but quiet exhalation is just that: quiet. On the other hand, vibrations cannot travel without some medium to travel through, which is where the air comes in. And finally, although vibration plus air can certainly make a sound, that sound will be singular and invariant, like the note produced by a tuning fork. A single sound would not offer a great deal of potential in terms of communication; and just think how irritated we get when we are exposed to one sound over and over again.

For human speech, these three characteristics are all part of what is known as the source-filter model of speech production. Most speech sounds are pulmonic (though pulmonic speech may be interspersed in some languages by glottalic or velaric sounds, produced using intrinsically limited airflow initiated elsewhere in the vocal tract); and here the initiator is the respiratory system, notably the lungs. It is in this sense that speaking is often described as *modified* breathing: whereas in normal respiration, breathing in and breathing out each take up about half the cycle, with expiration being an automatic consequence of relaxation at the end of the breath in, speech brings the breath out under muscular control so that it is more gradual and takes much longer. If quiet expiration takes around 2 seconds, breathing out for speech can take up to about 40 seconds (Lieberman 1975: 57). This gradual deflation of the lungs is managed by the action of the diaphragm and the intercostal muscles between the ribs, which work together to create steady pressure, and therefore ensure a consistent and controlled supply of outgoing air for speech.

The next step is to set this air in motion and therefore to change quiet breathing into sound. The main source for this transformation lies in the larynx (or voicebox), a cartilaginous box located at the top of the trachea, the tube through which air passes on its way to and from the lungs. Above the larynx, and with muscular attachments to it, is the hyoid bone, which is 'floating' in that it does not articulate directly with any other bone. Through the hyoid bone, the larynx is connected to the jaw and skull, and can be raised and lowered. Inside the larynx are the vocal folds, or cords, elastic bands of tissue which are abducted or opened back against the larynx walls during quiet breathing, but adducted, closing off the glottis (the space between them) during coughing and swallowing. While closed, the vocal folds also impede the passage of air outwards from the lungs. The pressure of outflowing air, however, is considerable, and will act to push the vocal folds apart; their inherent elasticity makes them snap back together, supported by the Bernouilli effect, a force pulling the cords together in response to negative pressure generated by the passage of air through the gap; and this cycle is repeated, setting up regular vibration which creates voicing in vowels and voiced consonants like English [m b n d l r j w].

These cycles of opening and closing of the vocal cords are the main source of vibration in the column of air issuing from the lungs, and they determine the fundamental frequency of the resulting sounds, which we hear as pitch: the characteristic pitch of any individual's voice is therefore created by the length and size of the vocal folds, though speakers can adjust this to a limited extent. For some sounds, audible sound energy is also created in the supralaryngeal vocal tract by the approximation of the vocal organs, such that air flowing through the oral cavity is forced through a narrow gap, creating turbulence. Such local audible friction is characteristic of both voiced fricatives, like English [v z], where turbulence co-exists with vocal fold vibration, and their voiceless equivalents like [f s]. For the most part, however, the action of the vocal organs above the larynx is that of a secondary filter: here we find not the creation of sound energy through vibration, but the modulation of that energy by reconfigurations of the space inside the vocal tract. In vowels, this contribution is perhaps most notable, since the vocal organs are never particularly closely approximated, so that the highly salient perceptual differences between [i] and [u], for instance, are created solely by movements of the tongue and lips which create resonating chambers of different dimensions. The vocal tract will have a different configuration for each position of the vocal organs; and each configuration will have its own formants, or 'gaps', which effectively allow sound energy at certain frequencies to pass through, while absorbing other frequencies. Overlaying these different formants on the consistent signal from the larynx will reveal different peaks and troughs of energy corresponding to different output spectra for each sound.

We have, then, three contributions to sound production. Air is supplied by gradual, muscularly controlled (though not complete) emptying of the lungs following inspiration. This air is set in motion primarily by vibration of the vocal folds in the larynx, though there may be further contributions here in the supralaryngeal vocal tract caused by close approximation of the articulators. For the most part, however, these articulators act to change the effective dimensions of the vocal tract itself, functioning as a series of filters which shape the output sound according to the frequencies they preferentially allow through as opposed to those they block. Together, these interacting systems and structures work to produce the range of sounds characteristic of human languages.

5.3   Uniquely human?

5.3.1   Present-day comparisons

If we take the biological comparative method as our guide, the next steps must be to determine which aspects of vocalisation we share with other species and may have shared with our hominid ancestors, and which, if any, are unique to modern humans. Certainly, we are not alone in using sound as part of our systems of communication: all primates have calls as part of their communicative repertoire, as we have seen in previous chapters for vervet monkeys, for example; and mammals in general can be noisy beasts, as witness various larger and smaller members of the cat and dog families. Outside our immediate set of relatives, there is clear evidence that birds communicate through a relatively highly structured system of song, while cetaceans like dolphins are also under active investigation in terms of their own partly vocal signalling system (see Hauser 1996, Fitch 2010: Ch. 4, and Herman 2010).

There are, however, two distinct issues here. One is what different species do with sounds when they produce them: do those sounds form a potentially open, expandable and flexible set of signs, or a closed class? In other words, is the system expressed by the sounds characterised by design features like displacement, creativity and structure-dependence, or is it intrinsically more limited? These questions have more to do with the characteristics of language at a more abstract level, though this may be instantiated in speech, and we shall not be pursuing them in this chapter. However, the second issue involves the sounds themselves: what range can be produced, and what properties of the vocal tract are required in order for this to be possible? This is our main concern for the present.

The main candidate for something special in human sound production is precisely what makes our vocal tract unlike a uniform single tube, namely the position of the larynx and the functional division of the vocal tract into two sub-tracts with a right-angle bend between. From the lungs, air passes up the trachea, through the glottis, continues up the pharynx, then essentially turns a corner into the forward-facing, much more horizontal oral (and/or nasal) tract. In this respect, modern adult humans are unlike any other species of primate, and indeed also unlike human babies, who have the mammalian norm of a tongue lying relatively flat in the oral cavity, which is relatively longer, and a higher larynx. 'Indeed, in most mammals, the larynx is located high enough in the throat to be engaged into the nasal passages, enabling simultaneous breathing and swallow-ing of fluids' (Fitch 2002: 31, and see also Crompton *et al.* 1997). It follows that human babies can drink while breathing through their noses, because liquid can pass around the larynx and down into the stomach while leaving the airway clear; but following the descent of the larynx from around 3 months of age over the first 3–4 years of life, the lowered adult larynx sits at the bottom of a shared single tube down which both air and food must travel, and the glottis consequently has to be closed during swallowing as an insurance policy against food or drink acci-dentally falling into the lungs. Obviously, while the glottis is closed, air cannot get into or out of the lungs either, so that the option of breathing while eating or drinking is removed. This does, however, confer on adult humans the dubious distinction of being able to choke if any of these interacting mechanisms should go wrong.

If the lowered larynx causes potential problems for modern adult humans in removing the flexibility to be able to eat and breathe at the same time, what, if anything, does it do *for* us? Here the answer seems to lie in flexibility of a differ-ent kind: the descended larynx elongates the total functional length of the vocal tract, and moreover allows this to be divided by a potential constriction of the tongue into two tubes of different lengths and dimensions. This might be expected to expand the total range of sounds which can be made compared to those pro-ducible by a single continuous tube, as the different dimensions created by vocal organ movements will create a wide range of combinations of formants.

How do we go about demonstrating that this kind of additional flexibility in terms of speech sounds does indeed follow from a descended larynx? We might compare the sounds adult humans make to those baby humans make; and it is certainly true that babies initially produce only a relatively small and undiffer-entiated range of sounds, without anything like the range of fine articulatory distinctions which characterise adult vowels and consonants. On the other hand, in developmental terms, baby humans operate only at a fraction of adult power in a range of other domains, so we cannot be sure whether the only relevant factor here is the undescended larynx. When we turn to other primates, and in particular to our closest living relatives, chimpanzees, we find again that only a rather limited range of sounds is produced. The chimp Viki, who was raised by Hayes and Hayes (1952) for more than six years, purportedly learned to produce only four discernible words, 'mama', 'papa', 'cup' and 'up', and even here her articulations were by no means clear. Otherwise, despite extensive training and prompting, she did not learn to speak. Clearly, there are also various complex factors at work here, in terms of whether non-human primates have a capacity for language learning regardless of modality, or an inclination to use such language-like elements as they do learn in social situations; but, as we saw in Chapter 2.4 above, subsequent more successful experiments on language learning in non-human primates have typically involved signing rather than speaking, suggesting that differences in the vocal tract might have something to do with Viki's struggles.

Lieberman *et al.* (1969) were the first to demonstrate that adult humans do have a descended larynx, while human babies and non-human primates do not. On the basis of acoustic modelling work based on vocal tract analogues of different dimensions, Lieberman (1975) suggests that non-human primates could physically produce the difference between voiced and voiceless sounds; variations in fundamental frequency, heard as dynamic shifts in pitch; stop sounds, at a range of places of articulation including labial, denti-alveolar and glottal; and at least a schwa-like vowel quality (see also Lieberman 2003). This reflects the fact that we find in humans of any age and in other primates the capacity to produce sound and to modify it using air produced in the course of respiration, so that we have in all these cases at least a rudimentary source-filter model. However, Lieberman's modelling indicated that the most important contribution made by the special supralaryngeal vocal tract anatomy in human adults, with the two-tube configuration and right-angle bend, is the capacity for steady-state vowels like [i] and [u]. As Fitch (2002: 31) summarises, 'Such vowels are highly distinctive, found in virtually all languages (Maddieson 1984), and play an important role in allowing rapid, efficient speech communication to take place. These vowels require extreme constriction in some vocal tract regions, and dilation in others, which the two-tube configuration allows.'

5.3.2   Reconstruction and the fossil record

If non-human primates today could, at least physically, produce a subset of the phonetic distinctions we find in modern human languages, but without the apparently crucial steady-state vowels, what can we say about the stage when these additional capacities might have developed in the human family tree? Lieberman and Crelin (1971, and see Lieberman 1984) attempted to model the likely speech production capacities of a range of hominid fossils, including Australopithecines, *Homo erectus*, *Homo heidelbergensis* and Neanderthals, using an extension of the same techniques they had previously used for modern species. For modern humans and other primates, silicon rubber casts of supralaryngeal vocal tracts were made; the coordinates of these were then fed into a computer, which modelled the possible articulations. For fossils, as we saw in the last chapter, the difficulty may be that we have rather sketchy coverage of earlier parts of the fossil record; and even in more recent times, complete specimens are hard to come by, while even a complete specimen would not show most of the crucial apparatus for speech production, since it involves mainly cartilage, muscle and other soft tissues, which do not fossilise at all. There is, it is true, the hyoid bone: but it is so small, and so embedded in soft tissue rather than being attached to other, larger bones, that it may well be lost or detached substantially from its original position in the specimen, especially where this dates from the period before deliberate burials.

Lieberman and Crelin suggested that reconstructions of fossil vocal anatomy could be achieved by two different techniques. First, it is possible to examine the marks left on skulls by muscle attachments, which indicate place of attachment, but also relative size and the direction the muscle is pulling: methods of this kind are familiar today from techniques of forensic facial reconstruction. Second, the right-angle bend in the pharynx in modern humans is associated with a change in skull shape. Comparing the vocal tract in modern humans and modern primates, we find that adult humans have a relatively higher oral cavity, and a much longer pharynx with a lower larynx; and these physical changes are accompanied by an increase in the bend in the basicranium, the base of the skull just in front of the spinal cord. These differences are shown in Figure 5.1, with a chimpanzee on top, and a modern adult human skull below.

There are various arguments about whether this change in the basicranial flection is cause or effect in terms of speech. On the one hand, increases in human brain size may have rotated the front part of the skull forward. On the other, the basicranium may have shifted to accommodate the new position of the tongue in the two-tube vocal tract, thereby allowing it more scope to articulate in an enlarged oral cavity. This may also be seen as a developmental consequence of shortening the muzzle, which would have contributed to the revised proportions of the modern vocal tract, and which has led to the distinctively flat face of modern humans as opposed to other primates. Either way, assessing the angle of the basicranial flection allows us to access information about soft tissues from aspects of the skull, which of course is much more likely to fossilise.

Lieberman and Crelin (1971) made casts on the basis of these reconstructions, and their dimensions were used to suggest the range of sounds that could have been produced by these vocal tracts. One of the best-known suggestions here (Lieberman and Crelin 1971, Lieberman 1984) is that Neanderthals must have had a rather high position for the larynx: 'The Neandertal vocal tract, though larger, is essentially the same single-tube vocal tract as that of newborn *Homo sapiens* and the chimpanzee' (Lieberman 1975: 138). According to these simulations, Neanderthals would have been able to produce labial and alveolar stops, some fricatives, and approximations to some vowels like [u], but not true steady-state vowels, or velar consonants, or in all probability the distinction between

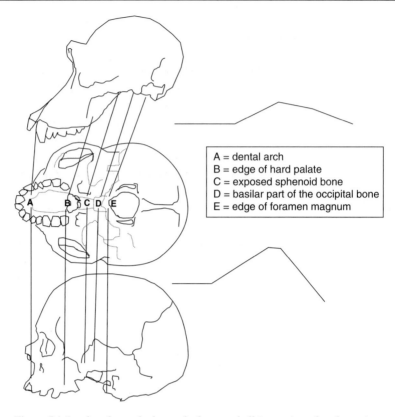

A = dental arch
B = edge of hard palate
C = exposed sphenoid bone
D = basilar part of the occipital bone
E = edge of foramen magnum

Figure 5.1 *Landmarks on the base of a human skull (centre) used to determine the shape of the basicranium of a chimpanzee skull (top) and a modern adult human skull (bottom). The degree of basicranial flection is indicated by lines drawn between the landmarks as shown to the right of the respective skulls (redrawn after Laitman, Heimbuch and Crelin 1979).*

nasal and non-nasal sounds, with all speech being slightly nasalised in character. Lieberman and Crelin propose that the first fossil which can be reconstructed with anything close to the modern human right-angle bend in the supralaryngeal vocal tract is the *Homo heidelbergensis* Broken Hill I fossil, which is now referred to as the Kabwe fossil, and taken to date from approximately 125 kyBP (McBrearty and Brooks 2000, Table 1).

5.4   Complexities and critiques

None of these claims can be taken at face value; certainly, none have been. There is no doubting that Lieberman and Crelin's contribution to work on comparative vocalisation is absolutely seminal; but that does not mean that aspects of it are not ripe for reconsideration. Some aspects of this work remain unassailable: adult humans really do have a significantly lower larynx, and a longer laryngopharynx, than human babies or other primates (see Kent and Murray 1982). But the story is not as simple as Lieberman and Crelin's early modelling work would suggest.

First, let us turn to the case of the Neanderthals. In Lieberman and Crelin's initial modelling work, the suggestion is that Neanderthals may well have been different from their early *Homo sapiens* contemporaries in their capacity for articulation. This is not to say that Neanderthals could not produce speech-like sounds at all, but that the range available to them would have been more rudimentary than in modern human languages. Nonetheless, we must constantly remain aware that any work based on modelling depends crucially on the information fed into the system in the first place. Lieberman and Crelin draw their conclusions on the basis of assumed vocal tract dimensions: the predictions they make are absolutely accurate given their overall system, but their assumptions on the vocal tract characteristics of Neanderthals are essentially hypothetical. As we have seen, soft tissue does not fossilise, and because the hyoid bone is 'floating' and not attached directly to any other bone, its presumed location in a fossil can only be approximate; yet this will determine the coordinates of the reconstructed tract. Modelling can be absolutely accurate but give the wrong information for a particular fossil type, because it is so dependent on the decisions made about tract parameters in the first place.

There is a lengthy and ongoing debate in the literature over Lieberman and Crelin's conclusions with respect to Neanderthals (Morris 1974, Falk 1975, Arensburg *et al.* 1989, 1990, Arensburg 1994, Lieberman 2007, Fitch 2000). Boë *et al.* (2002) argue, on the basis of different assumptions and therefore different modelling results, that Neanderthals would have had approximately the same vocal tract space as modern humans, and that control of the tongue body, lips and jaw could have compensated for any limitations caused by a higher larynx. Lieberman and McCarthy (1999) argue, meanwhile, that there is no reliable correlation, and therefore probably no causal connection, between the angle of the basicranial flection and the position of the larynx in modern humans. If we cannot predict the height of the larynx from the shape of the basicranium in modern humans, this casts doubt on the use of precisely the same feature in fossils (especially from only a few individual examples) to draw conclusions about the descent of the larynx and therefore the capacity for articulations of particular kinds.

Of course, as Boë *et al.* (2002: 465) note, their research 'is strictly limited to the morphological and acoustic aspects of the vocal tract, and we cannot therefore offer any definitive answer to the question of whether Neandertals spoke or not'. It might seem that we can come closer to such an answer by considering evidence suggesting that Neanderthals could bring their breathing under active muscular control. The muscles responsible for expiration are controlled by thoracic motor neurons, which pass through the thoracic region of the spinal cord. This is much larger in modern humans than in other primates, and although this enlargement is not present in early *Homo ergaster*, for instance, it is clearly discernible in Neanderthals (MacLarnon and Hewitt 1999). However, although controlled expiration is important for speech, we cannot be sure that it evolved for this purpose initially; and in any case, knowing that Neanderthals were capable of regulating their outbreaths to allow continuous speech does not tell us anything about the range of segmental sounds they might have been capable of producing, or the complexity of the grammatical structures in which they may have embedded them. The same goes for parallel evidence on the size of the hypoglossal canal, the channel through which nerves activating the tongue pass. As Corballis (2003: 206) suggests, 'The hypoglossal canal is much larger relative to the size of the oral cavity in humans than it is in the great apes, presumably because there are relatively more motor units required to produce speech.' Kay *et al.* (1998) note that *Australopithecus africanus* and/or *Homo habilis* fossils (there is some uncertainly about the grouping the relevant specimens belong to) have a smaller hypoglossal canal than modern humans, and indeed than modern gorillas and chimpanzees. The two Neanderthal specimens in Kay *et al.*'s study, however, have hypoglossal canals within the size range of those for modern humans, as do late *Homo erectus* and early *Homo sapiens* fossils. This is not conclusive, however, since the *Australopithecus africanus* and *Homo habilis* individuals appear generally to have been small in stature, so that their nerves and vocal apparatus might reasonably be expected to have been on the small side too. Jungers *et al.* (2003), on the basis of a substantial data set of 298 skulls, show that siamang gibbons, for instance, significantly exceed humans in terms of relative hypoglossal canal size, and that the range of values for *Australopithecus africanus* fossils overlaps with the ranges for humans and gibbons when these are expressed relative to body size. In addition the cross-sectional area of the canal is poorly correlated with the density of nerve fibres passing through it (de Gusta *et al.* 1999). Jungers *et al.* (2003: 474) conclude that 'the relative size of the hypoglossal canal is neither a reliable nor sufficient predictor of human-like speech capabilities, and palaeoanthropology still lacks a quantifiable, morphological diagnostic for when this capability finally emerged in the human career'.

The one thing we have been taking for granted in this discussion so far is the descended larynx – both the fact that in adult humans the larynx is typically lower, with further anatomical consequences, than in other primates and human babies, and the connection of this lowering with the capacity for articulatory diversity in speech sounds. However, even this apparently straightforward evidence turns out to conceal further complexity when seen in the light of more recent comparative research. First, it would appear that there may have been two separate stages of laryngeal lowering in human evolutionary history, and that one of these is shared with chimpanzees. Nishimura *et al.* (2003) and Nishimura (2005) demonstrate (on the basis of magnetic resonance imaging) that chimpanzee larynges also descend during infancy, perhaps reflecting changes in the swallowing mechanism after the initial stages of infancy. However, whereas in humans there is then a further descent of the hyoid relative to the mandible and skull base, there is no evidence of this second phase in chimpanzees.

This does not challenge the relevance of the lowered larynx for speech, but suggests that the first phase of lowering may have evolved in earlier pre-hominins, with the second phase being characteristic only of the human line. Fitch (2002) argues strongly in favour of exactly this kind of investigative, comparative work involving live animals rather than dead ones or fossils: as he points out, it is very difficult to generalise reliably from the anatomical conformation of a dissected specimen to the on-line physical activities of a living beast. This turns out to be particularly relevant to the operation of the larynx in vocal communication, since Fitch (2000) reports that animals which have high larynges in the resting state may nonetheless lower them when they are vocalising. This appears to be true of pigs, goats and monkeys, and particularly of dogs, which during barking will lower the larynx, cut off the nasal cavity by raising the velum, and therefore produce voiced, non-nasalised sound; when vocalising stops, the larynx is raised again. Fitch (2002: 34) suggests that this 'may represent the typical mammalian vocal gesture'. Quieter vocalisations tend to be nasal, suggesting that the laryngeal lowering might allow louder calls, and providing a possible rationale for its development.

This hypothesis of increased loudness links indirectly to the additional fact that some animals have a permanently lowered larynx: these include koalas, red and fallow deer, and probably lions and Arabian camels (though as Fitch (2002: 35) points out, obtaining close-up X-ray pictures of roaring lions is neither easy nor necessarily recommendable). So, not only do other species lower their larynges while vocalising; it appears that humans are not alone in having the larynx low in adults in its resting state. Fitch and Reby (2001) studied red deer, and found that males had a descended larynx, though the hyoid remained high, with the linkage between the two maintained by a particularly elastic ligament. Moreover, during roaring, the larynx is retracted even further, creating a much longer vocal tract (nearly twice as long as in the resting position), and therefore much lower formants. The length of the vocal tract is important in that it is positively correlated with overall body size, giving an indication of likely body size and consequently likely success in combat; red deer stags roar mainly when there is another mature male close at hand. As Fitch and Reby (2001: 1673) point out, 'Effective size exaggeration would be highly adaptive for red and fallow deer, since both body size and vocalizations play key roles in deer behaviour, influencing aggressive interactions and mating success.' That is, red deer males may be elongating their vocal tracts by laryngeal lowering to make themselves sound bigger, as this has the dual advantage of attracting female deer and repelling other stags. Human males have a further laryngeal lowering at puberty, too, when their voices break.

Of course, none of this means that laryngeal lowering evolved in the mammalian common ancestor of deer and humans; we are dealing with analogy (or convergent evolution) rather than homology (or shared ancestry). Nonetheless, forces that act as powerful motives for a particular evolutionary development in one species might be invoked for a similar development in another. It is nonetheless still true that the vocal repertoire of the red deer is restricted to roaring, with a far smaller set of possible vocalisations than we find in human languages today. Whatever might have provoked the initial descent of the larynx, then, it may well have been a preadaptation for language – in other words, a prior physical development which, once in place, turned out to be a useful starting point for subsequent linguistically relevant innovations in the vocal apparatus. It is not necessarily the case that the larynx lowered because we were already talking and wanted to talk better or more clearly; it may well be that earlier laryngeal lowering set the stage for the initial phases of human speech.

5.5   Adaptations and complications

As we noted in Chapter 2 above, there is a natural human tendency to look at the functions that a particular organ, or mechanism, or system has, figure out which ones it is central to or contributes to in an important way, and then assert that the organ, or mechanism, or system evolved in order to facilitate that particular function. In evolutionary terms, such arguments are adaptationist: we argue that a particular structure has been adapted, by the forces of mutation and natural selection, to make a species better at something than its ancestors. This tendency is even stronger when we are faced with a structure that seems to be good for something, but bad for something else, as can be illustrated by this quote from Lieberman (2003: 262), on the human SVT (= supralaryngeal vocal tract):

> The primary life supporting functions of the mouth, pharynx, throat and anatomical components of the SVT are eating, swallowing and breathing. These functions are … impeded by the human SVT. The low position of the adult human larynx and the shape and position of the human tongue results in food being propelled past the opening of the larynx. Food lodged in the larynx can result in death. Chewing is also less efficient in the shorter human mouth; our teeth are crowded and molars can become impacted and infected, resulting in death in the absence of dental intervention. The only selective advantage that the human SVT yields is to enhance the neural process that estimates SVT length by making it possible to speak the vowel [i]. This must yield the conclusion that speech was already the mode of linguistic communication well before the appearance of anatomically modern *Homo sapiens*. In other words, the neural capacity to produce speech must have been in place before the evolution of the modern human SVT.

Lieberman's argument is that earlier hominins must have evolved the neural circuitry for language; this was expressed in speech, but only in a relatively rudimentary way, because that was all the vocal tract at the time would allow. The drive towards improvement in speech production then motivated a gradual evolutionary change culminating in the second-step lowering of the human larynx, the shortening of the mid-face projection, and the general reconfiguration of the vocal tract into its current two-tube shape, with the capacity to produce an increased range of consonants and steady-state vowels. There would be no point in working towards this additional complexity if other members of the species were not already equipped to understand the output, hence the assumption that the vocal tract changes must have postdated at least some of the necessary neural structures for language production and processing.

These adaptationist arguments are a key part of evolutionary theorising, and will recur frequently in the following chapters. We must, however, take some care with our formulations; otherwise, as we noted in Chapter 3, they are apt to shade into 'just-so stories' (Gould and Lewontin 1979); and we return to these issues and to non-adaptive alternative explanations, with their own attendant problems, in Chapters 8 and 9. It is certainly the case that the vocal tract anatomy modern human adults have allows us to produce a wide range of sounds, which we could not produce if we had different vocal tracts; but we must be careful to remember that this does not necessarily mean these anatomical features evolved specifically and only to expand the range of available sounds. For instance, as we have seen, at least part of the laryngeal descent might be motivated by considerations of loudness in vocal signalling, or by the possibilities it offers for signalling body size (perhaps not altogether truthfully). Once those factors have started vocal tract remodelling on its way, then it might well be found that this happens also to increase our capacity for making different noises. But that is a very different hypothesis from one that cites steady-state vowels as the initiating reason for the anatomical change, let alone as its sole goal. We must beware teleology here, the tendency to see a goal when evolution does not, but rather muddles along adding little bits and pieces to the raw material it starts with. The latter view (paraphrasing Vincent 1978 on explanations for language change) allows us to recognise only termini, where cumulative changes happen to end up, but not goals. Come to that, we might not be at the terminus yet.

On the other hand, we must equally beware of ascribing negative consequences for other species or groups to a lack of a particular adaptation. If the two-tube vocal tract is good for speech, and modern humans have it and are thriving (setting aside for the moment the damage we do to the planet and each other), and if Neanderthals arguably did not have the same two-tube vocal tract and are no longer with us, then surely the two factors in each case go together? Adaptationist reasoning can easily, if overdone, lead us to assume that the lack of steady-state vowels (to oversimplify somewhat) was the Neanderthals' curse and downfall, and inexorably led to their demise. As we have seen, it is already controversial whether Neanderthals would have been seriously disadvantaged in their vocal production; and they were in fact an extremely successful species, and lasted a good deal longer than modern humans have been in existence. They were, however, arguably adapted (that idea again) to a particularly cold environment, and as the world warmed up after the last Ice Age, they were unable to adapt to the new conditions and to use their new context to its best advantage. We can combine these suggestions with the possible contemporary expansion of *Homo sapiens*, who may have hogged all the best localities and resources. Zubrow (1989) demonstrates using computer modelling that, if our ancestors had only a marginal initial advantage in terms of use of resources and mortality rate, the replacement of Neanderthals across Europe would have taken only about 1,000 years, which appears to match what can be hypothesised from the archaeological record. This might reflect the neural capacities of Neanderthals in general, but it is not so easily connected to their potential inability to produce the cardinal vowels.

Even if human vocal tract anatomy is adaptive, and even if the main motivating force behind the component evolutionary changes was the development of a wider repertoire of speech sounds, we must bear in mind that language still has to be balanced against a range of other priorities; as Laitman *et al.* (1996, quoted in Davidson *et al.* 2005: 497) observe, 'The acquisition and processing of oxygen and its byproducts is the primary mission of any air-breathing vertebrate. Chewing, walking, reproducing, thinking are all fine, but first one has to breathe. Anthropologists sometimes seem to forget this; evolution never does.' It is therefore possible to evolve risky situations, like the possibility of choking in modern adult humans, if those relatively rarely come into play, and perhaps if the opposing advantages are particularly salient; but mutations which are necessarily fatal will be unlikely to spread through reproduction and will not therefore become the norm for a species. In the case of potentially adaptive anatomical developments for speech, it is also possible to overshoot and create maladaptive situations. For instance, in individuals with Apert syndrome, a skeletal condition resulting in malformations of the skull and hands, the palate moves further back during development than its usual position, causing a constriction of the pharynx and compromising the capacity to produce vowels like [i u], with the result that the phonetic output of these individuals is often misinterpreted (see Shipster *et al.* 2002 and OMIM number #101200, accessed 14 May 2011).

The discussion so far seems predicated on the assumption that language from its earliest stages was expressed or realised vocally, though clearly the range of vocalisations would have been more restricted initially; we will return in Chapter 8 to the question of what kinds of linguistic structures we might hypothesise for these earliest utterances, and how those became differentiated and structured into the complex phonological and syntactic systems we find in human languages today. However, we must address the underlying assumption that sounds came first in linguistic terms. True, other primates give vocal signals; but they also have a considerable inventory of manual ones. Could language as we know it have originated in gesture rather than vocalisation?

The gestural hypothesis is put forward with considerable conviction and aplomb by Michael Corballis (1991, 2002, 2003), though it has a much longer history, going back to Condillac (1746), with a much more recent defence in Hewes (1973). Corballis argues that, although non-human primates do vocalise, human language may well be built on a completely different set of principles from their largely involuntary, situationally primed and holistic calls. He sees manual gestures as a far better precursor to language as we know it. It is certainly true that the great apes have complex systems of gesture: these have been observed for captive lowland gorillas by Tanner and Byrne (1996) and Tanner, Patterson and Byrne (2006), and for free-ranging captive chimpanzees by Tomasello *et al.* (1997), for instance; and Corballis (2003) stresses that such gestures are most often directed to another individual, as opposed to calls, which are often given regardless of whether any conspecifics are watching or listening. Corballis notes that gestures have by no means disappeared from the modern human communicative repertoire, and remain as an accompaniment to speech in all human societies; moreover, sign languages are full linguistic systems which operate without speech.

Corballis suggests that the gradual evolution of bipedality through the early hominin line would have left the hands free for communicative gesturing; these might well have been accompanied, as they are in other primates, by vocalisations. However, a great many more evolutionary modifications of both the brain and the vocal organs (recall the flattening of the face and the formation of the two-tube vocal tract, and bear in mind also the increases in brain size and specialisation to which we turn in the next chapter) would have been necessary before speech could have been used as the primary modality for language, leaving a lengthy period when gesture would have been the main communicative mechanism in hominins. Nonetheless, spoken language may have certain advantages over gestural communication which led to its gradual dominance: for one thing, it leaves the hands free for other activities, and for another, we do not need to see the person speaking, or to pay visual attention to them, in order to receive a spoken message.

It is certainly the case that speech is, more often than not, still accompanied by manual and facial gestures; it is also true that non-human primates both vocalise and gesture. This means there is no major conceptual or evolutionary barrier to the two having developed together in humans. However, Corballis's argument about the clear sufficiency of gesture in the case of contemporary sign languages almost works against him here: if humans had the capacity to evolve a purely gestural system of such structural complexity (albeit accompanied by the occasional grunt by way of emphasis or embellishment), why should this have given way to a spoken system? What evolutionary pressure could have led to the wholesale changes necessary for spoken language to develop, with its attendant physical dangers and compromises, when a flexible gestural system was already within our grasp? In order to accept Corballis's arguments here, we would need a very clear idea of how and why humans would have shifted towards speech as a carrier for language; and to pursue those questions, we first need to know a great deal more about the human brain. That is our destination in the next chapter.

5.6   Summary

In this chapter, we have described the modern human vocal tract, and the role it plays in speech production. Our vocal tract is very different from those of human babies, non-human primates and, as far as we can tell, many of our immediate and more distant hominin ancestors. In particular, it has a two-tube conformation, which provides very considerable flexibility in the range of sounds we can produce; this goes along with a shortening and flattening of the face and a lowering of the larynx. Although laryngeal lowering is not unique to humans, being found either in the resting state or as a special setting for vocalisation in a range of other species, it is not found to anything like the same extent in other primates. However, although these vocal tract differences certainly appear to be correlated strongly with our capacity to produce a wide range of speech sounds, it is frustratingly difficult to reach any absolute conclusions about when and where they might have evolved from the fossil record, since so many of the structures involved are non-fossilising soft tissue.

This leaves us with two main open questions. First, even if we knew exactly when the vocal tract attained its current form in the history of our species, we could not know for sure what our ancestors were doing with it. Having the apparatus to control expiration does not mean we know that potential was used for speech; and knowing there was the capacity to speak does not mean hominins had language. Hominins might have developed aspects of language in a more abstract sense, perhaps externalised mainly through gestures, and then undergone progressive vocal tract changes to allow more flexibility in sound production; or the vocal tract anatomy may have changed first, for whatever reason, and turned out happily to be useful for language. In short, there are always understandable temptations in historical work to home in on what is apparently the most immediately accessible evidence, which would probably be the vocal tract for modern humans and language. But sometimes the most obvious structure is the hardest to actually engage with, and the evidence base turns out to be limited in various crucial respects. We must recall, however, that language involves a whole range of abilities, structures and systems, not just a single one: our vocal tract story might begin to make much more sense when we put it together with developments in other systems. Of course, there is also no point in speaking if nobody can hear or interpret what you say, so that obviously we need to engage with the largely neurological basis of speech perception. We therefore turn now to the role of the brain and the genes in language: these are strongly interconnected, and although they might initially appear less accessible and less specialised for language than the more obvious vocal tract, we shall find that detective work here can tell us a great deal about the evolution of language.

Further reading

Good general introductions to phonetics are provided in Ladefoged and Johnson (2010) and Catford (2001), with an overview of the sounds of the world's languages in Ladefoged and Maddieson (1996), and a much more detailed overall framework for phonetics in the authoritative Laver (1994). There is more detail on acoustics, anatomy and physiology in Raphael, Borden and Harris (2006), Ladefoged (1996), Kent (1997), and Johnson (2002). For information on vocalisations in other primates, see Hauser (1996), Fitch (2010) and Goodall (1986) specifically for chimpanzees. Fitch (2000) provides a helpful comparative review of the evolution of speech; Fitch (2002, 2009) are accessible overviews of the issues surrounding the descended larynx and fossil markers for the evolution of speech. Lieberman and Crelin (1971) is the classic reference on the modelling of speech production capacities through the fossil record, with a helpful summary in Lieberman (1989) and a book-length survey in Lieberman (1984). References on the alleged language capacities of Neanderthals are given in the text, but include Boë *et al.* (2002), Lieberman (2007) and MacLarnon and Hewitt (1999), and see also the recent computer modelling work of de Boer (2010) and Carré (2009) for discussion concerning the optimality of the modern human vocal tract. Fitch (2002) and Fitch and Reby (2001) outline the evidence for descended larynges outside modern humans, and the possible evolutionary motivation in terms of body size. Arguments for a gestural origin for modern human language can be found in Corballis (1991, 2002, 2003), Armstrong *et al.* (1995), Armstrong (1999), Hewes (1973), and Condillac (1746).

Points for discussion

1. Using the references in the further reading section above, find out more about the detailed anatomy and physiology of the larynx and associated structures. For example, what are the muscles that suspend the larynx? How is the hyoid attached, and what is it attached to? Where does the tongue fit in?

2. This chapter has not engaged at all with the mechanisms of speech perception. Although the brain is the topic of the next chapter, it would be helpful now to consider the ear in more detail. What is uniquely human about our ears? Find out what you can about the comparative anatomy of the ear, and therefore by extension about sound perception, in adult humans, baby humans, other primates, other mammals and non-mammals. Is as much information available about ears as about vocal tracts? If there is a discrepancy, why might it exist?

3. How do parrots make human-like sounds, and how far this can be said to be associated with human-like language capacities or language use? What does this tell us about the relationship between and relative importance of speech and language? What does Pepperberg (2005) mean when she talks about homologues and analogues, or homologies and analogies, in the comparison between humans and vocally communicative non-primates like African Grey

parrots? For a state-of-the-art review of communication in parrots, and a range of references, see Pepperberg (2010).

4. Read some of Corballis's work on gesture, and some earlier proponents of a gestural origin for language. What are the main arguments for the gestural hypothesis, and what arguments can you think of, or find in the literature, against this idea? How similar are human gestures in different language communities, and how similar to the gestures of non-human primates are our human gestures today?

6   Language and the brain

6.1   Overview

Having considered the vocal tract, we now turn to another organ with a level of apparent specialisation for language, namely the human brain. There are two essential types of difference between the human brain and brains in other species: first, it is just vastly bigger than we would expect given human body size; and second, its internal structure and connectivity (as far as we can currently tell) are distinctive. In this chapter we shall provide a brief overview of the shape and function of the brain, and consider the evidence we have for linguistic specialisations and for brain evolution. As with the vocal tract, much of this evidence is indirect and interpretations vary; furthermore, much of the relevant neurolinguistic data is very new, so for the most part we shall be reviewing ongoing debates rather than coming to definite conclusions. In addition, we proposed right back in Chapter 1 that the human capacity for language cannot be understood as a single object of enquiry; rather, we have to engage with it on three levels at once if we hope to cast any light on how it emerged and how it works. Those three levels are the controlling *genes*; the physical or neurological *structures* they build in particular environments; and finally the phenotypic, surface *systems* or *behaviours* which these structures permit. It follows that evolved aspects of the human brain must have genetic underpinnings, and we turn to these in the next chapter.

6.2   Brains and genes: one topic, not two

In the last chapter, we looked for evidence on the evolution of language in what is arguably the most obvious place – the vocal tract. After all, the human vocal tract has physical characteristics which are clearly different from those we find in any other species, including our closest primate relatives; and those differences seem to be linked with our capacity for production of a diverse range of sounds. What we found, however, was a common though irksome fact of scientific investigation: sometimes you look in the obvious place, but you can't see very much. Although it is true that we can find some clues from the fossil record about the approximate time in the hominin lineage when the crucial right-angle bend in the vocal tract evolved, that evidence is suggestive rather than conclusive for a number of reasons. The fossil record, as we have seen, has more gaps than evidence; and even if we could pinpoint the emergence of the modern-type vocal tract precisely, we could not conclude that this marks the emergence of language too. Many of the requirements for spoken language (like being able to use symbols) might equally hold for gestures, and might consequently pre-date the development of the vocal tract, since we know that other primates gesture as part of their communication systems. Conversely, although probably less plausibly, the vocal tract might have developed some of its current characteristics for reasons quite separate from spoken language. And, of course, 'spoken language' might mean many different things, from the utterance of individual, indivisible vocal signs, to the complex combination and manipulation of vowels, consonants and suprasegmental features into lexical items we find in human languages today.

Let us turn, then, to an alternative candidate, the human brain. Initially, the brain might seem less promising than the vocal tract. Although both are multifunctional, it is clear that the brain is vastly more complex; and while we might reasonably argue that spoken language represents the major function of (at least part of) the vocal tract, this would be a harder position to maintain for the brain. Furthermore, there might seem on the face of it to be less of a gap between human and other primate brains than between human and other primate vocal tracts. On closer inspection, however, this is very far from the whole truth. Although we shall find, again, that there are conspicuous gaps in the fossil record, and encounter, as with the vocal tract, the difficulty that most of the structures we are interested in are soft tissue and therefore would not fossilise anyway, there is plenty of evidence that the human brain really is unique, both in its absolute size and its internal configuration.

It has to be confessed at the outset, however, that we cannot talk sensibly about the brain without talking about genes too. As we noted right at the beginning of the book, trying to explain a behavioural system like language in evolutionary terms requires reference to the physical and neurological structures that enable that system, but also to the genetic instructions that cause the structures to be built. Of course, that is true of the vocal tract as a system, just as it is for the brain; but perhaps we have rather more direct evidence for the way the vocal tract works. The brain is a structure of such exceptional complexity, with so many and varied internal interconnections, that ascribing particular effects to its individual subcomponents is simplistic at best, impossible at worst. We can often best appreciate the workings of parts of the brain by observing the results when things go wrong; sometimes this reflects illness or injury, but in other cases, problems arise because genes are damaged. In the next chapter, then, we shall invoke genetic differences between humans and other species, and between humans with certain kinds of inherited conditions and those without, in trying to understand how the brain is wired up, what consequences the wiring has for systems of behaviour like language and how it got to be that way through the processes of evolution. It follows that the brain and associated genes really represent a single topic which, for ease of explication, we have had to split in two.

6.3   Elementary brain geography

Anything we can say in a section this short, in a single book with other agendas to pursue, about an organ as complex as the human brain, is bound to be inadequate and over-simplistic. It is absolutely vital that readers who are interested in finding out more about the brain and in pursuing its links with language further should move on to the more detailed reading suggested at the end of the chapter – and even that barely scratches the surface of a huge, interdisciplinary literature. However, what we can provide at this stage is a very general road-map of the human brain as it is in our species now, to provide a focus for discussion of its evolutionary development.

In terms of gross anatomy, the average adult human brain weighs around 1330 grams, which is approximately 2 per cent of average human bodyweight; on the other hand, it is a spectacularly expensive organ to run, since it consumes around 20 per cent of human energy when at rest (Deacon 1992: 115). Using the biological comparative method, we can establish that certain parts of the brain are much older than others (Nolte 2002, Loritz 1999). The oldest of all is the brain stem, which mainly controls actions like breathing that are not under conscious control: damage to the brain stem results in deep coma, and brain stem tests are used to determine whether a person is brain dead or has a chance of recovery of higher functions. The brain stem is conventionally divided into the hindbrain and the midbrain. The former contains subparts like the medulla, which maintains breathing, heart rate and blood pressure; the pons, which is involved in motor control and monitoring of sensory information from the environment; and the cerebellum. The cerebellum is attached to the back of the brain stem, and is interesting since its structure is a miniature version of the whole of the brain, with two highly convoluted hemispheres. It seems to be implicated in controlling and coordinating movement, posture and balance, and in birds, for instance, controls the various elements of flying. The midbrain, in turn, includes the tectum and tegmentum, and is involved in vision, hearing and movement.

Some creatures, like fish, fill their environmental and behavioural niches perfectly well with brains that strongly resemble a slightly expanded human brain stem, but in higher vertebrates and notably in mammals there are other, more recently evolved structures. In other mammals, this newer **fore**brain is aptly named, since it has developed in front of the older, pre-existing brain structures. However, human bipedality has led to an effective rotation of the older brain below the new forebrain, which in modern humans therefore effectively covers the older structures, as well as protruding out in front. The forebrain can itself be divided into more archaic elements such as the limbic system, sometimes thought of as the 'emotional brain': this includes the thalamus, which is essentially the earliest processing centre of the mammalian brain, receiving information from the environment and relaying it elsewhere in the brain for action. Interacting with the thalamus is the hypothalamus, which specifically monitors information about the organism's internal biochemical state, relating, for instance, to body temperature, thirst and hunger, and releases hormones when particular actions are required. Also involved in the limbic system are the hippocampus, with connections to learning, memory and spatial awareness; and the amygdala, active in memory and emotion. Above the limbic system lies the newest and uppermost part of the forebrain, the cerebrum, which is most important for us since it is involved in coordinating the voluntary functions of the animal, including what we might call thoughts and actions.

In mammals, the most salient aspect of the cerebrum is its outermost part, the neocortex (also known in the literature as the isocortex), which is organised into six layers of specialised neural cells (see Figure 6.4 below).

The nervous systems of all animals are made up of only two types of cell: neurons, the information-processing and signalling components, and glial cells that act in supporting roles. A neuron conveys information from one part of itself to another via electrical impulses, and communicates with other neurons and somatic cells using chemical signals at special junctions called synapses. As shown in Figure 6.1, a typical neuron consists of a metabolically active cell body; single or multiple hair- or tree-like projections called dendrites that accept incoming (afferent) signals from other neurons; and a single axon which looks like a long cable and connects output (efferent) signals from the neuron to other cells. Although most nerve cell bodies are nearly invisible without a microscope, the axons can be really quite long, as in the primary sensory cells of the giraffe's foot, which travel around 5 metres from the cell body in the spinal cord to the toes. Longer axons tend to be surrounded by a fatty sheath of myelin laid down by specialised glial cells (which are called Schwann cells in the peripheral nerves and oligodendrocytes in the central nervous system). The myelin sheath acts as electrical insulation and vastly speeds up the rate at which signals can be transmitted along the longer axons. Although only a single axon usually leaves any particular cell, this can subsequently branch and connect to many other neurons, both locally and in distant parts of the brain; in addition, synapses can be either stimulatory or inhibitory, so that an activated cell might stimulate a distant neuron while repressing the activity of its immediate neighbours.

There are around 25 billion of these neurons in the neocortex, and they are interconnected in particularly complex ways by over 100,000 km of axons, with an average of 7,000 connections per cell – this gives an estimated 3×10^{14} connections overall (Nolte 2002: 525, Drachman 2005). The six outer layers of tissue of the neocortex are known as the 'grey matter', since the neural cell bodies have a grey appearance in newly dead tissue. Below these outer layers run the myelinated axons; and the white appearance of the fatty myelin provides the term 'white matter' for the interior parts of the brain containing many axons. The relative geography of neural cell bodies and axons is illustrated in Figure 6.2, which shows a giant pyramidal cell (so called because the early anatomists, who had no idea of individual cells' functions, labelled them according to the pyramid-like shape of the cell body as seen under the microscope). This type of cell is common throughout the cerebral cortex, although by no means the only type of cell

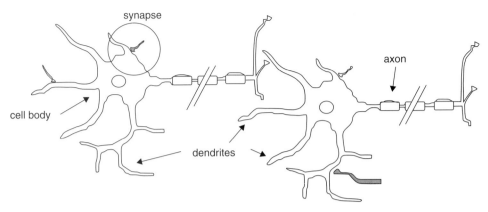

Figure 6.1 *A pair of typical neurons. The break indicated in the myelinated axons shows that these may be up to several metres long. The synapses, where dendrites make external connections with axons from other cells, can have an excitatory or inhibitory effect.*

present: the specific example in Figure 6.2 comes from the third layer (see, for example, Nolte 2002: 4 and Ch. 22 for a more detailed discussion of neural cell types and functions).

In humans, the neocortex has assumed many of the functions of coordination and control performed by the older parts of the brain in lower vertebrates; as a result, many of the structures in the more archaic brain in mammals now act effectively as relay stations which seem to coordinate and distribute information from the body and pass it to the forebrain for action, and likewise to pass instructions from the higher 'executive brain' (Goldberg 2001) back to the body.

The neocortex consists of two parts, or hemispheres, joined to each other by extensive neural connections in the corpus collosum below the outer layers. In lower mammals, the cerebrum (that is, the outer, visible surface of the neocortex) is often smooth and unconvoluted, but in more advanced mammals the increasing surface area, associated with greater processing power, requires increasingly deep wrinkles to form in order for the brain to fit in the skull (the total surface area of the human cerebral cortex is approximately 2,500 cm^2, http://faculty.washington.edu/chudler/facts.html, accessed 14 May 2011, so you can imagine how much space it would occupy if it were flattened out). The resultant pattern of foldings and wrinkles is species specific: the 'hills' are known as *gyri* (from the Latin for 'circle', because the elevations are typically rounded), and the 'valleys' are *sulci* (from the Latin for 'groove' or 'trench') – so, we find regions of the brain called the angular gyrus, for instance. Although there are individual differences in the location, height and depth of the gyri and sulci, some fall into recurrent patterns with limited variation, so that these can act as landmarks in the road-map of the human brain. Sulci in particular are vital in the conventional division of the cerebrum into four lobes. For example, the deep central sulcus (also called the central fissure) forms an S-shaped groove

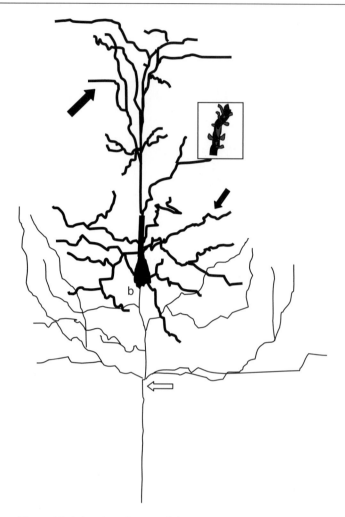

Figure 6.2 *A drawing of a typical, large pyramidal neuron from the cerebral cortex, showing the cell body (b), dendrites (filled arrows) with an increased magnification of one dendrite (boxed) to show the 'dendritic spurs' where other neurons synapse with this one, and the single thin axon (open arrow) leaving from the base of the cell to send signals to distant sites. Note that the axon also branches to synapse with adjacent cells. Dendrites and axons are collectively called cell processes and only a small proportion are drawn in this diagram for clarity. Each pyramidal cell synapses with around one thousand other cells via the dendritic spines. This is one of the largest cells in the brain at around 3 mm from the top of the dendritic 'tree' to the point indicated by the open arrow. The surface of the brain is at the top of the diagram; hence the 'grey matter' of cell bodies lies above the 'white matter' of axons. (Adapted from Nolte 2002: 4, Figure 1.4e.)*

running from the top of both hemispheres about halfway down on each side, delimiting the frontal lobe from the rest of the cerebrum. The central sulcus terminates in a second deep sulcus, the sylvian fissure, which runs from front to back on both sides and is taken to be the boundary between the temporal lobe

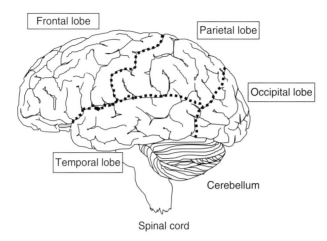

Figure 6.3 *Surface of a human brain in lateral view from the left-hand side. The front of the head would be to the left. Dotted lines demarcate the major sulci used in mapping the surface layout of the cerebral cortex (see text for details).*

below, and the parietal lobe above. The back of the brain is the occipital lobe, which is divided from the parietal and temporal lobes by the parieto-occipital fissure; unfortunately, this is clearly visible only from underneath, so that on the surface this particular boundary is a good deal more notional than the others. The lobes are shown in Figure 6.3.

These four lobes are not purely used for physical delineation of areas of the brain, but appear to map onto a series of functional specialisations: the occipital lobe is primarily responsible for processing information on vision; the temporal lobe deals with sound; the parietal lobe with touch; and the frontal lobe with movement planning. Since most of these types of sensory information have some connection with human language, it is not surprising that language has frequently been associated with the cerebral cortex; but that does not mean we can pinpoint a specific area and argue for a simplistic localisation of language functions there. It is particularly important to bear this in mind if we accept that functional specialisations may have shifted over evolutionary time. For example, Lynch and Granger (2008) suggest that the associative neocortex may have its origins in the interpretation of olfactory information; certainly, a large proportion of the mouse genome is dedicated to the production of nasal receptor proteins, and hence linked to navigation of the world by smell. It may be that in our primate ancestors, an improvement in vision and a shift away from smell as the major sense has left some of these regions free to develop connections to other sensory input and storage regions.

We consequently need to be prepared to find that aspects of language seem to relate to different, possibly spatially distant parts of the brain. In discussing these we can use the gyri and sulci to navigate; but we have already seen that these form a largely species-specific pattern; and of course when we are discussing the evolution of the current human brain, we also need a way of applying the

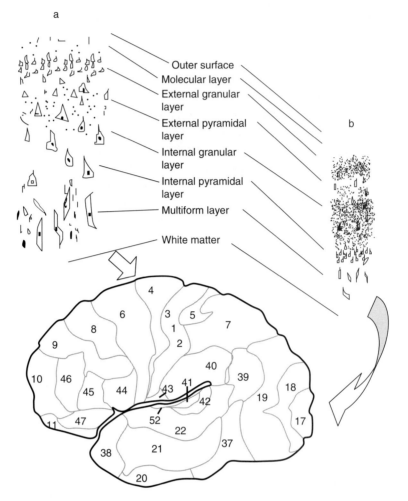

Figure 6.4 *Cytoarchitectonic map of the human left cerebral hemisphere. An example of so-called agranular cortex dominated by pyramidal cells is shown in (a) (from Area 4; see text) and an example of granular cortex (from Area 17) dominated by non-pyramidal cells is shown in (b). Note that only cell bodies are shown in these diagrams, and that the names of the layers reflect the shape of the cell bodies visible in each. (Redrawn from Roberts, Leigh and Weinberger 1993, in Goldberg 2001: 34.)*

biological comparative method and hence of matching up areas of human and other primate brains, for instance. One approach to this problem involves the structure of the neocortex. As we have seen, this is uniformly six layers thick in humans; but the relative density of the layers varies across different regions of the brain, and it is presumed that this is related to the function of these different areas (Nolte 2002). The areas of different cellular concentration can then be used to construct a cytoarchitectonic map of the sort shown in Figure 6.4.

Maps like the one in Figure 6.4 have been developed over the past century by microscopic examination of the different cellular layers in the neocortex, and the numbers conventionally used derive from the work of Brodmann (1909). For instance, Area 17, in the occipital lobe (the source of the magnified example (b) in Figure 6.4), is involved in the primary processing of visual information from the retina; while Areas 18 and 19, just in front of this, are the secondary or associative visual cortex areas and appear to be involved in more complex analysis of the visual signal. Areas 22, 41 and 42 in the temporal lobe, meanwhile, seem to be involved in processing auditory inputs. Surrounding the central sulcus, we find two mirrored strips: Areas 1, 2 and 3 behind the central sulcus constitute the sensory strip, which is essentially composed of a series of subregions, each receiving sensory input from a specific area of the body. These areas are arranged in an order matching the physical structure of the body, so that data from the toes and feet are received by areas deep down within the fissure between the two hemispheres, while regions corresponding to the leg, hip, trunk, neck, head and shoulders, then arm and hand, and finally face, follow down the side of each hemisphere. Likewise, Areas 4 (see Figure 6.4, magnified area (a)) and 6 to the front of the central sulcus contain areas controlling voluntary muscular movements in each of these same regions of the body, in exactly the same order as for sensory input in Areas 1–3.

These Brodmann maps have gained considerable further support from more recent investigations involving stimulation of the brain in patients undergoing certain types of surgery (starting with Penfield and Rasmussen 1950, Penfield and Roberts 1959). As we saw in the previous chapter, there are very good reasons to want to study vocal tracts in action in living animals, rather than drawing all our conclusions from dissection of dead ones; and in exactly the same way, there is a strong case for using emerging data on brain function from living humans where it is available, to supplement what we know from anatomical investigations of dead brains. The importance of this kind of data will become clear in the next section.

6.4   Specialisation of the brain for language

6.4.1   Brain asymmetry and analysis of naturally occurring lesions

Studying brain structure and function relies on the general assumption that activity within certain neurons will either lead to a mental perception of particular characteristics of the external world (such as colour, depth of field or motion of an object), or initiate some behavioural characteristic in the animal being studied. Such assumptions led the Victorian phrenologists to search for external signs of (over)development of particular faculties of mind, which they envisaged as giving rise to bumps of particular shapes and sizes on the surface of the skull (Greenblatt 1995). While we would no longer expect differences in functionality between different brains to translate into tangible bumps on the head, the relationship between aspects of mind and brain still holds at a range of more subtle levels. Relevant brain activities could be the activation of a single neuron (Parker and Newsome 1998, Barlow 1972), or of groups of neurons either close together or in diffuse locations (Parker, Cumming and Dodd 2000), or the temporal sequence of incoming action potentials (Singer and Gray 1995). On the basis of experimental evidence, a certain perception or behaviour can be assigned to a particular group of neurons, under a certain, testable set of circumstances: either loss of neuronal activity should lead to some change in behaviour or perception in the individual; or artificial chemical or electrical stimulation should generate or block the appropriate perception or action; or measured variation in neuronal activity should have a relationship to some perception or action manifested in behavioural changes of some kind.

Of course, when we talk about 'experimental evidence', we do not always mean that the relevant set of circumstances is artificially created by an experimenter – especially when we are dealing with human subjects, many such situations would be wholly unethical. However, certain changes in brain activity can occur naturally, with results which can subsequently be observed and correlated with changes in behaviour or perceptions on the part of the affected individual. For example, when blood vessels are blocked through infarctions, or ruptured through strokes, the areas of the brain affected will be determined by the perfusion patterns of the particular vessels involved, which are often relatively extensive and include damage to grey and white matter, and often also to subcortical brain structures. Indeed, the correlation of the region of brain damage in aphasic patients with the degree and nature of cognitive deficit was the traditional method used to determine Broca's and Wernicke's areas (see Broca 1861, 1865, Wernicke 1874, and the discussion in Chapter 2.6 above).

On the Brodmann map in Figure 6.4 above, Area 44 is marked in the perisylvian region (that is, near the sylvian fissure) of the frontal lobe. Area 44 is typically identified with Broca's area; as we saw in Chapter 2.6, damage to this area (usually in the left hemisphere) has, since the mid nineteenth century, been recognised as resulting in a specific type of aphasia characterised by slow speech, lacking in fluency, and often without affixes and function words. Broca's aphasics are typically able to understand speech, and their main problems lie in production, as we might anticipate since the frontal lobe is involved with action and planning.

However, if we expected to find recurrent, one-to-one mappings between the numbered sectors on the Brodmann map and specific linguistic consequences of brain damage, we would be very much mistaken. For one thing, although Area 44 of Figure 6.4 corresponds closely to Broca's area, there is no single Brodmann region mapping onto Wernicke's area; rather, this covers parts of Areas 22, 41 and 42. Furthermore, although these regions fall within the temporal lobe and would therefore be expected to be involved with processing auditory inputs, Wernicke's aphasia does not uniquely involve problems of comprehension. Wernicke's aphasics do have trouble with processing language, and on the face of it their speech is often fluent; but they can nonetheless experience great difficulties in making themselves understood, since they 'make frequent lexical errors, substituting semantically related but inappropriate words (e.g. *table* for *chair*) or incorrect sounds (e.g. *bekan* for *began*), or, in more severe cases, unidentifiable phonological forms such as *taenz*' (Dąbrowska 2004: 40). Likewise, although it is often claimed that Broca's aphasics retain their capacity for comprehension, they may struggle with grammatical comprehension; on the other hand, while Broca's aphasia has often been related to an absence of inflections, Dąbrowska (2004: 46) reports that this appears to be true for English but not for more highly inflecting languages like Polish or Turkish, where inflections are typically preserved (though they may be misused).

There is little evidence, in short, for a simplistic association of production problems (only) with Broca's area, and comprehension problems (only) with Wernicke's area. Although there are obvious links of Broca's area with speech production, and of Wernicke's area with reception and perception, grammar clearly does not 'live' in one of these regions and vocabulary in another. As Dąbrowska (2004: 48) puts it, 'It is undeniable that some regions of the brain are more involved in linguistic, and specifically grammatical, processing than others. However, the strongest version of the anatomical specialisation hypothesis – that grammar resides in the pattern of connections in Broca's area – is clearly false.'

The difficulties we have encountered with these attempts at mapping areas of damage to specific deficits also reflect profound asymmetries and differences in brain structure. Perhaps most importantly, the patterns of sulci and gyri are consistently asymmetrical between the hemispheres of the cortex. Data for twenty-five out of thirty-six human brains (from Rubens, Mahowlad and Hutton 1976) is summarised in Figure 6.5, which compares the majority position of the central sulcus and the sylvian fissure (called the lateral sulcus in this study) for the left and right hemispheres. Recall from Figure 6.3 that the central sulcus divides the frontal lobe from the remainder of the neocortex, while the sylvian fissure is the boundary between the temporal and parietal lobes. It is clear from Figure 6.5 that the sylvian fissure extends further backwards on the left, as shown by the solid line, as opposed to the dotted line for the right hemisphere. This discrepancy is usually taken to indicate an expansion of the superior temporal gyrus on the left, in the region immediately behind the primary auditory cortex. This expanded region is called the planum temporale (temporal plane), and corresponds broadly to Wernicke's area. The argument would then be that, as language evolved, the apparatus in the brain necessary for reception and perception of language would also have to increase, consequently altering the normal shape of the left hemisphere for most individuals, though by no means all.

This highlights yet another issue, since brain shape, size and conformation are not uniform across all humans: frequently, we observe differences that do not correspond to any deficit or damage, but simply represent individual-specific

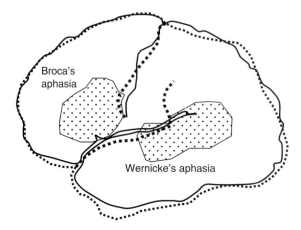

Figure 6.5 *Outline of the left hemisphere (solid line), with an inverted outline of the right hemisphere (dotted line) superimposed. The relative positions of the central and lateral sulci are shown, for the majority pattern in Rubens, Mahowlad and Hutton 1976. Stippled areas show regions of cortical damage for patients diagnosed as having Broca's or Wernicke's aphasia, in Kertesz, Lesk and McCabe (1977). (Diagram drawn from data in Rubens, Mahowlad and Hutton 1976 and Kertesz, Lesk and McCabe 1977.)*

patterns. Figure 6.5 also shows the results of an approach to aphasia which takes such differences into account. Rather than defining a group of patients with brain lesions affecting particular regions, then testing them for particular suites of linguistic competences (Chao and Knight 1998, Friedrich *et al.* 1998), it is possible to begin with a group of patients who exhibit a particular deficit, then overlay their lesions on a brain map to identify the commonly affected areas (Dronkers 1996). For example, Kertesz, Lesk and McCabe (1977) detected areas of cortical damage in 27 patients with aphasia; these areas were detected using radioisotope scanning to detect leakage of an injected radioactive tracer from small blood vessels damaged by an infarction. The stippled areas of Figure 6.5 indicate the overlap regions for areas of cortical damage in the two subgroups of these 27 patients who had been characterised as having Broca's (14) or Wernicke's (13) aphasia.

Matters are, in fact, even more complex. Strong links between lexical and grammatical difficulties appear to characterise a wide range of aphasias, and also emerge in normal ageing and Alzheimer's disease (Bates and Goodman 1997); and more perplexingly, there can be aphasic symptoms even when the classic language areas in the brain do not appear to be damaged, as well as damage to those areas without the expected type of aphasic symptoms (Dronkers *et al.* 2000, 2004). On the basis of this evidence from aphasia, then, it seems highly likely that language functions are not localised to precise areas of the brain, but rather that they correspond to diffuse networks of neurons across the brain, as illustrated in Figure 6.6, which shows the far-reaching networks formed by recurrent bundles (or fasciculi) of axons in the white matter. Indeed, lesions in the arcuate fasciculus, which connects Broca's and Wernicke's areas, are associated with

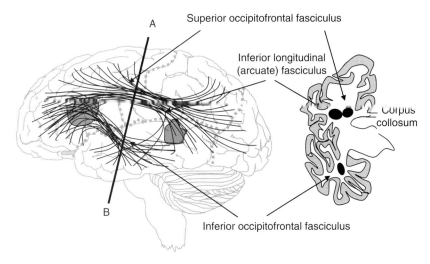

Figure 6.6 *The paths of bundles of axons connecting distant parts of the cerebral cortex. A lateral view of the left hemisphere only is shown on the left; on the right is a section cut through this hemisphere along the line indicated by A to B to show how the 'grey' matter is folded and lies above the connections forming the 'white matter'. The bundles, called fasciculi, are indicated in black. Grey stippled areas on the left diagram represent the location of cerebral damage associated with Broca's (left) and Wernicke's (right) aphasias. (Adapted from Nolte 2002, Figure 22–10: 532.)*

conduction aphasia (Obler and Gjerlow 1999: 62–3), which leads to difficulties in repeating words.

Given the contributions of these fasciculi, and their role in linking apparently distant parts of the cortex, we might well anticipate that damage in a range of different areas could have the same functional results. If the brain is also configured (even slightly) differently in different individuals, we might also predict a range of consequences from damage in apparently much the same region. This, however, is hard to investigate or demonstrate conclusively using naturally occurring deficits: brain damage is, almost by definition, uncontrolled, and consequent disruption to language functions may be associated with a range of other symptoms. The particular causes and effects which we might be most interested in can therefore be hard to disentangle from other factors; and we could wait for a long time before we were able to identify even a single patient whose situation would allow us to test a particular hypothesis. One solution lies in the exploration of techniques for stimulation of working, undamaged brains.

6.4.2   Electrocortical mapping and newer methods

In functional lesion studies, we move from the investigation of naturally occurring damage and consequent deficits to the temporary, experimental inducing of localised disruptions and the study of their effects. For example, electrocortical mapping involves electrical stimulation of the exposed brains of conscious patients; any effect on various linguistic tasks can then be determined. Although this particular technique will not be explored further below, chemicals like sodium amytal can also be used to anaesthetise parts of the brain, often a single hemisphere, to determine the effect on behaviour (Wada 1997, Wada and Rasmussen 1960).

Electrocortical mapping is used in a clinical setting to identify (in order to preserve, if possible) speech, language and motor functions in patients undergoing exposed brain surgery to treat seizure disorders, tumours or vascular malformations (Ojemann 1983, Penfield and Roberts 1959). The process involves applying a low-level electrical current of around 10–15 mA between two electrodes a few millimetres apart for 5–10 seconds. This current disrupts the normal electrical signals of the cortical cells lying immediately below and around the electrodes, effectively creating a localised brain lesion; any resultant deficit can then be physically assigned to the location of the electrical stimulation. The technique appears to generate no long-term detrimental effects, and the brain returns to normal activity immediately after the current is switched off. This provides a great advantage over investigations using clinical stroke patients, as a whole range of sites can be tested in the same individual. Since the brain has no pain receptors, the electrical stimulation itself causes no pain to the patient, although the necessary prior removal of part of the skull means the process as a whole is seriously invasive, limiting the subjects on which it can be performed, often to those with neurological deficits or a need for surgery (though care is typically taken to ensure that results with a general bearing on language are collected from individuals whose condition would not have an adverse impact on language functions in itself). In addition, the superficial physical nature of the electro-stimulation means that deep sulci whose relevance to language might be identified through some other scanning techniques (e.g. Binder *et al.* 2000) cannot be easily tested.

Electrocortical mapping can be employed, for instance, to identify the areas of the brain involved in speech perception. Frequently, individuals will be presented with a so-called 'forced-choice' discrimination task while the current is switched on; where performance deviates from control conditions, this is taken to indicate a critical importance for the region being stimulated in the particular process involved. For acoustic-phonetic processing, listeners may be asked to categorise as 'same' or 'different' the members of pairs of sounds, which will sometimes be identical and sometimes form minimal pairs differing in one or more features (as with /ba/ ~ /pa/ for consonant voicing, /ba/ ~ /da/ for place of articulation, or /bi/ ~ /ba/ for vowel quality). Such forced-choice protocols show that acoustic-phonetic discrimination of speech sounds, regardless of the specific features involved, is blocked in most individuals by stimulation of a small area of the middle part of the superior temporal gyrus lateral to the primary auditory cortex (Micelli *et al.* 1978, Boatman 2004, Boatman *et al.* 1994). Interestingly, however, stimulation of this site does not block discrimination between other auditory signals, such as pure or frequency-modulated tones, suggesting that these induced lesions interfere only with aspects of hearing specific to language processing, not with acoustic discrimination as a whole. Boatman *et al.* (1997) also show that some listeners retained the ability to distinguish vowels even when their ability to discriminate syllables on the basis of consonants was disrupted, depending on the very specific region stimulated.

When listeners were asked a more complex question about a specific part of the syllable (not whether *pit* and *pat*, say, were the same or different as wholes, but whether their initial consonants were the same or different), studies have revealed more variability and a much wider distribution of sites where electrical stimulation can induce errors. Nonetheless, these areas usually include and surround that identified for acoustic-phonetic decoding above (Ojemann and Mateer 1979). It would appear, then, that dividing the stream of speech and comparing its elements to internal representations to determine contrast is a more variable, individual and diffuse process in neurolinguistic terms. On the other hand, the initial detection of 'same' or 'different' in phonetic terms is less variable, more localised, and closer to the primary auditory area of the cortex.

In all mammals, vibrations in the air caused by sound waves are collected by the outer ear, and transferred and amplified by the bones of the middle ear to set up vibrations in the fluid of the inner ear, the cochlea. The base of the cochlea consists of a membrane lined with hairs, each of which is connected to a neuron. Different hairs are structured to move in response to different frequencies of incoming sound waves, so that higher-pitched sounds are detected at one end of the membrane, and increasingly lower pitches along its length to the other end. As the hairs move, they activate the attached neurons, generating a 'tonal map' of incoming sound frequency in the activity pattern of the neurons, which in humans transmit their signals via the 8th cranial nerve to the brainstem, and thence in particular to the primary auditory cortex on the transverse temporal gyrus extending down into the sylvian fissure. Remarkably, the activity pattern of the neurons in the cochlea is mimicked in the auditory cortex. The cells here are arranged in a tonotropic map, such that higher-frequency sounds evoke activity in those cells deep in the fissure, with increasingly lower-pitched noise generating activity in cells closer to the surface (see, for example, Romani *et al.* 1982, Cansino *et al.* 1994). This kind of tropic map, where areas of the brain seem to reflect geographically certain aspects of a given sensory organ, which itself is detecting aspects of the external world, is fairly common: neurons in a range of Brodmann areas in the brain seem to be organised in an order that reflects the location of organs in the body, or receptors within an organ.

The problem is that for humans, the relevant areas of the primary auditory cortex are mainly buried deep down in the sylvian fissure. This rather inauspicious and inaccessible location means the precise workings of hearing and perception in human brains are difficult to determine directly. However, the relevant region in humans does correspond to areas that have been associated with processing species-specific sounds in macaques (Kaas, Hackett and Tramo 1999); and because the macaque brain is physically much less convoluted, examination of the more superficial matching areas in macaques allows us to determine the situation in a relatively closely related primate species. Furthermore, the findings for macaques are generally from bilateral lesion studies or direct recording of individual neuron activity, which cannot be replicated easily in humans.

In macaques, auditory information is received from the cochlea by three core regions of the cortex; again, these primary auditory regions form tropic maps, responding to signals from the cochlea in a pattern that matches the distribution of frequency-sensitive cells in the cochlea itself. These primary areas then send out connections to a surrounding belt of around seven areas, at least four of which appear to be loosely cochleatropic in organisation, with the others mapping the cochlea much less directly, and showing stimulation in response to more complex aspects of the auditory signal. In turn, these belt regions send out connections to an as yet less well-characterised para-belt region. Both the belt and para-belt regions are believed to have an integrative function concerned with initial pattern recognition and discrimination of incoming auditory signals. They in turn connect to diffuse sites in the temporal, parietal and frontal cortex where 'higher-order' processing of features such as space perception and auditory memory are believed to lie. In other words, the primary area extracts spectral and temporal features of the incoming auditory stream; these are then integrated in the belt and para-belt regions to form the 'mental image' of incoming sounds which are interpreted in time, space and meaning in the higher associative cortex.

Aspects of lexical-semantic processing have also been examined using electro-stimulation during linguistic tasks such as the Token Test (de Renzi and Vignolo 1962), in which patients are asked to perform manipulations of coloured tokens using simple instructions that require interpretation, such as 'move the red square'. Brain areas where stimulation causes disruption of these functions show even more diffuse localisations within both left and right hemispheres, probably because the tasks involved require a complex interaction of a range of functions, including long- and short-term memory, categorisation, and lexical and grammatical interpretation. Similar results have been confirmed outside the context of invasive surgical procedures using techniques such as PET (positron emission tomography) and fMRI (functional magnetic resonance imaging), which provide indirect ways of detecting the electrical activation of individual neurons (see Poeppel and Hickok 2004, and also Chapter 2.6 above). These techniques are relatively non-invasive, and allow on-line mapping of areas of brain activity while specific tasks are being performed; they do not rely on the artificial stimulation of particular areas of the brain, and therefore allow us to observe much more realistically the interaction of different regions. In PET, a radioactive tracer is injected into the subject, and sensors are arranged around the head to detect the *emission* of *positrons*, which cause the release of gamma rays during radioactive decay. Snapshots of activity are taken for 'slices' through the brain (*tomography* comes from the Greek *tomos*, 'slice'), and these are combined computationally to provide an overall, three-dimensional reconstruction of activity in the whole brain. fMRI is even less invasive: individuals have to lie down surrounded by immensely powerful magnets (and have to be very careful to remove anything metallic, like jewellery, lest the magnets do it for them); they are then exposed to pulses of high-energy electromagnetic radiation at the frequency of radio waves. The nuclei of hydrogen atoms in water, or in blood within the human brain, behave like tiny magnets, and are therefore aligned within the magnetic field, absorbing energy from the radio waves and emitting it in patterns that can be detected and measured. Again, the chemical processes and patterns of blood flow in the brain are then reconstructed computationally.

While results obtained from PET and fMRI so far replicate many of the findings from earlier and more invasive techniques, they also provide additional and more accurate findings on areas like the relationship between the physical asymmetry seen in the pattern of brain sulci and gyri and the differential geography of the hemispheres, and the behavioural asymmetry determined by aphasia studies and electrocortical mapping. Evidence from fMRI studies do not suggest, for example, that the perisylvian region in Wernicke's area, the planum temporale, is actually larger on the left than the right, but rather that it may merely have a different shape (Westbury, Zatorre and Evans 1999). In addition, although the majority of individuals show a left–right hemispherical asymmetry in shape and left dominance for language processing, there are some individuals who have right dominance for language processing with the left–right asymmetry, and others with right–left asymmetry and left dominant language. For instance, Knecht *et al.* (2000) used transcranial Doppler tests, based on ultrasound, to investigate lateralisation for language in right- and left-handed individuals. Knecht asked for passive recall of as many words as possible beginning with a specific letter in a given time while monitoring blood flow in the different hemispheres, and found that 12/204 right-handers and 29/122 left-handers had language functions centred in the right hemisphere.

These newer methods seem likely to revolutionise the future study of the human brain, allowing for more individuals and activities to be investigated. Moreover, the technology involved will inevitably become gradually more flexible, less restrictive, and more direct, as signalled in the current development of EEG (electro-encephalograms) and MEG (multi-channel encephalograms), both of which allow observation of electrical activity directly, not through any kind of secondary emitted signal. We can therefore hope to come progressively closer to a naturalistic observation of the brain in action, though the costs of the equipment and expertise necessary to make and interpret these observations will ensure that these techniques will remain, for the foreseeable future, restricted to a relatively small number of research centres.

6.5   Evolution and the human brain

6.5.1   Stressing similarities

'Given what we know about the human brain, two facts stand out as astonishing: (1) We know very little about what distinguishes the human brain from that of other species; and (2) apparently, few neuroscientists regard fact 1 as much of a problem' (Preuss 2000: 1219). Of course, to understand differences, we have to be looking for them, and to be able to see them when we look; and from Huxley (1863) and Darwin (1871) until relatively recently, neuroscientists have tended to be preoccupied by similarities between the brains of different species rather than by differences. The emphasis on commonality seemed to be vindicated in the 1950s with the discovery of parallel functional organisation, with groups of similar cells appearing to act together in localised units within the neocortex of all species that *have* a neocortex. Such groups are composed of columns (or more correctly 'mini-columns', Mountcastle 1997), units of vertically arranged cells running through the cortex. Each column is typically 40–50 micrometres across, and consists of around 80–100 pyramidal cells and other neurons spanning the six layers of the cortex and interconnected within the column and to other columns. Columns may constitute the fundamental unit of computation in the cortex: they seem to be information-handling units that receive a series of incoming signals, assess the series and transform it into an outgoing signal (Mountcastle 1997, 2003), rather in the manner of an on–off switch which sends a yes or no signal depending on whether the sum of the input signals crosses a particular threshold.

Across species, columns vary in diameter by a factor of only one to two, though the brains in which they reside may differ by three orders of magnitude (Mountcastle 1997); this might suggest that columns perform the same types of information transformation wherever they appear in the cortex. In that case, evolution could then operate mainly by altering the number of columns in particular regions, or the function or interconnections of certain columns, not the internal organisation of the individual columns. This doctrine of continuity was useful in that it allowed, amongst other things, for model animals such as rats to be used to interpret the human system; but this assumed that humans do not have brain specialisations not found in rats (and conversely that rats do not have specialisations not found in humans). However, this is untenable in modern evolutionary thinking, where each species can be anticipated as having its own particular specialisms, as it will have adapted to its own particular environment. It follows that highly complex behaviours such as language cannot be assumed to be simply extensions of similar but less-developed functions found in other species, but must at least be examined as possible specialisations found only in humans. But the same is true of any highly developed trait found in any species. McAlpine (2005) notes that for at least 50 years, it has been hypothesised that animals are able to locate sounds in space around them because columns receive the sound in each ear, then effectively calculate the time delay between the nearer ear and the further away one. This model was successfully tested and confirmed for owls (Knudsen and Konishi 1978), where the relevant neurons form a tropic map of external three-dimensional space. The initial assumption was that this model could therefore be accepted also for mammals; but recent work suggests that this happy map-based knack is in fact a special property of owls (and note the fact that owl ears are offset, with one raised higher than the other, in an apparent adaptation maximising the efficiency of these spatial calculations). Mammals do something ostensibly similar, but using different strategies in different parts of the brain, reflecting the fact that they are solving different problems using different resources. Continuity or discontinuity, in other words, have to be considered and interrogated for each behaviour and for each underlying system.

In keeping with this more recent openness to discontinuity, much evidence has emerged that, in addition to behavioural specialisation across species, there is variability of cell shape within the columns (Glezer *et al.* 1993), and in the organisation of the columns within homologous areas in different species (see Buxhoeveden and Casanova 2002 or Mountcastle 2003 for a more detailed review). One particularly relevant example is the organisation of the mini-columns within certain areas of the left and right auditory association cortex. In humans, there appear to be wider columns with more space between them in the left hemisphere compared to the right, in contrast to similar regions in chimps and rhesus macaques, which appear not to be asymmetrical (Seldon 1981a,b, Buxhoeveden *et al.* 2001) – though the significance of these findings for language processing is as yet unclear (Hutsler and Galuske 2003). The important point here, however, is that if we anticipate possible differences between the brains of modern humans and other species, and if we have techniques like those reviewed in the last section for identifying these and their behavioural implications, we can begin to ask how, when and why these differences arose through the processes of evolution. In the subsections below, we shall consider such differences under the headings of changes in absolute size, relative size of particular brain regions, and specific alterations in the functional connections between and within particular brain regions.

6.5.2   Size matters?

'For two centuries, popular and scientific opinion assumed that intelligence is determined by brain size. As a result, studies of the brain have focused on this measure, at the expense of others ... There is still no proof that bigger means smarter' (Jones, Martin and Pilbeam 1992: 115). It seems, on the face of it, natural to assume that more means better; that having evolved more brain means more can be done with it. However, this ostensibly sensible starting point turns out to be both simplistic and problematic.

It is certainly true that there is a general trend for larger animals to have larger brains; so, for example, an elephant has a brain of around 4.6 kg, around three times the size of a human brain. But this simply means that using absolute brain size as a marker for increased cortex (which Goldberg 2001 sees as the 'executive brain', the control room overseeing processing power) will be confounded by body size. Taking this into account requires the use of a scaling calculation that controls for the effects of body size, such as the encephalisation quotient (EQ; see Jerison 1973). EQ involves an allometric equation, which assumes that there is a correlation between two factors (here, body size and brain size), but that they are not related in a straightforward, linear manner. Essentially, there appears to be a certain, basic requirement in terms of brain size to manage a body of a particular size; where we find brain sizes in excess of this expectation, we assume a capacity for additional brain processing or integrative functioning beyond the servicing of sensory input and body maintenance. There is, however, considerable argument over the 'correct' scaling when comparing across species, and even within species (Deacon 1990). Brain size can, for instance, be scaled according to the surface area of the organism, since this is argued to reflect the amount of sensory input to the brain, on the basis of a relatively consistent concentration of nerve cells across the skin of animals; but this is not uniformly true, since 'in primates the tiny retina is represented by more cortical surface than the entire surface of the body' (Deacon 1990: 202). Scaling can alternatively be based on energy requirement, or a range of other factors, but in each case there are caveats and exceptions which need to be taken into account.

These allometric equations, assuming we can agree which one to use, allow a non-linear relationship between two factors to be converted into a linear one, leading us to expect a straight line correlating brain and body size, and hence to predict an expected brain size on the basis of a particular body size. Where we find deviations from this straight line in one direction or the other, we have an indication that the species in question has either more or less brain than we would expect. However, it is not straightforward to extrapolate that a larger EQ will always mean more advanced behavioural ability. For example, Deacon (1997) notes that chihuahuas have a larger EQ than we would expect; that is, they lie above the normal range for carnivores of their weight. However, chihuahuas are not generally recognised as particularly clever dogs; their level of intelligence, as expressed in behavioural terms, is normal for dogs (at best). What has happened here is that some people (for reasons unclear to us; if you want a small animal, why not have a cat?) like little dogs, and consequently body size in dogs has been selected for smallness by human intervention in breeding in the ancestors of modern chihuahuas, whose brains are therefore (Deacon 1997: 166)

> fairly typical dog brains (possibly even deformed by miniaturization), with fairly typical dog abilities. They are just in very small dog bodies. Differences in dog encephalization are the result of breeding for body size effects ... There are numerous ways to alter the ratio between brain and body size in dogs that have nothing to do with selection for cognitive traits per se.

It follows that we must be very cautious in cases where EQ is used in attempts to explain apparent increases in cognitive functions in cross-species comparisons. Clutton-Brock and Harvey (1977, 1980), for instance, examined the differences in EQ among primates; they show that monkeys with a fruit-based diet are more highly encephalised than those that eat leaves. The potential problem arises when they try to explain these observations in an adaptationist paradigm, using a classic 'just-so story'. Clutton-Brock and Harvey suggest that the increased

EVOLUTION AND THE HUMAN BRAIN

memory, forethought and planning required to exploit patchy food resources such as fruit, which will provide the required resources only if you know which trees are nearly ripe and where they are in your territory, are greater than those needed to recognise an edible tree and strip it of its leaves. For leaf-eating, that is, you just turn up and go; but for fruit, you need the itinerary all worked out days or even months in advance. These hypotheses have been very influential, leading to putative connections between increases in human brain size and dietary changes, perhaps from fruit to meat (Dunbar 1992, 1993, and see further Chapter 9 below). But this argument only follows if an increase in EQ is always associated with an increase in intelligence; and since we have seen that this is debated, Clutton-Brock and Harvey may have fallen into Deacon's 'chihuahua fallacy'. The converse of the actual situation in chihuahuas and other small dogs would be a context where selection for *larger* bodies results in *lower* EQs within species, without obvious or necessary changes in intelligence. Since there is plenty of evidence that leaf-based diets require large digestive tracts, and thus select for bigger bodies, this might equally explain the higher EQs associated with habitual fruit eaters; and this could be exacerbated by the physical constraints of obtaining fruit from high in trees, which might also favour smaller body size (Fleagle 1985). Paraphrasing Deacon, we might then see leaf-eating primates as having fairly typical primate brains, with fairly typical primate abilities; they are just in rather large primate bodies. It follows that an apparent change in EQ must be taken as a marker for behavioural difference rather than as proof of increased intelligence or behavioural complexity; any further consequences would have to be probed by testing behaviour and other markers of intelligence more directly.

Of course, these criticisms can be circumvented to some extent by comparing EQ and body size simultaneously with other species from the same taxonomic group, which is precisely what Deacon (1997) does in showing that domestic dogs at the extremes for EQ are also at the extremes for body size. Consequently, Figure 6.7 shows a comparison of humans with the other extant great apes and a range of hominin fossils, based for the fossils on calculations of likely brain volume derived from cranial size, since brains of course do not fossilise directly. (Note that brain weight is readily translatable into brain volume terms, as 1 ml of brain tissue weighs roughly 1 g.) These comparisons indicate that although our body weights are well within the range of the other living apes, our EQ is well outside. Human brains really are the 'largest primate brains that have ever existed, both in absolute terms and relative to body size' (Deacon 1992: 115).

Figure 6.7 clearly shows that although early Australopithecines were close to the expectation from living non-human apes, there is an obvious increase in the size of the cranium relative to estimated body weight throughout the history of *Homo*. From the origin of *Homo*, there is a clear increase in absolute brain size without a dramatic change in body size. If we take a change of body size without a corresponding change in brain size (as for the chihuahuas) to indicate the likely operation of selection on body size, it follows equally that a significant change in

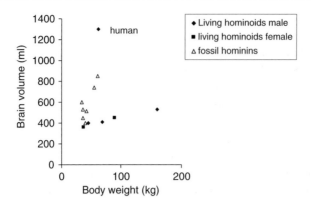

Figure 6.7 *Graph of brain volume in ml against body weight in kg for a range of extant hominoids, with estimated values for fossils derived from cranial size. (Adapted from Lewin 2005, using data from McHenry and Coffing 2000 and Holloway 1996.)*

brain size without an equivalent effect on body size might suggest that selection is acting on some aspect of the brain. The question is whether this is connected, directly or indirectly, to the emergence of or the capacity for language. We return to these issues in Chapters 8 and 9 below.

The difference in brain volume from Australopithecus to modern humans is around 1000 ml, and although this is a threefold increase, it is comparable to the maximum 'normal' range seen amongst modern individuals (Holloway 1996: 87 opposes Anatole France, winner of the Nobel Prize for Literature in 1921, with a brain weight close to 1000 ml, to Jonathan Swift, the rather earlier but at least equally famous author of *Gulliver's Travels*, whose brain was much closer to 2000 ml). There are numerous documented cases of individuals with microcephaly or severe hydrocephalus, whose amounts of cortical material fall well down the scale, within the range of modern chimpanzees; yet these individuals may appear almost functionally normal, and are certainly capable of speech (Giedd *et al.* 1996, Dorman 1991; OMIM 251200 for microcephaly, OMIM 236600 for hydrocephalus). This has been observed for individuals with both primary and secondary microcephaly (where the primary type involves a congenitally small brain for genetic reasons, and the secondary type is caused after birth, for instance by infection or a premature fusing of the skull bones which restricts brain growth). On the other hand, there is evidence of a correlation between IQ and brain size even within humans (Andreason *et al.* 1993), though methods of measuring IQ, let alone the issue of exactly what is being measured, are hotly debated (Gould 1981). So the significance of a particular degree of brain-size increase for the evolution of language is not clear or simple. It seems likely that this increase had some part to play, but this alone cannot tell us when, how or why spoken language arose. In other words, size does matter, but it is not the only thing that matters, and it may not matter in its own right as much as we initially think (Lewin 1980).

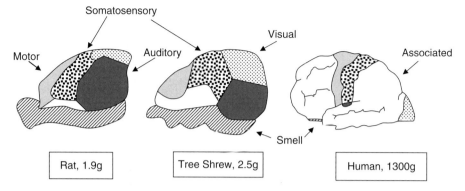

Figure 6.8 *Left lateral view of one cerebral hemisphere of three mammals drawn to approximately the same size, even though the human brain is much larger, as indicated by approximate brain weight shown. Note the relative reduction of the neocortex given over to primary sensory input and increase in association cortex between rat and human. (Redrawn after Nolte 2002 Figure 22–14: 538.)*

6.5.3   Asymmetry and development of specialised structures

The cellular structures of brains and their underlying connections do not fossilise, so it may at first seem that we can say little about the structural changes which may have emerged during the evolution of language. If we compare cytoarchitectonic maps for different species of mammals, we find within the cerebral hemispheres of the neocortex the same broad types of tissue: there are areas that seem to receive primary sensory input; others controlling motor functions; limbic areas which are involved with emotion and control of memory; and associative areas, which seem to be largely integrative in function, and are generally connected with higher-level processing and intelligence. However, Figure 6.8 shows that the human brain has a higher relative proportion of association cortex than other mammals, with a resultant apparent reduction in tissue dedicated to primary sensory functions.

As we have seen, in absolute terms, human brains are also very much increased in size; and it is unlikely that such increases will have occurred without some restructuring of the internal layout and organisation of the cortex and its connections.

There are several reasons for assuming that this increased complexity and size will have resulted in difficulties in maintaining the integration of functions; for example, processing that relies on the timing of incoming signals from different parts of a brain may become more difficult as brains grow larger and interconnecting axons longer, resulting in slower communication between neurons (Deacon 1990, Kaas 2000).

One common solution in species with larger brains is that more processing happens within each hemisphere in isolation before integration, which results in a higher degree of lateralisation of functions and/or a greater subdivision of processing tasks into local subroutines. This can be seen in the size of the human corpus collosum, the tract of axons connecting the two cerebral hemispheres, which is relatively much smaller than would be expected on the basis of animals with smaller brains if we assumed the same degree of connectivity between the hemispheres (Kaas 2000). This pattern of reduced connectivity within larger brains has been tested in other species where individual axons can be identified; for instance, connections between areas of the primary visual cortex across the two hemispheres are extensive in small mammals, reduced but numerous in prosimians, and practically absent in macaques (Kaas 1995, cited in Kaas 2000: 20). If this decrease in connectivity results in an increase in local subroutines within each hemisphere, we might also expect an increase in the tendency for new functions to be concentrated in one hemisphere, and hence for an increase in the basic level of asymmetry in larger brains (Gazzaniga 1995). The consequences of maintaining small-brain levels of internal connectivity would be severe; if the human brain were a scaled-up version of a rat brain in terms of connections between the hemispheres, the corpus collosum would have to be several feet long, and the human brain around the size of a bathtub.

Division into subroutines would lead to extremely slow processing of information from different sources if each one had to be handled serially; consequently, big but less well interconnected brains favour parallel processing. For example, in speech perception, incoming auditory signals appear to undergo initial decoding bilaterally in the superior temporal gyri, but subsequent processing is split, so that certain aspects of the signal such as slow changes of pitch associated with prosody are concentrated in one hemisphere, usually the right, while other aspects of linguistic processing relating to more rapid signal fluctuations like the spectral changes involved in formants occur generally on the left (Zatorre, Belin and Penhune 2002). In addition, left-hemisphere processing may in turn be split into two parallel streams that may correspond to a similar division of 'what' and 'where' visual processing streams (Milner and Goodale 1995), the 'what' being related to identification of an image, and the 'where' being the processing of the image's position in space relative to the brain and the activation of necessary motor patterns required to orientate the body towards the object. In hearing language a similar division, into 'dorsal' and 'ventral' streams (so called because of the relative direction of the processing waves across the cortex centred on the initial input in the primary auditory region) has recently been proposed (Rauschecker 1998, Hickok and Poeppel 2004). In this model, processing proceeds from the initial auditory decoding in two principal directions: the ventral stream (the 'what' stream) mapping sound to meaning runs from the temporal sulcus to the rear of the inferior temporal lobe, where the auditory signals are matched to conceptual representations that are widely stored throughout both left and right hemispheres; and the dorsal stream (the 'where' stream) runs via the parietal-temporal boundary to frontal regions which may be involved in tracking changes in incoming signal frequencies with time, and in localising the external source of the incoming signal. The dorsal stream may also be involved in mapping the incoming auditory signals to their corresponding motor patterns in speech perception and production. This division of analytical labour may reflect ancestral acoustic systems similar to those designed to process non-speech sounds, as in macaques, which have subsequently become specialised and highly integrated with memory and higher cognitive functions during the evolution of language in humans.

As we have seen, increased specialisation and decreased interhemispheric connectivity predispose to lateralisation of functions in the brain. Lateralisation in itself is not unique to humans; on the contrary, it is very common in mammals performing complex behavioural activities (see McManus 2002: Ch. 9 for a review of 'handedness' in animals, from whales that slap the water preferentially with one flipper, to chimps who are more successful if they show handedness in termite fishing). However, in many species the actual hemisphere that is dominant in controlling a particular task in a particular individual appears to be more or less random, whereas in humans a consistent pattern of right-hand dominance associated with left-hemisphere dominance for language is common, if not exceptionless.

Structures in the human brain do show clear physical asymmetry related (at least in part) to the functions of language, and it might be suggested that we should look for evidence of those in the fossil record. However, most of the functions of language considered above have indicated dispersed localisation through the brain, or at least through the left hemisphere; this means there are problems in deciding exactly how a particular difference in human brain morphology is related to language, and therefore what to look for in the fossils. As we saw in 6.4.1 above, for example, the posterior sylvian fissure extends further back in the left hemisphere than in the right, a fact ascribed to the expansion of the temporal gyrus into the planum temporale (aka Wernicke's area, or at least part of it); but the significance of even this well-attested asymmetry for language processing is unclear (Binder *et al.* 1996).

Even if we could agree on where to look, and what to look for, the fossil evidence is rather fragmentary and indirect. Our best information comes from endocasts, which are formed from marks left on the inner surface of the brain case by the developing brain during life and therefore partially reflect the structures of the living brain that was held within the skull. However, the living brain is covered by three membranes, the dura, arachnoid and pia mater, which intervene between the brain surface and the skull, obscuring the visible detail on the internal bone structure. In particular, the extra convolutions present in the human inferior frontal lobe (Broca's area) and posterior temporal and parietal lobes (Wernicke's area) 'are seldom well preserved on fossil endocasts ... and are areas of considerable interpretative controversy among paleoneurologists' (Holloway 1996: 84). Under very rare conditions, endocasts can occur naturally as the fossilising skull is filled with the surrounding material, such as mud, which subsequently becomes the rock matrix in which the fossil is embedded; but they are more normally created by flooding an empty skull with latex to take an impression of the marks left on the skull's inner surfaces. These casts give some indication of major features, but since fossil skulls are often partial reconstructions from fragments, and since the impressions are only slight in the first place, there are many opportunities for disagreement on what is represented.

One example is the position of the lunate sulcus, a curved groove that marks the edge of the occipital and parietal lobes in ape brains. This is set much further back in the human cerebrum, presumably as a result of the relatively larger amount of association cortex in the parietal lobe where major cross-modal integrative processing is believed to occur (Geschwind 1965), and the consequent relative reduction of primary visual processing cortex in the human brain. There has been a lengthy debate in the literature over the position of this lunate sulcus in the endocast of the 'Taung child' (*Australopithecus africanus*; Holloway 1979, Falk 1980, Holloway 1983, Falk 1985, Holloway 1996, and many others – for a flavour of the discussion, see the title of Holloway 1991, 'On Falk's 1989 accusation regarding Holloway's study of the Taung endocast: a reply'). These studies clearly indicate the problem in interpreting the bumps and ridges on a few preserved skull casts. However, the general outcome of this debate suggests that the lunate sulcus is not in a completely ape-like position. This is important because it indicates that *A. africanus* specimens more than 2 million years BP may already have had some brain reorganisation towards the modern human pattern, even though their brains had not yet radically expanded in size from the EQ ape norm. One early skull (KNM-ER 1470, usually attributed to *Homo habilis*, and with a cranial capacity of approximately 750 ml) is well enough preserved to indicate that by this stage, around 1.9 million years BP, there was already a relatively complex human-like pattern in the inferior frontal lobe corresponding to Broca's area (Holloway 1996, Falk 1983). None of these findings in themselves, however, translate in a direct or predictable way into consequences for the behaviour of the species; knowing that the brain was reorganising at a particular evolutionary point does not tell us why, or what the reorganised bits of brain were used for.

Moreover, Holloway and Lacoste-Lareymondie (1982) demonstrate a high degree of asymmetry in brain form among apes. For instance, 28/40 gorillas and 20/41 bonobos (*Pan paniscus*) had an expanded occipital pole on the left hemisphere, compared to 12/14 humans. However, although 11/14 modern humans in their study showed this left hemisphere occipital expansion in conjunction with right-hand frontal expansion, this correlation was not so clear in other species (although they **did** observe such a pattern in 13/40 gorillas, 4/34 common chimpanzees (*Pan troglodytes*), 14/41 bonobos (*Pan paniscus*) and 3/20 orangutans (*Pongo pygmaeus*). Among fossils, they detected this pattern in 2/3 early hominids (Australopithecines) (with an additional 4/4 showing left occipital expansion but with skulls too poorly preserved to see whether the right frontal expansion was in place), 11/12 *Homo erectus* and 7/9 Neanderthals. These findings are supported by other studies noting asymmetry in a range of other species, in particular chimps and rhesus monkeys (Yeni-Komishian and Benson 1976) and other primates (Heilbroner and Holloway 1988).

We shall return to the issue of lateralisation in Chapter 8 below, pursuing one further line of evidence not dealt with here, namely the potential connection of lateralisation for language with the dominant right-handedness for tasks shown by modern humans. It is possible that evidence of handedness in the making and use of tools through the hominin lineage might tell us something about the evolutionary trajectory of the brain. It is clear, however, that we cannot hope for immediate or conclusive answers about the evolution of language and the role of the brain from either the fossil record, or even from the increasingly fine-grained and naturalistic studies within cognitive imaging. These sources of evidence can help us build up a picture of what the modern human brain does in the context of language production and perception, and when approximately it would have had the size, structure and organisation to allow those language functions to take place. However, this leaves outstanding questions on how brains develop, and consequently how language functions are acquired or alternatively emerge, if they are genetically prespecified, in individuals. In the next chapter, therefore, we will start to look at how genes can be implicated in the development and evolution of the brain, and what genetic evidence and genetic defects can and cannot tell us about language.

6.6   Summary

This chapter has outlined the physical and cellular structure of the human brain, with a strong emphasis on the cerebral cortex. We considered the shape, size and considerable convolution of the modern human brain, and its internal workings in terms of neurons and axons. We also showed that the neocortex is conventionally divided into four lobes, each of which is associated with certain functions. Although we emphasised that the areas of the brain are not differentiated in any absolute way, and furthermore that certain functions (like those associated with language, indeed) are widely distributed in the brain, we also introduced cytoarchitectonic maps, which allow approximate areas to be distinguished for the purposes of inter- and intra-species comparison. These Brodmann areas are correlated with functions on the basis of a whole range of techniques, from studies of naturally occurring lesions, through electrocortical stimulation in which stimulation of the brain surface effectively generates short-term lesions, to the more recent development of less-invasive cognitive imaging methods like PET, fMRI and MEG, which identify blood flow or electrical activity associated with specific functions in the normal and normally functioning brain.

All these types of evidence, though they do not always by any means lead to clear and generally accepted conclusions, can be used in comparisons with the brains of other species. We considered issues of continuity and discontinuity between the modern human brain and the brains of other mammals, and specifically other primates. The salient features of the human brain from such comparisons involve its substantially greater size than would be expected for a great ape on the basis of calculations like the encephalisation quotient; and the substantial reorganisation (relative to other living primates and many fossil hominins) indicated by the asymmetry in size and function between the two hemispheres in the neocortex. Lateralisation per se is not unique to humans, but there is a particularly strong tendency in modern human brains towards a particular pattern of left–right asymmetry, which will be explored in more detail in later chapters. In short, the older picture of individual, physically bounded areas of the brain (like Broca's and Wernicke's areas), each responsible for a single language function and nothing more, is gradually being replaced by a more complex, dynamic and individually variable picture of diffuse, interacting networks of activity. Understanding more about these patterns of subroutines requires more information about how the brain is wired up; and this requires crucial reference to genetics.

Further reading

Good and accessible introductions to the relationship between language and the brain are to be found in Obler and Gjerlow (1999, especially Chapters 2 and 3), and in Dąbrowska (2004, especially Chapters 4 and 5); Dąbrowska also introduces the vexed question of the relationship between the mind and the brain. There is a good review of the discordance between the classical ideas of Broca and Wernicke and modern neurology in a special issue of *Cognition* Volume 92 (2004), summarised by the editors, Poeppel and Hickok in their introduction (pp. 1–12). Goldberg (2001: Ch. 4) provides a clear outline of essential brain architecture, and Nolte (2002) is an extremely detailed and plentifully illustrated manual of brain anatomy. Deacon (1990) is a clear critique of allometric scaling equations (like the EQ approach), and the difficulties of applying these in an evolutionary setting; and Deacon (1997) is an exceptionally detailed discussion of symbolic representation in the brain and how this may have evolved (though it is fair to say that your authors disagree on this one, with one finding it extremely approachable and the other putting it in the Very Difficult category; we do agree on the readability and clarity of Chapters 5 and 6, which cover the chihuahua fallacy and the influence of genes on brain growth, to be discussed further in the next chapter). Gould (1981) is an impassioned plea for not believing everything you read about IQ.

McManus (2002) provides an intriguing and detailed view of handedness in humans and its relationship to issues of brain asymmetry; Chapters 8 and 9 are particularly relevant to language. Gazzaniga (2000) is an Enquire Within Upon Everything volume on cognitive neuroscience, with a whole section on language; and Boatman (2004) provides an excellent review of the literature on electrocortical mapping and its relevance to speech perception. Issues of auditory processing in humans and other primates are also dealt with in Kaas, Hackett and Tramo (1999). We have deliberately not given any overview references on the newer techniques like PET, fMRI, EEG and MEG, because the field is simply moving too fast; but we recommend a regular scan through journals like *Behavior and Brain Sciences*, *Cortex* and *Nature*, and Ross (2010) might be a particularly good place to start with a useful historical overview. It is worth noting that many of the references given in the text for this chapter are to review articles, though there are lots of them because the specialist literatures are so extensive; these have not also been collected in this section, but can be found throughout the chapter.

Points for discussion

1. Read Dronkers *et al.* (2004). Their findings suggest that individuals with components of Broca's aphasia, though not the whole, classical condition, may have small, specific lesions which do not involve Broca's area at all, but other areas surrounding it. How do these specific lesions relate to particular deficits, and what might this tell us about language in the brain? Map the findings from Dronkers *et al.* on a Brodmann-type map of the left hemisphere. Do these findings challenge the existence of Broca's area, or of Broca's aphasia?

2. What can the various techniques described in this chapter tell us about the way words are stored in the brain? Do the findings all agree?

3. Find out whatever you can about the use of endocasts in discussions of the evolution of the brain for any specific hominin lineage (you might find it interesting to focus on *Homo erectus*, or the 'hobbit', *Homo florensiensis*). What controversies do you find from your reading? Are these inevitable, given the paucity of evidence from endocast studies? Can you think of other ways of resolving the questions raised?

4. Are the relationships between the cerebral hemispheres of left-handed individuals simply mirror-image versions of those for right-handers? If not, why not?

7 Language and genes

7.1   Overview

Through the last two chapters, we have built up a picture of the physical and neurological systems underpinning language in modern humans. We
have tried wherever possible to make suggestions about the stages in the human
family tree where innovations in vocal tract structure or brain size and conformation may have arisen; and perhaps even more importantly, we have introduced
all the multifarious reasons why fossil evidence does not lend itself to absolute or conclusive interpretation. The biological comparative method does allow
these human systems to be compared with those in other living species, and
reveals both continuities and discontinuities; this helps us begin to catalogue
those aspects of human anatomy and brain structure and function which may
be special, and may be linked with language. Investigations of modern humans,
both those with specific deficits and, increasingly, given the advent of new and
less-invasive technologies, those functioning normally, can tell us more about
how the brain contributes to language learning, production and perception.

However, we agreed at the outset that we would consider behaviours like language as reflexes of structures like the brain and the vocal tract; and in turn,
that those structures must be seen in an evolutionary context, as the outcomes
of inheritable modifications in the genetic code. Understanding the structures,
therefore, crucially means understanding the forces that modify genes in populations, and the mechanisms by which those genes interact with the environment to
build individuals. In the two main sections of this chapter, after introducing some
central ideas and concepts from genetics, we will therefore turn to the techniques
of modern population genetics and molecular genetics respectively. For population genetics, we shall concentrate on the dynamics of gene frequencies over
time, developing in more detail processes like mutation, natural selection and
drift which were introduced briefly in Chapters 1 and 2 above. If we understand
these forces, we may be able to reverse them to reconstruct the timescale for
certain changes among populations, even where direct evidence is not available.
Turning to molecular genetics, we shall consider how genes influence the construction of particular systems both during embryological development and later,
in association with learning. As ever, we shall find that understanding how systems work normally mainly follows from observing and explaining cases where
they go wrong. This section will focus on the brain, allowing us to reopen some

questions on the evolution of brain size and structure and their relevance to language which we could not fully answer in the last chapter.

7.2 What is a gene, and how does it work?

What do we mean when we say 'gene'? The technical meaning of this word has undergone radical change during the course of the twentieth century, and as the results from the Human Genome Project start to be examined, the twenty-first-century definition is becoming even harder to pin down in a simple way. At a meeting of the Sequence Ontology Consortium intended to determine what computer algorithms should look for in the mass of data from the genome projects, 25 scientists took nearly two days of sometimes acrimonious discussion to arrive at a working definition they could all accept (Pearson 2006). Although for our purposes much of the complexity can and will be ignored in the following discussion, it is important to bear in mind that this is consequently oversimplified in places.

Modern genetics has its origins in a monastery garden in the middle years of the nineteenth century where an Augustinian monk, Gregor Mendel, spent a decade counting and recording the characteristics of over 28,000 pea plants. Mendel's work, which was published in 1865 and completely ignored until 'rediscovered' by researchers at the beginning of the twentieth century, demonstrated that inheritance was controlled by packets of information that he concluded were passed from one generation to the next as effectively indivisible units (or 'genetic factors'). These information packets are what we now call genes. It is possible that Mendel's readiness to accept this idea of packets of transmissible information was encouraged by his two years of leave from the monastery to study botany and mathematics at the University of Vienna, at just the time when experimental physicists like Doppler, whose lectures Mendel attended, were setting the scene for work on atoms and elements. But his success in finding them lay in identifying the simple mathematical patterns revealed by breeding experiments using clearly defined characters.

Mendel studied only a few characters, such as seed colour and seed-coat texture, which he had established came in two clearly identifiable forms in his garden peas. For example, the seed coat of the garden pea can appear either wrinkly or smooth, and these are inherited characteristics that are passed from one generation to the next. Mendel crossed plants with different forms of the character together to form a hybrid generation (often called F1, for filial generation one). In many cases, the hybrids looked like only one of the parents – so, in our example of seed-coat pattern, the hybrids all had smooth, round peas. However, when Mendel then went on to cross the F1 individuals together, the next (F2) generation contained both wrinkly and smooth peas again, and always in the ratio of one plant of the type not found in the F1 generation to three of the plants with the F1 form.

Mendel deduced that each trait he studied was controlled by a single gene that existed in his population of peas in at least two forms (for the seed-coat example, these two forms would be wrinkly and smooth). In addition, each individual pea plant had two copies (now known as alleles) of each gene, which could be the same (both wrinkly or both smooth) or different (one wrinkly and one smooth). When a pea plant produces either pollen or egg cells, the alleles are partitioned at random into these sex cells (zygotes or gametes), so that each contains a single form of the gene. If both alleles in the parent plant are the same, then all the zygotes from that plant will be identical for that trait, and the plant is described as being *homozygous*. On the other hand, if the parent plant contains different alleles, then two forms of zygotes would result and the plant would be *heterozygous*.

In the next generation, the same principles apply, so that a new plant formed by fusion of two parental zygotes will in turn have two alleles, one inherited from each parent. The observable appearance or phenotype of a plant will therefore be determined by the alleles inherited for that gene. However, Mendel encountered cases where only one phenotype was found in the F1 generation, in our case peas with smooth seed coats, but where this could not be explained by absolute loss of the wrinkly allele at F1, because both smooth and wrinkly phenotypes then appeared in the following F2 generation. Mendel instead suggested that one allele (in this case, smooth) could be *dominant*, masking the presence of the *recessive* allele (wrinkly). The F1 individuals (which are all heterozygotes, remember, because of Mendel's breeding protocols) each produce zygotes, approximately half of which contain one smooth and half one wrinkly allele, by random chance. When these F1 individuals are mated, we predict that the offspring will fall into four classes in terms of their genetic characteristics, namely smooth/smooth, wrinkly/smooth, smooth/wrinkly and wrinkly/wrinkly. Since smooth is dominant to wrinkly, the first three classes will all have smooth peas at the phenotypic level, while only the last will have wrinkly peas.

Although Mendel had no idea about the physical structures underlying his observations, he was absolutely correct in his interpretations. By the time of the rediscovery of Mendel's work, microscopists using dyes to visualise the internal structures of plant and animal cells had discovered a central compartment which they called the nucleus. In most cells, this lacked obvious structure and was therefore homogeneously stained, but in actively dividing cells the staining was concentrated in paired structures (known as *chromosomes*, from the Greek 'colour' plus 'body') that differed in number from species to species. Each daughter cell following cell division will contain one member of each chromosome pair; this pattern makes chromosomes good candidates for Mendel's rather vaguely conceived 'genetic factors'. This chromosome theory of inheritance was developed independently in the early 1900s by Walter Sutton and Theodor Boveri (see Sutton 1903, Boveri 1904).

The physical nature of the inherited material itself was much debated through the early twentieth century, but was finally confirmed as DeoxyriboNucleic

Acid (DNA) by a series of experiments in the 1940s involving bacteria (Avery, MacLeod and McCarty 1944). Watson, Crick and Franklin (Watson and Crick 1953, Maddox 2002) determined the three-dimensional structure of this molecule, proposing the now iconic double helix, which can perhaps be slightly demystified by comparing it to a metal spiral staircase. Figure 7.1 is a diagram of DNA, showing the substructures to be discussed in more detail below.

DNA molecules are very long polymers, or complex chemicals, composed of strings of relatively simple subunits called nucleotides. Each nucleotide consists of three parts. The first is one of four different nitrogen-rich molecules called bases, which are either double-ring structures (the purines, adenine = A and guanine = G) or single rings (the pyrmidines, cytosine = C and thymine = T). The bases are like the treads of our spiral staircase (to continue the analogy introduced above), and they protrude from a backbone or handrail composed of the other two DNA elements, namely molecules of the sugar deoxyribose, and of phosphate. Deoxyribose has a basic ring shape consisting of an oxygen atom and four carbon atoms, conventionally numbered from 1′ to 4′. Next to the oxygen atom and attached to carbon 4′ is a further carbon atom, number 5′, which protrudes from the central ring and connects the sugar molecule to an adjacent phosphate group. The phosphate group in turn is linked to the next deoxyribose molecule in the chain by an additional protruding oxygen atom attached to the 3′ carbon. In other words, each deoxyribose molecule has its five-part ring (one oxygen and four carbon atoms), plus an extra carbon atom reaching out to the next phosphate molecule up, and an extra oxygen atom connecting to the next phosphate molecule down. This means that the linkage in the backbone is not symmetrical, and DNA is referred to as having a 5′ to 3′ orientation, referring to the carbons linking deoxyribose to phosphate (mediated by an oxygen atom in the case of carbon 3′). Carbon 1′ of each of the deoxyribose sugar molecules forms the attachment to one of the A, C, G and T bases, and it is common to refer to a complex of phosphate, sugar and base as a nucleotide named according to the base.

Watson and Crick's insight was to suggest that molecules of DNA actually consist of two of these strands of nucleotides wound round each other in opposite orientation to form an anti-parallel helix, with the two sets of bases pointing towards each other and held together, not by the strong covalent bonds that link the atoms within molecules, but by weaker and reversible interactions between hydrogen atoms in the base of one chain and either oxygen or nitrogen atoms in the other. (Sticking with the staircase analogy, each tread is composed of two bases, each protruding from one of the handrails, and connected by hydrogen bonds.) In fact, the pyrmidine thymine (T) always pairs with the purine adenine (A), and the pyrmidine cytosine (C) always pairs with the purine guanine (G), so that the presence of an A or a G on one strand will always predict the presence of a T or a C at the corresponding position on the other. This relationship means that either strand can act as a template for replication of the other when parent cells split during growth or prior to the production of sex cells (as Watson and

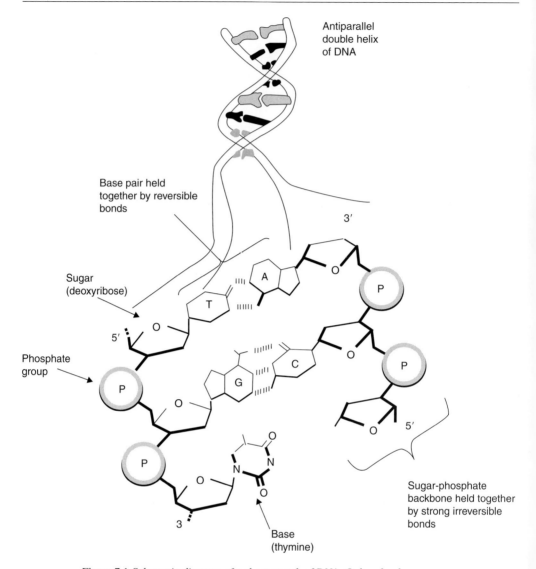

Figure 7.1 *Schematic diagram of a short stretch of DNA. Only a few bases are shown from a molecule many millions of base pairs long; for further details see text. In the diagram solid lines represent covalent bonds involving carbon atoms. For clarity, hydrogen atoms and carbon atoms are not shown. The hydrogen bonds that keep the two anti-parallel molecules together are shown for two base pairs as hatched lines. A purine (A = adenine or G = guanine) always pairs with a pyrmidine (T = thymine and C = cytosine), A with T or C with G.*

Crick (1953: 737) rather understatedly put it, 'It has not escaped our notice that the specific pairing we have postulated immediately suggests a possible copying mechanism for the genetic material'). DNA replication is therefore a semi-conservative process: during synthesis of new DNA, each copy of the parental molecule starts from one of the 'old' strands, which remains intact to act as a template and as a means of proof-reading for correction of errors induced in the 'new' strand in the process of replication. This means that DNA replication is tremendously accurate, and makes it possible, though highly unlikely, that half of the DNA present in any one of your genes may actually have been created in one of your dinosaur or even single-celled ancestors.

DNA, with its complex internal structure as described above, makes up the chromosomes within cell nuclei. The nucleus is rather like the library of the cell, and the chromosomes are the library shelves, each containing many different genes (while humans have only twenty-two pairs of chromosomes plus an additional two chromosomes: a pair of X chromosomes in females or an X and a Y in males, the Human Genome Project has revealed that we have approximately 30,000 to 35,000 genes). Each chromosome is a physically contiguous length of DNA, stretching for many thousands of nucleotides. Geneticists measure the length of DNA in base pairs, meaning the number of pairings of A or C on one strand with T or G on the other, and each gene has a characteristic length and composition. The average length for a gene is around 1,000 base pairs (or one kilobase), but we find much shorter and much longer cases: while the Histone H4 gene (which makes one of the proteins responsible for packing DNA into the chromosomes) is only around 500 base pairs long, the Dystrophin gene, which is damaged in cases of Duchenne Muscular Dystrophy (OMIM 310200), is around 2,500 kilobases long, or around 2,500,000 base pairs.

Pursuing our library analogy, the genes can be considered to be the instruction manuals of the cell, each containing the information needed to build a chemical called a protein, that in turn forms either a structure or a functional component of the cell – though we might argue that this genetic information is more like a magazine article than a book, with different parts of each article being physically interspersed with other articles, and adverts and other assorted bits and pieces of junk. For instance, though the Dystrophin gene mentioned above is around 2,500,000 base pairs long in total, the important information for protein-building purposes takes up only 0.6 per cent of this, distributed over 79 different, discontinuous sites along the gene (Strachan and Read 2004: 291). These meaningful instruction sequences are called exons, and are interspersed with introns, the intervening strings which may in some cases be implicated in timing and level of protein production, but in others appear to be parasitic, self-replicating chunks of DNA, or just plain meaningless junk arising from accidental but faithful repetitions of parts of meaningful sequences.

Genetic information is coded in a 64-word language, each word (or codon) being three letters long, with our now-familiar As, Cs, Gs and Ts as the letters. The linear arrangement of codons is read and converted by cellular machinery, producing an intermediate version called RNA (RiboNucleic Acid), minus the non-information-bearing introns. This is then moved from the nucleus to the factory floor of the cell where it is translated into a linear arrangement of amino acids which fold up according to their chemical properties into proteins: thus, ATG produces methionine (the universal signal for the start of any protein); CGG means the amino acid arginine; and TGA is one of the three STOP codons, which indicates the end of a protein sequence. There are sixty-four possible codons, which code for the twenty amino acids found in normal proteins, with three ways of saying 'stop'; and a sequence of around twenty or more amino acids will make a protein, though many proteins are much more complex. Virtually all the work of a cell and many of the structural characteristics of cells and organisms are determined by proteins. These can form long rigid rods, such as the collagens that give our skin and bones strength and flexibility; large globular synthesis machines, like the DNA polymerases that copy the DNA; or multifunctional chemical scissors, such as the caspases that are involved in programmed cell death (apoptosis). This occurs, for example, during embryonic development, where the hand begins as a paddle-like structure, and cell death leads to the separations between the fingers; and on a day-to-day basis in the destruction of damaged cells that would otherwise give rise to cancer (Earnshaw *et al.* 1999). We shall return to the properties of proteins and the interaction between development, genes and their products below.

First, however, it is worth tackling one common misconception, namely the idea that every gene is 'the gene for' some trait, and conversely that each and every trait has a corresponding gene. This one-to-one assumption is relatively easy to debunk: if there was a gene, for instance, for each and every one of the 3×10^{14} synapses estimated to connect the neurons in the human brain, we would quite rapidly encounter problems of 'poverty of the genetic code' (Müller 1996: 626), where there simply weren't enough genes to go round. Even in the case of a reasonably well-characterised genetic condition like Cystic Fibrosis (OMIM 219700), which primarily affects the lungs and digestive function, we find the same phenotype arising from different genotypes. In the UK, Cystic Fibrosis most commonly arises from cases where a child inherits two copies of the p.delF508 variant of the *CFTR* gene on chromosome 7, one from each parent: around 1 in 25 of us have one copy, which has no adverse phenotypic effects at all, because only a few molecules of the CFTR protein are needed for normal functioning. At the last count, however, in addition to p.delF508, there were 1,871 recorded variants in this gene associated with Cystic Fibrosis (Cystic Fibrosis database, www.genet.sickkids.on.ca/cftr/, accessed 14 May 2011); and in other parts of the world, different mutations from this set are more common causes of Cystic Fibrosis. At least in this case all the mutations are in the same gene; but the same phenotype may also arise from different genes – as Strachan and Read (2004: 479) put it, 'locus heterogeneity ... the situation where the same disease can be caused by mutations in several different genes ... is the rule rather than the exception'. For example, Osteogenesis Imperfecta (OMIM 166200, 166210, 166220, 166230), or brittle-bone disease, can result from mutations in two collagen genes, *COL1A1* on chromosome 17, and *COL1A2* on chromosome 7. On the other hand, different phenotypes may arise from the same genotype: the p.R117H mutation in the *CFTR* gene may be associated with full-blown Cystic Fibrosis, but can also result in male infertility with no other symptoms.

The (now rejected) one-to-one view leads, however, to the much broader and potentially more dangerous question of exactly what might count as a trait under genetic control, whether by one gene or several. We have already seen, and argued against, the rather mechanistic view that physical phenotypic traits arise inevitably from one particular genetic source; but worryingly, there is evidence that behavioural traits are increasingly being seen in this way too. We looked (in a seriously poorly controlled experiment) at whatever we were reading outside work for a week in May 2005 and the matching week in May 2006, and found, for instance, a discussion in the *Times Higher Education Supplement* (14 May 2005) about whether the predisposition to murder is genetic, with Steven Rose pointing out that, contrary to what one might then expect in terms of evolutionary theory, murderers are not necessarily known for leaving more children than the average, non-murdering member of the population. *Scotland on Sunday* (16 May 2005) suggested that anxiety may be 'in the genes', and a year later (21 May 2006: 5) led a story with the headline 'Doomed to failure by "poverty gene"'; while Miranda McMinn in *Red* magazine (June 2005: 87) claims she has 'the Overdraft Gene and the Can't-Stop-Watching-Reality-TV Gene. Don't blame me, it's my biological heritage'.

David Weatherall, in a *Times Higher Education Supplement* book review a year on (19 May 2006: 26), agrees that 'it is unusual to open a newspaper these days without reading an account of the discovery of a "new" gene for something or other. The list seems to be never-ending: heart disease, diabetes, divorce, homosexuality and many more'. He continues, however (2006: 27):

> Except in rare cases, we do not have 'genes' for schizophrenia or heart disease; rather we have a number of genes the structure of which may make us more or less susceptible to particular environmental insults and hence make us more or less likely to contract these diseases.

If this is true of disease, where in some cases we now have unassailable connections between a certain condition and a range of genetic variants, links between essentially societal and behavioural issues and possible genetic underpinnings must surely be treated extremely tentatively, if they are countenanced at all. Take, for instance, sickle cell anaemia (OMIM 141900), which has one of the more strictly localisable causes for a genetic disease, being caused by a single amino-acid substitution of valine for glutamic acid in the β-globin gene, which makes part of the haemoglobin protein responsible for oxygen transport in the blood. Sickle cell anaemia is a classic recessive condition: individuals show symptoms of this life-threatening disease when they have inherited two damaged copies of the gene. But although all sufferers, to a greater or lesser extent, have chronic anaemia, there are other symptoms, including stroke, renal dysfunction and pain crises, which are brought on by environmental circumstances, such as stress, bacterial or viral infections, and low-oxygen conditions, such as those encountered when flying, exercising or climbing. In fact, even non-affected parents of such affected individuals, who will have only one damaged gene, can be observed to have a small proportion of damaged, sickle-shaped red blood cells. These carriers of the disease will normally show no symptoms of anaemia, but again the environment plays a crucial role: if carriers exercise in low-oxygen conditions, for instance at high altitude, a larger proportion of their red cells will become sickle-shaped, and these can block blood vessels, which may lead to strokes (Ashley-Koch, Yang and Olney 2000, Kark *et al.* 1987).

If symptoms of the sickle-cell trait depend on the environment, then, can we say this condition is truly genetically determined? This takes us back to Mendel's wrinkled and smooth peas: again, we find a clear genetic change underlying the phenotype, but environment plays a part: we might not like the look of wrinkly peas so much, but they survive because they are particularly sweet, since the wrinkled appearance results from sugars not being converted into starch (Bhattacharyya, Martin and Smith 1993). Although in the case of sweeter peas (or smaller dogs, returning to the chihuahua story from the last chapter) we see deliberate intervention by humans over a relatively short period, there is also evidence of selection over far longer time-spans and without deliberate breeding programmes. Thus, the sickle-cell variant of the β-globin gene has survived in populations in spite of its catastrophic effects on affected individuals unfortunate enough to inherit two copies, because a single copy helps to protect against malaria. In many such cases, we are dealing with complex interactions of genes and environment, which are only observable and understandable at the level of the population, where we can see patterns developing over many generations because of the interacting forces acting on allele frequencies; and this is the topic of the next section.

7.3   Genes in populations

7.3.1   Mutation

There are three key forces acting on genes in populations, and these could in principle be divided and grouped in a number of different ways. We could lump together selection and drift as forces which tend to remove or reduce variation, as against mutation, which introduces it; or we could combine mutation and drift as stochastic or non-directional forces, as opposed to the directional activity of selection. We shall consider each of the forces in its own section; but it is important to remember that they tend to work together, or at least to operate alongside one another, though they may sometimes be pulling in opposite directions.

Mutation is the primary process by which genetic variation is introduced. So far, we have been suggesting that DNA copying is very conservative, if not error-free; and even Mendel's pre-DNA idea of 'genetic factors' seemed to rely on unchanging packets of information being passed on, like a baton in a relay race, from one generation to the next. In post-Mendelian terms, this means that in the absence of other factors, the allele frequency in one generation is predicted by the allele frequency in the previous generation, and the two will be identical. The mathematical formula relating the two generations in a random-mating and infinite population is the Hardy-Weinberg Law. (If you're especially interested, this says that the frequency of the genotypes A1A1, A1A2, A2A2 for a gene A with two alleles, 1 or 2, is p^2, $2pq$, q^2 where p and q are the frequencies of A1 and A2 in the previous generation and $p = (1-q)$. This predicts the allele frequencies of A1 and A2 in the next generation as p and q, where $p = (1-q)$, which is, of course, exactly the same as in the previous generation. See Hardy 1908, Weinberg 1908, Falconer and Mackay 1996.)

However, the main use of the Hardy-Weinberg Law in population genetics today involves attempts to explain observed deviations from it, which means there must indeed be interfering factors of one kind or another. Some of these follow naturally if we substitute real-world assumptions for the idealisations of the Hardy-Weinberg Law: populations do not tend to be of infinite size, and they are frequently geographically divided, leading to non-random mating on grounds of distance, even setting aside any kind of mate selection. Hermann Muller, working in the 1920s, demonstrated that mutations in genes could be induced using radiation, debunking the view of genes as indivisible and unchangeable packets; and now, of course, we know that DNA has a rather complex internal structure, so that even given the inherent conservatism of the replication process in the double helix, we would not be surprised if changes sometimes occurred naturally.

In fact, new variation can be introduced all the time via mutation during replication of DNA. For instance, when one strand is being used to reconstruct its opposite number, the wrong base can be incorporated by chance, so that instead of the intended T, a C may be added to the chain. The detailed proof-reading scheme of DNA will note a mismatch of A to C rather than the correct A to T, and in many cases this will then be corrected. However, even the best proof-reader can make mistakes, for instance by replacing the right base (here, the original A) rather than the wrongly inserted C. This is a more serious error: DNA will not replicate a mismatched base pair like A to C, but if the A is replaced by G to match the C, then replication can proceed, and we have an inheritable mutation. Because each base will be read as part of a set of instructions to make proteins, having a different base at a certain point can lead to the production of a different protein, which may mean a necessary job relying on the protein that would have been produced is not done, or that the new protein does something inappropriate for its current context. Alternatively, the new sequence may mean 'stop' so that no protein is produced.

Proof-reading error rates in DNA replication are estimated at around 10^{-7}, which is low, but not insignificant. And, as Muller's work suggests, naturally occurring errors in replication are not the only source of mutation: environmental factors like radiation and chemicals (such as the free radicals in smoke) can also damage DNA. Cells have two major repair mechanisms to deal with different types of error, removing and replacing either the mismatched bases, or a longer string of damaged DNA. But these correction mechanisms are much less efficient than the proof-reading process active during replication, and therefore allow more mutations through. Furthermore, the repair mechanism itself may delete or introduce a small number of bases, leading to further errors in reading protein-making instructions. For instance, a sequence of four three-base codons TAT CCT CAT AAA would make tyrosine, proline, histidine and lysine. Deleting the second T and therefore shifting the boundaries between codons gives TAC (tyrosine, so no change here), CTC (leucine), ATA (isoleucine), AA (uninterpretable). If you prefer a linguistic analogy, think of a sequence of four three-letter words, such as 'let ton eat ear'. Again, deleting the last letter of the first word and rearranging the results into three-letter words gives 'let (no change) one ate ar', where the last word is fragmentary and meaningless. (We couldn't get our four-word sequence to make sense, but are happy to offer a small prize for the best example sent to us that does.)

The examples above involve local changes, typically in single base pairs, which result in the production of a different protein, or in the failure of protein production. However, similarly local genetic changes can leave protein structure unchanged but alter the sequences of bases in the introns, between the codons which translate into particular amino acids. These introns, which are sometimes disparagingly seen as simply 'junk', also include strings responsible for controlling when, how much, and in which tissues of the body a particular protein is produced. So, mutations in the introns upstream of the β-globin gene can cause foetal haemoglobin to persist into adulthood, resulting in a mild anaemia (Collins *et al.* 1984); while *PAX6* (one of the Homeobox genes involved in gene regulation), which is usually switched on during embryological development in a group of cells which will become the future eye, can be affected by mutations which cause switching-off in humans (leading to blindness) and inappropriate switching-on in fruit flies, which may result in additional eyes on the abdomen or legs.

Mutations sometimes extend over more than a single base, and can involve substantial chunks of chromosomes, or even whole chromosomes: human chromosome 2 represents a fusion of two ancestral chromosomes, which are still distinct in chimpanzees and gorillas. Although this might seem rather dramatic (it changes, after all, the number of chromosomes characteristic of the species), it appears not to have affected any genes, since the fusion took place in the 'junk' area between genes towards the ends of the primate chromosomes. However, other major developments of this kind are found only in certain individuals, and do have dramatic effects, as with the Philadelphia chromosome, formed by a fusion of parts of chromosomes 9 and 22, and causing chronic myeloid leukaemia.

Chromosomal translocations of this kind have been used to identify the underlying genetic cause of diseases; the Philadelphia chromosome, discovered in 1960 (guess where?) was the first chromosome abnormality found to be consistently associated with any form of cancer. As another example, Sotos syndrome (or cerebral gigantism; OMIM 117550) is an overgrowth syndrome associated with several dysmorphic features, including a prominent jaw and large hands and feet, and mental retardation. Usually people with Sotos syndrome have no children, and the disease is sporadic, making it difficult to track the gene(s) implicated. However, one affected individual was found to have part of the tip of chromosome 8 transferred to chromosome 5; the breakpoint lay in a piece of DNA whose sequence matched that of a known gene in mice (*NSD1*). The function of this mouse gene is not entirely clear, but its protein is present in the nucleus of cells where it appears to be implicated in regulating the activity of many different genes – hence, perhaps, its involvement in an overgrowth disorder affecting many different systems. This example usefully illustrates the conservatism of DNA change (mice, after all, are some distance from humans in the average family tree); the usefulness of genes in animal models, here mice, which can tell us more about the operation of parallel systems in humans; and the advances being made in medical genetics in tracking familial conditions.

7.3.2   Selection

In contrast to mutation, which can be seen as the introduction of random changes into the genetic code, selection is a directional process that acts to remove some and increase other variants, decreasing the overall diversity in a population in the process. Natural selection is often seen as survival of the 'best' individuals in a generation through conflict and competition for limited resources such as food and mates, but really it is more like the survival of the genetic alleles best able to leave descendants, which need not be the same thing, as we shall discuss below. At its simplest, selection is the differential survival and therefore increase in frequency over several generations of one allele at the expense of another allele or group of alleles present initially in the population at the same locus. Eventually this process will result in an initially rare variant (remember any new allele has to start as a mutation in a single individual, no matter how large the population) spreading to 'fixation', which is the point at which all members of the population are homozygous for that variant. The rate of this process is particularly dependent on the degree of dominance shown by an allele. In the earlier discussion of Mendel's work on peas, we encountered the dominant smooth and recessive wrinkly traits; but although these characters are absolutely dominant and recessive respectively, it is also possible to find every degree of dominance in between, so that a heterozygous individual may lie precisely midway between the two homozygote phenotypes, or may be indistinguishable superficially, if not genetically, from one or the other. If the allele under selection is dominant in effect and beneficial, then selection will quickly favour it until it is very frequent; however, as this dominant allele approaches fixation, the majority of the remaining recessive alleles of lower fitness will be present only in heterozygotes who will in general mate only with the more numerous homozygote-dominant individuals – so that recessive allele can be passed on without discernible effect, and can persist at low frequency. For a beneficial but recessive allele in a large population, it may take hundreds of generations for a sufficiently large proportion of the population to be heterozygous, enabling homozygotes of the beneficial genotype to arise and be selected. Only in the so-called additive case where the heterozygotes demonstrate an intermediate phenotype, therefore providing a discernible surface signal of the presence of that allele, can selection act quickly on a new variant; where heterozygotes are completely intermediate between the two homozygotes, selection can take a new allele to fixation in a few tens of generations.

In most populations at most times, however, directional selection of the kind discussed above will not be operating. Rather, under constant environmental conditions it is likely that the vast majority of phenotypic selection acting on a population will be purifying or stabilising: most organisms are the product of many generations of evolution and have relatively good genetic conditions for being alive, so the less those are disrupted, the better. The corollary is that any change major enough to be visible in the phenotype is likely to have a detrimental effect, and this is even more likely for mutations of larger effect. Dawkins (1986) develops R. A. Fisher's analogy of a microscope that is close to being optimally focused: random very small adjustments in the mechanism might improve the focus, but they are at least as likely to make it worse, while adjustments with a hammer are very unlikely to improve optical clarity. So in a relatively constant environment, selection will usually act to remove most new mutations, as these will typically move the phenotype away from the optimal norm (as has been observed in humans, where neonatal infant mortality is higher for larger and smaller babies: Karn and Penrose 1951). On the other hand, when the environmental conditions faced by a species change, selection can rapidly increase the frequency of mutations or variants in a population that are beneficial in the new environment, resulting in a rapid change of phenotype in a relatively short time.

In the fossil record, it sometimes seems that periods of phenotypic stasis are interspersed with rapid replacements of one phenotype with another (although it must be remembered that a 'short' time in evolutionary terms might be in the order of 200 to 1,000 generations for a large population). This has come to be known as the punctuated theory of evolution (Eldredge and Gould 1972), and has been considered by some to argue against the slow accumulation of variation predicted by Darwinian natural selection. However, it should be clear from the above that even in times of stasis, mutation and selection are still acting on the population at the same rates as during the 'punctuations': it is only the environment and therefore the directionality of the selection that is different, in one case favouring the status quo and in the other favouring alleles that result in a change in phenotype towards a new optimum. Under this interpretation, apparent periods of greater or speedier activity in the fossil record would be completely reconcilable with a Darwinian perspective, making punctuated equilibrium a description of speciation rather than a specific type of evolution requiring its own, independent model or refuting the Darwinian gradualist paradigm (see Chapter 8 and points for discussion below; see also Dawkins 1986).

Mutation and selection, then, can work together in particular environments to develop complex phenotypes which appear to be designed for, or especially fit for a particular purpose; and this is unsurprising, as long as we remember always to interpret 'fitness' as meaning 'fitness for a specific way of life', not in any abstract and global sense. Everything in evolution, in other words, has to be interpreted in a strictly local fashion. Dawkins (1976), for instance, suggests in his 'selfish gene' proposal that the ultimate unit of selection is the gene, and that bodies, or phenotypes, simply represent a way for genes to make more genes. Likewise, Ridley (2003: Ch. 9) identifies seven different meanings commonly used for 'gene', and takes the example of the gene *SRY* to illustrate how selection cannot straightforwardly be construed as survival of the fittest at the phenotypic level – 'fitness' must instead be understood in terms of allele survival. This example also demonstrates how the timing of particular expression patterns can result in alterations and repercussions throughout the life of an individual.

The *SRY* gene is found at the top end of the small arm of the human Y chromosome. It is one of the smallest in the genome at 612 bases long, with a single coding exon, and is found in all male mammals from rats and mice to humans and chimpanzees. This gene gives rise to a protein of 204 amino acids which is produced during a relatively short period of embryonic life both in the brain and in a small strip of tissue called the primitive streak; in mice, this usually happens on day 11 of gestation. If the SRY protein is undamaged and functioning normally, a mammal carrying this gene will develop into a male of its own species, so mice engineered to have no Y chromosome but to carry the *human SRY* gene will develop nonetheless into male mice. The SRY protein, called the testis-determining factor, is in fact a gene regulator or transcription factor, which switches on at least one other gene, *SOX9*, in the embryonic brain and primitive streak; the latter will then develop into the testes, whereas in the absence of *SRY* it would become the ovaries. Once *SOX9* is activated, the *SRY* gene plays no further part in development (as far as we know at the moment: never say never, given the current state of genetic investigation). *SOX9* in turn switches on and turns off a whole cascade of different 'downstream genes' (with names such as *Wnt4* and *Dhh*), with many different interacting changes in other genes. These in turn alter hormone levels in the developing embryo, resulting in the specific growth characteristics that distinguish males from females. It should be stressed that for the developing body tissues, for example in the abdominal fat depots, the presence of male hormone levels is only one of many 'environmental factors' affecting the expression levels of different genes and hence the development of the tissue. Many of these environmentally induced changes in gene expression continue to be influential throughout life: to these have been ascribed effects as different as male-pattern baldness, and the tendencies to tinker with gadgets and to die earlier of heart disease, but the environment will also have its continuing interactive effect, as in the influence of diet on obesity and male heart attacks. Ridley points out that we could therefore see the aggression and excessive risk taking of adolescent males as resulting from the activity of the *SRY* gene, which we might then think of almost as (paraphrasing Ridley) a 'getting-yourself-killed-young gene'. And yet the highly conserved nature of *SRY* across species indicates that selection has been working very hard to maintain the functioning of this gene, a statement that may at first appear to go against the Darwinian idea of natural selection, where a gene for earlier death should be rapidly heading for extinction itself.

Sexual selection may come into play here as a way of resolving this apparent paradox. If males who take risks and survive get more sexual partners and produce more children than males who do not take risks, for whatever reasons, then the inclusive fitness of an allele that encourages moderate risk taking may be higher than one that makes its bearer risk averse, where inclusive fitness is the probability of survival to reproduction multiplied by the number of descendants likely to be produced. Of course, in humans the nature of the risks taken and the likely outcomes will depend on the cultural environment of the individual. Many aspects of human culture that do not seem to be obviously 'advantageous' to the individual such as art, dance, humour, singing and ritual are pursued with considerable energy by human males in their late teenage years at around the age of mate selection. From a direct survival-of-the-individual point of view, these behaviours may seem wasteful; we might not see them quite so readily as downright dangerous, though story telling, for instance, might easily attract a hungry predator. So, these behaviours have a survival 'cost', whether directly (the hungry predator scenario), or because the story-telling time could be spent doing something more immediately productive, like hunting; but when seen in the light of female sexual selection of the 'best' mate they could make a lot of sense.

Sexual selection has been seen to be one of the more powerful forces for evolutionary change. However, there is still discussion as to whether or not the process is merely a runaway feedback loop between alleles slightly favouring mate choice for a particular characteristic resulting in offspring who have either that characteristic (male) or a preference for it (female), or whether there is truly an underlying 'good genotype' benefit of mate choice related to the calculation that only a potential mate with the best underlying total genetic make-up can take risks or invest the time required in learning to sing well, for instance, and still survive to reproductive age (Ridley 1994, Hamilton and Zuk 1982). Either way, if taking risks and surviving is seen as an attractive character for mate choice and a genetic variant, or group of genes, make it more likely that the body they are in behaves in this way, then those genes are more likely to survive and increase in frequency in subsequent generations until they dominate, even if they do not have a direct survival value to the individual. Indeed, the same result will be predicted even if they are mildly detrimental, for instance by increasing the risk of being eaten by a predator alerted to your presence as you practise scales and arpeggios.

7.3.3   Drift

In the sections above we may have given the impression that most mutational changes in DNA have some major effect on a phenotype, but this is far from the truth. The Human Genome Project has shown that less than 5 per cent of our DNA actually codes for protein structures; and while some of the intervening material is implicated in regulating the tissue specificity and timing of protein production, as in the cases of foetal haemoglobin or extra eyes in fruit flies discussed in 7.3.1 above, much has no known function. Changes in this material are consequently unlikely to have dramatic effects on phenotypes, and variation in certain classes of junk DNA, in particular sequences of two, three or four nucleotide repeats that lie next to each other throughout the genome in blocks of different length, are so variable that each individual has effectively a unique pattern that can be used to identify or exclude them in forensic investigations (Jeffreys, Wilson and Thein 1985, Evett and Weir 1998).

Even mutations affecting the protein-coding region itself may not have any effect, because of the redundant characteristics of the genetic code. Each of the twenty possible amino acids used in proteins, with the exception of methionine, is coded for by more than one codon (see Strachan and Read 2004, inside front cover, for the whole codon usage table, although note that this substitutes U for T, as the table is based on the RNA copy of a gene, where uracil replaces the thymine present in DNA). As an example, the codons TCT, TCC, TCA, TCG, AGT and AGC all code for serine, and GTT, GTC, GTA and GTG for valine. In general, then, mutations that change the third base in the codons tend not to have any effect on the amino acid added to the protein chain at that point and are said to be synonymous. The major factor affecting the survival and eventual fixation of such variants in a population is chance.

Indeed, even for mutations capable of generating different amino acids (non-synonymous substitutions), chance plays a major part in initial survival, since every mutation has to occur in a single DNA strand, so there is a 50 per cent chance that the new variant will not be included in any particular sex cell and hence will be lost immediately after creation.

Chance factors can also dramatically affect the frequency of alleles in small or subdivided populations. Sampling a finite population of alleles from a previously finite population of alleles generates a stochastic, directionless 'wander' in allele frequency from one generation to another, a process known as random genetic drift (Wright 1931). This process affects both selected and neutral alleles and although the direction of drift cannot be determined except with hindsight, the amount of change from generation to generation can be predicted from the size of the breeding population. In real-life populations, the number of breeding individuals is not simple to determine and is often very different from the census population size (the number of individuals alive in any one generation), so that an idealised value N_e (the effective population size) has to be used instead (Wright 1931). Many different factors determine the value of N_e, including the past history of the population size, non-random mate choice, or differences in the number of offspring per individual, and it is often much smaller than the apparent number of individuals in a population (see Jobling, Hurles and Tyler-Smith 2004, section 5.3: 131–8).

In a population with two neutral alleles at a locus, drift alone predicts that one allele will eventually become fixed and the other lost, but it does not predict which one will actually be lost for any particular population. So, in a subdivided population originally containing two alleles at 50 per cent, drift may result in individual groups diverging, or drifting apart, in allele frequency until eventually all are fixed for one or other of the originally variable alleles. The neutral theory of evolution (Kimura 1968, 1983) argues that the majority of DNA differences between populations and species results from this process, and predicts that any new mutation has a probability of becoming fixed of $1/2N_e$, and it will take $4N_e$ generations to get there. In smaller populations, new alleles will therefore have a higher chance of becoming fixed by chance drift alone, and will take fewer generations to reach that point. It follows that speciation is more likely to occur in peripheral, isolated populations, as these are typically smaller.

The neutral theory alone might seem to predict that all proteins in the same species should change at the same rate; but in fact we can explain differences across species in the observed rate of substitutions (a measure of evolutionary change) by putting the neutral theory together with the argument in 7.3.2 above that selection is often purifying or stabilising at the phenotypic level, removing detrimental changes. Variation in rate will then reflect the extent of balancing or purifying selection, and this in turn depends on the inherent flexibility of the structure of a particular protein, or how far it can tolerate internal substitutions in amino acids without compromising or changing function (Nei 1987). Haemoglobin, which we encountered earlier in the context of sickle cell disease, is a globular blood protein that binds oxygen in the lungs and swaps this with carbon dioxide in the tissues. The oxygen-binding reaction takes place in a functional pocket or indentation in the protein, with the rest of the molecular surface effectively acting as a framework for the pocket. It has been shown that changes to *any* of the nineteen amino acids within the pocket involved in oxygen binding are associated with a reduction in the oxygen-binding function, and that these amino acids are relatively invariant in haemoglobins across many different species. Thus, while the rest of the haemoglobin molecule is relatively free to accumulate changes in amino acids by the drift of effectively neutral changes to fixation, very little change if any can be tolerated in this specific area of the molecule. At the other end of the spectrum, neutral drift reaches a maximal rate of 9.0 amino-acid changes per site per 1,000,000,000 years in the fibrinopeptides, whose amino-acid sequence is known to have little effect on function, compared to 1.2 for haemoglobin alpha and 0.01 for Histone H4 (whose structural shape is absolutely critical to its function of binding and packing DNA) (Kimura 1983).

This suggests that where we see low relative frequency of amino-acid substitutions between species, we can interpret this as a marker for constrained evolution, and assume preservation of an important function across evolutionary time. At the same time, this trend establishes one vital aspect of evolution, namely that proteins which perform basic functions of cellular activity will be highly conserved across distantly related species. This is not to say that selective changes do not occur, but only that they are rarer than neutral changes; sticking with haemoglobin, the crocodile and human versions differ by 123 amino acids, since crocodile haemoglobin has specific ways of binding oxygen and carbon dioxide that allow it to function in the highly acidic blood conditions that develop during the reptile's prolonged dives under water. These changes can be thought of as adaptations, and therefore the results of active selection in the crocodile's ancestors; but only 5 of the 123 differences between the crocodile and human haemoglobin genes are materially involved in explaining these changes in function.

By comparing the frequency of substitutions between species, it is therefore also possible to suggest when a gene or part of a gene has changed function or at least come under different selection pressures. In blind mole rats (Hendricks *et al.* 1987) the alpha A-crystallin gene, which makes the lens of the eye, no longer has to produce a protein that is completely transparent to light, since these animals live their entire lives underground in the dark. Although hamsters and mice do not have a single difference in amino-acid sequence for this gene, mole rats differ from both by nine amino-acid changes; on the other hand, all three species do differ in the sequence of the synonymous 'third-base wobble' changes. The observed rate of synonymous substitution averaged over all sites where a synonymous change *could* happen gives a value K_S, and the corresponding rate for non-synonymous sites gives a value K_A. The ratio of K_A to K_S can then be used to detect the strength of any relevant evolutionary constraint.

This is only one of a whole series of tests which geneticists have developed, based on the theoretical interaction of drift, selection and mutation, to examine observed levels of diversity within and between populations in order to estimate or detect the action of selection, for example (see Jobling, Hurles and Tyler-Smith 2004: Ch. 6, and Sabeti *et al.* 2006 for reviews and examples of recent results). In the rest of this chapter and later in the book we will consider how these arguments about selection and drift, along with comparisons of synonymous and non-synonymous substitution rates, have been used to identify genes implicated in the development of the human mind and specifically associated with language.

7.3.4   Genes and population histories

If we understand how the forces of mutation, selection and drift interact to generate diversity between populations and eventually between species, we might anticipate being able to use present-day diversity to reconstruct the action of those same forces in the past. Moreover, because those forces are affected by parameters like population size or time since the last common ancestor, knowing the present-day outcome in terms of patterns of diversity should allow us to estimate values for those parameters through the histories of the relevant populations. As we already know, however, the human genome consists of 46 DNA molecules, the chromosomes; and each molecule can be inherited independently into subsequent generations. Every chromosome can therefore have its own history and tell a different story about the population in which it is found. It follows that thinking in terms of a single history for each population, even for each individual, considerably oversimplifies the real and complex situation.

We also have to take into account the possibility of recombination, or crossing over of parts of paired chromosomes during the production of sex cells; this occurs at chiasmata, or special crossing points, which are more or less randomly positioned along all the paired chromosomes. There is an approximately 1 per cent chance of recombination per 1,000 kilobase pairs (= 1 Megabase) per generation, and since the average length of a chromosome is 140 Megabases, this means we should expect to find at least one recombination event for every chromosome in every generation (Strachan and Read 2004). This random assortment means that each locus effectively gives an independent history of the influences on the population; but recombination means that each chromosome is also potentially a patchwork of ancestral molecules (Pääbo 2003).

One hopeful possibility involves the study of individual genes and the diversity in surrounding areas to detect 'selective sweeps': in other words, in cases where a particular allele has been under intensive selection pressure, we might expect to see virtual homogeneity across the population under selection for that locus, which will therefore stand out from the normal variability found in the surrounding areas of the genome, indicating a locus of likely evolutionary importance. Again, the confounding factor is that recombination can occur within the selected region, disrupting this pattern of homogeneity; because this is more likely as the time depth increases, the power of this technique is greatest in diagnosing relatively recent cases of selection.

In addition to the paired chromosomes, which contain the autosomal gene loci, cells also contain mitochondrial DNA; this is inherited only through the female line but found in both males and females. In males, we also find a Y chromosome, which we encountered in 7.3.2 above in our discussion of the *SRY* gene. The mitochondria and Y chromosome are effectively each a single large non-recombining locus; their size relative to any locus on the paired chromosomes means there are more potential sites for mutation, while the absence of recombination makes it more straightforward to reconstruct the history of the whole molecule. The mitochondria for women, and Y chromosome for men, are then preferentially used to construct gene trees using cladistic methods. Each mutation is treated as a unique event, and shared derived mutations (or synapomorphies) can be clustered hierarchically to visualise patterns of inheritance and split. Moreover, counting the number of mutations on each branch (and assuming a more or less constant mutation rate) allows approximate dating of split points, and hence calculation of the length of each phase of a population's history (see Figure 7.2, and Page and Holmes 1998).

However, any gene tree can provide only a single picture for a single molecule. As we have seen, this may have entered the population only very recently and, more generally, different molecules may not share all their history with other molecules in the population of interest (as shown in Figure 7.2b). Constructing

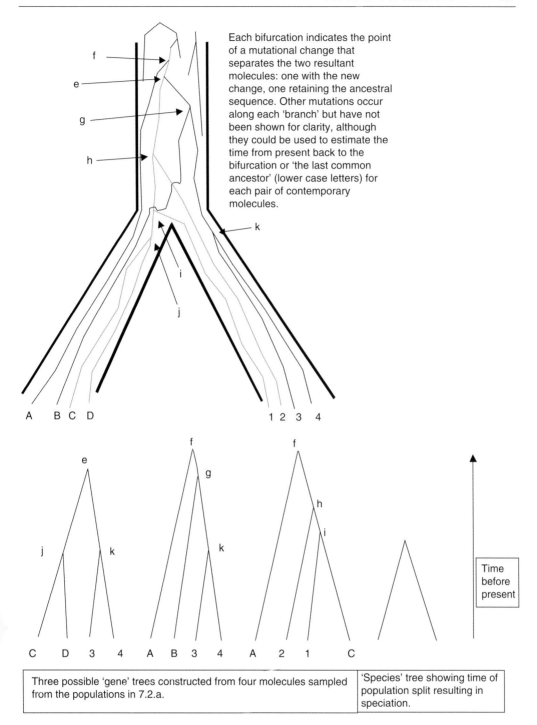

Each bifurcation indicates the point of a mutational change that separates the two resultant molecules: one with the new change, one retaining the ancestral sequence. Other mutations occur along each 'branch' but have not been shown for clarity, although they could be used to estimate the time from present back to the bifurcation or 'the last common ancestor' (lower case letters) for each pair of contemporary molecules.

Three possible 'gene' trees constructed from four molecules sampled from the populations in 7.2.a.	'Species' tree showing time of population split resulting in speciation.

Figure 7.2 *Diagram of species tree and gene trees*

an overall population history would require summing across many molecules sampled from a representative range of individuals. This requirement brings its own problems. In the case of more distant population histories, we might wish to employ new techniques for extracting DNA from fossils (see Krings *et al.* 1997, Höss 2000, Richards and Macaulay 2000, and Schmitz *et al.* 2002 for work on Neanderthals); but even when these are successful, we should not expect to find many individuals from a particular population or time period to apply them to. On the other hand, when we turn to the present-day situation, how are we to determine what a population is or how it is to be sampled? Is it sensible, for instance, to talk about the population of a small, rural community in 2011; or a major city like Edinburgh, London or New York; or a country like France, Wales or Ecuador? Might we need to develop different sampling protocols for these and, if so, will the results from one kind of community be strictly comparable with those for another?

7.3.4.1   Distant histories

Chromosomal comparison of great apes and humans indicates a close similarity, particularly between humans and chimps, although there has been significant reordering of chromosomal regions by inversion (where areas of a chromosome are turned upside down in one species relative to the other), translocation (where part of one chromosome moves to another) and, in the case of human chromosome 2, fusion (Yunis and Prakash 1982). As for populations within a species, molecular analyses can be used to build individual gene trees with accompanying dates; these can then be averaged to give the best overall species tree. The overall conclusion is that humans and chimps are the closest relatives, having shared a common ancestor 4–7 million years ago, with gorillas as a sister group at 6–9 myBP and orang-utans as outliers at around 12–15 myBP. These composite dates are based on whole genome hybridisation profiles, involving heating, mixing and cooling whole genomes from the different species, and counting the overall differences as they anneal (Sibley and Ahlquist 1984), as well as on data from individual loci (e.g. Chen and Li 2001). These ranges of dates reconcile well with the assumptions from fossil evidence which we have used in previous chapters (Stauffer *et al.* 2001), though there are some apparent inconsistencies with respect to earlier hominin groups. To take one example we have not mentioned before, *Orrorin tugenensis* has been claimed as a direct human ancestor (Senut *et al.* 2001); since fossils from this group are dated around 6 myBP, this would argue for a human–ape split around 8 myBP (Pickford and Senut 2001). However, if we agree that the genetic evidence now conclusively dates this split between 4 and 7 myBP, we must necessarily review the status of *Orrorin*, which would now have to be interpreted as an ancestor of the chimp–human line, or a sister species of *Homo* which has not been maintained.

The chimp and human genome projects are beginning to allow whole genome comparisons to establish what genetic differences underpin observable phenotypic differences. True, we are almost identical to chimps in our coding DNA sequences, differing on average by only around 1.6 per cent of our nucleotides; but we must remember that all living things share basic cellular functions based on genes which evolved early and have effectively remained unchanged (in other words, being 98% chimpanzee appears in a slightly different perspective when we consider that both chimpanzees and humans are more than 30% identical with daffodils and parsnips). On the other hand, there is considerable difference between humans and chimpanzees in the non-coding genetic material, with extensive and different patterns of insertion and deletion of genetic sequences, correlating with a range of quite striking differences in the tissue-specific levels and timings of gene expression (Enard *et al.* 2002a, Chimp Sequencing and Analysis Consortium 2005, Gilad *et al.* 2006).

7.3.4.2   More recent population histories

One significant result to emerge from the comparison of human and ape population variation is the striking lack of diversity within the human species. Studies of mitochondria (Gagneux *et al.* 1999) and of the X chromosome locus, Xq13.3 (Kaessmann *et al.* 1999, 2001) indicate that compared to the other great apes, humans have a considerably lower level of within-species diversity, with variability across the whole human species being on a par with the level present in two local populations of chimps living a few hundred miles apart (Gagneux *et al.* 1999).

This observation flatly contradicts the expectation that the largest and geographically most spread populations (which in the current cross-species comparative context would definitely be us) should support the highest levels of diversity, and has been taken to indicate that the whole current human population has derived from a relatively small group of individuals in the recent past, as proposed by the out-of-Africa model of recent human origins (for discussion and references, see Chapter 4.4.2 above). Using calculations based on the theory of drift, a minimal population size associated with the speciation of *Homo sapiens* (the speciation bottleneck) has been estimated as having an N_e (effective population size) of 10,000 individuals. If, on the other hand, we adopted the multiregional hypothesis, whereby *H. sapiens* populations evolved from a widespread *H. erectus* population, reasonable assumptions on relevant parameters predict an N_e of 40,00–300,000 (Harpending and Rogers 2000). However, global analysis of polymorphisms (alternative patterns in the genetic conformation of the same protein) at many loci within human populations also indicates a large number of low-frequency polymorphisms suggestive of a population expansion worldwide from a single, rather small and recent ancestral population. In addition, most genetic systems that have been examined by phylogenetic reconstruction indicate trees rooted in Africa with very recent coalescent ages (the date of the last shared ancestor at the root of the tree); so, for example, three most recent common ancestor estimates are 122–222 kyBP for fifty-three complete mitochondrial genome sequences (Ingman *et al.* 2000); 40–140 kyBP for 64kb of Y chromosome from forty-three individuals (Thomson *et al.* 2000); and 400–1300 kyBP for a

2,670 base-pair region of the β-globin gene on chromosome 11 tested from 349 chromosomes (Harding *et al.* 1997).

On the face of it, these types of genetic evidence argue strongly against the multi-regional hypothesis, though it might be objected that the coalescent age dates above are rather discrepant, with the third range of dates both wider and substantially earlier than the other two. However, autosomal loci are expected to give older dates than mitochondria or Y-chromosome trees because they tend to involve larger effective population sizes (having two copies per person), and recombination also allows sampling of a higher proportion of polymorphisms present in the last common ancestral population. Even after a population split, a particular polymorphism may be present in both the new species, though in fact it arose much earlier in the shared history of their common ancestor; this means that gene trees need not match species trees. If we then based our coalescent dates on this particular polymorphism, we might significantly overestimate the time since population divergence (see again Figure 7.2). In fact, the neutral theory predicts that a polymorphism will persist for an average of $4N_e$ generations (that is, 4 times the effective population size), so even assuming the smallest estimate of 10,000 for the ancestral human population, this represents 40,000 generations × 20 years, or a minimum potential error of 800,000 years per locus. A discrepancy of 800,000 years between the coalescent date and the true age of the population split is an insignificant error when comparing speciation events that occurred 4–10 million years ago, but is highly relevant when dealing with the origin of *Homo sapiens* a few hundred thousand years ago. Moreover, a particular polymophism may persist by chance for much longer than this average, and selection can favour maintenance of particular polymorphisms for very long periods indeed, as in the DR locus, where chimpanzee and human populations share common class II Major Histocompatability Complex (MHC) polymorphisms involved in immune responses (Fan *et al.* 1989).

Templeton (for example in Jobling, Hurles and Tyler-Smith 2004: 261) has argued that far from being a problem, the longer-lived nature of autosomal polymorphisms means that these data must be given priority when testing the recent out-of-Africa versus multi-regional models of human origins. Combining eight autosomal loci (with gene trees rooted at 670,000 to 8.5 million years ago) with Y-chromosome and mitochondrial data in a nested cladistic analysis, Templeton (2002) rejects both the simple out-of-Africa and the extreme version of the multi-regional hypothesis, since his analysis suggests several expansions out of Africa. He proposes instead a revised model, Out of Africa Again and Again, with a weak signal of initial expansion dated at around 2 myBP and a subsequent minor expansion at 420–840 kyBP, corresponding perhaps to the *Homo erectus* and *Homo heidelbergensis* fossils. Templeton also recognises a later set of major expansions around 80–150 kyBP, corresponding to an increase in the range of the species. This revised model probably represents the developing consensus among geneticists, which is that all modern human populations share a significant genetic component originating in Africa about 200 kyBP; all non-African populations derive from a subsample of the African groups present around 100 kyBP, and these expanding populations from Africa have interbred with the local (*H. erectus* or archaic *H. sapiens*) groups they encountered, though the extent of admixture was not great in comparison to the inherited genetic component from the African population.

Recently, it has been suggested that an interactive group of genes giving rise to 'modern' behavioural characteristics may have driven this demic expansion, and that contraction of population numbers during recent glaciations and subsequent expansions of local populations has given rise to the variability that we see today (Eswaran 2002, Eswaran, Harpending and Rogers 2005; for discussion see Mellars 2006). From the point of view of language evolution, we might speculate that if modern humans were breeding with derived *Homo erectus* populations across part of their range, then they were probably also communicating with them; this suggests that at least *Homo heidelbergensis*, if not earlier *Homo erectus* populations may have had elements of language, which would therefore pre-date the ultimate origin of modern humans. Quite what we mean by 'elements of language', of course, is another matter, and one we shall discuss further in the next two chapters.

7.4   Genes in individuals

7.4.1   Genes and embryogenesis

In the last section we were mainly concerned with investigations of genes in populations, variously defined; but of course, populations are composed of individuals, and genes play particular roles in building the bodies, brains and behaviours of each of us. In embryological terms, we each start as a single cell, which divides into daughters; these in turn divide into daughters themselves, and the iterative process of cell division eventually results in a complete organism. Cells do not only become more numerous, but also more specialised, in the developing body. The first egg cell and all subsequent daughters contain an identical set of genetic instructions encoded in the sequence of bases along the DNA of the chromosomes (ignoring the small, acquired or somatic mutations that accumulate throughout your life and are not inheritable); yet an adult cortical neuron looks nothing like a liver cell and the proteins present in these two cell types are completely different, reflecting very different patterns of gene expression. How the interactions of genes in time and space contribute to this process is only now beginning to be unravelled by a combination of developmental biology and molecular genetics (see Gilbert 2003).

Although every cell has identical genes, from the moment of the first cell division the actual genes that are active in every subsequent cell start to diverge according to an evolved program of development. Which genes are switched on in a particular cell depends both on the position of the cell in the developing embryo/foetus and the cell line to which it belongs (that is, the active genes present in its immediate ancestor cells). A developing embryo can therefore be seen as a collection of cell lines or lineages. In the early stages of development most cells are pluripotent, with the potential to form the ancestor for many different types of cells; but as development progresses many cell lines become increasingly specialised in their active genes and lose the potential to change type. The eventual function of these cells in the adult becomes fixed, or, in developmental biological terms, the cells become committed to a particular fate; and in many cases, such as the neurons of the brain, the specialisation is associated with loss of the ability to divide, and they are described as being terminally differentiated. A small proportion of cells in most tissues remain relatively unspecialised, retaining the ability to divide and to develop into a range of different cell types, and these stem cells allow the adult to maintain tissues such as the skin and blood, and to repair accidental damage to a wide range of tissue types.

However, not even specialised cells have only a constant, fixed set of active genes. On the contrary, in response to signals from the environment even fully differentiated cells can alter their detailed expression patterns, as for example in the process of learning outlined in 7.4.2 below. The final adult phenotype is not completely dictated by genes either; so, for instance, your two index fingers will show genetically determined similarities in the patterns of ridges and furrows, but also minor differences between left and right. We know there is a genetic component here: individuals in different human populations have inherited tendencies for particular ridge types (Jantz 1987), and chimpanzees and koalas also have finger ridges, presumably indicating convergent evolution for gripping (South American spider monkeys, by the way, use their tails as a fifth hand and have 'fingerprints' on the pads of their tails). But although the cells of both your hands have inherited the same genes, and indeed have arisen from the same ancestral cells during the same period of your development, the patterns on the two fingers will *not* be identical. During development and growth the exact environment and fate of every cell that formed your finger skin varied in small ways: perhaps a cell on one side developed a lethal mutation and left no progeny, or perhaps as a child you cut one finger tip. Neither eventuality leaves a hole where the cells died; instead, the neighbouring cells have adapted to the change in their local environment and grown into the 'empty' space, altering their genetic programming in response to unpredictable changes in their environment. Developmental plasticity is a common feature of cellular systems, and reflects a general tendency to evolve a degree of buffering to cope with environmental variability. So, genetic programs might be seen as having more malleable instructions, like 'divide until you touch another cell', or 'leave a particular concentration of a signal protein', rather than 'divide until you generate 64 descendants', or 'grow an axon exactly 2.333 mm long'. What this means is that chance plays a significant part in the precise networks of cells that emerge as the brain develops. A particular cell has a range of fates open to it, depending on what happens around it; so although the process of brain growth in outline might seem like a well-oiled and totally predictable machine, in fact there are many cases where the genes alone do not determine precise outcomes.

During human brain development, neurons must proliferate, move to the appropriate parts of the emerging nervous system, and grow axons to make correct connections with other regions, which may be a long way away. Embryological development of the nervous system is largely under genetic control, though subsequent remodellings during childhood are highly influenced by environment. In fact, the primary development of the central nervous system, starting with an elementary neural tube which expands and specialises into the spinal cord and preliminary brain, takes place with extraordinary rapidity between the third and sixth weeks of embryological development (for more details see Levitt 2000 or Lillien 1998). Between the third and fifth month of embryogenesis the developing neural system is dominated by massive cell division and migration of cortex cells; continued growth gives rise to an increasingly convoluted surface (Nolte 2002, Ch. 2). The number and position of gyri and sulci can be modelled as the result of different growth rates of different parts of the cortex in this phase, in conjunction with stresses created by the underlying white matter, determining the match of functional domains with surface physical patterns characterised by the Brodmann maps discussed in the last chapter (Toro and Burnod 2005).

This initial, exuberant expansion of the cortex involves the growth of immature neurons from a region known as the ventricular zone, where they are created from the progenitor cells by asymmetric cell division: one daughter cell stays in the same position in the ventricular zone and continues to divide while the other migrates away towards the outside of the developing cortical surface, eventually forming the six layers of the cortex. These migrating cells are formed into columns, which are believed to be the basis of the functional columns forming the minimal units of brain function discussed in 6.5.1 above (see Rakic 1988). The neurons in each column are clones, and preserve cell-line information present in the progenitor cell from which they derive, including any genes being expressed. Some of these genes will produce signalling proteins which attract incoming axons from thalamic neurons (the relay sites from the peripheral nerves and other brain areas). If the proteins on the surface of these axons 'match' the proteins on the surface of the neurons, then both will live; and if not, both will die (Levitt, Barbe and Eagleson 1997). After the initial expansion phase of cell division, the brain continues to grow and develop but, perhaps surprisingly, this involves considerable cell *loss* in the immediate perinatal period (Katz and Shatz 1996), as if the brain had initially vastly overgenerated cells to compensate for those which will die as they fail to make the right connections. However, axons and dendrites will continue to grow and mature throughout the neonatal period and well into puberty, with continued reorganisation of dendritic spines (the information transfer points between cells) throughout life, apparently related to periodic transfer of material to longer-term memory. The overall pattern of brain growth in humans is shown in Figure 7.3.

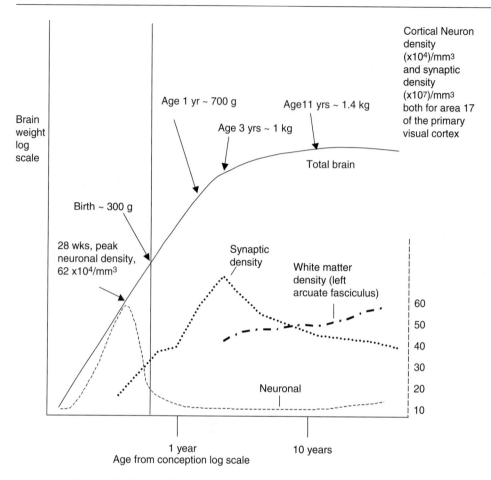

Figure 7.3 *Graph of brain weight in embryological and post-partum growth in humans.*

What this means is that brain size in itself is not the only relevant factor for functionality in the adult brain; the way connections are made and remodelled is vital too. It has even been suggested that certain behavioural characteristics such as general intelligence can be best predicted by dynamic aspects of the rate of brain reorganisation during post-natal development rather than by the size of brain structures at any particular time point (Shaw *et al.* 2006). Of course, to an extent, the amount of neuronal tissue devoted to a particular function in the adult brain will depend on the number of cells expressing the relevant receptors; and this is supported by, for example, post-mortem correlation of cortical thickness and cell density in Brodmann's area 44/45 with language skills (Amunts, Schleicher and Zillesa 2004), and from functional MRI (Paus 2005). Remember, though, that the number of active cells in an area is not completely under genetic control, since there is an element of chance in whether all the available cells have made appropriate connections to keep them alive. The developmental experience of the individual is also relevant here, since, for instance, disruption may be caused by the introduction of certain harmful drugs at particular embryonic stages. As for the relevance of these observations to language, Deacon (1997: 248–50) suggests that the increased size of the human forebrain has resulted in more cortical cells sending connecting axons into other brain regions, including the skeletal motor regions controlling face and tongue muscles and the visceral control nuclei that coordinate the larynx and muscles involved in breathing; these projecting axons displace local connections between and within these regions, resulting in greater cortical and therefore voluntary control of the associated structures. Hence, although some human newborns make noises resembling babbling at birth, canonical babbling emerges a few months later, just at the time when the axonic projections from the motor cortex are starting to develop their myelin sheaths. This allows rapid transmission of signals, and productive 'testing' of motor pathways, which are then gradually narrowed down through progressive learning and experience until only those relevant for the individual's language(s) are retained – exactly the kind of growth and then refinement and reduction we met above in the embryonic development of the brain. Tantalisingly, this seems to have an impact also on human non-linguistic vocalisations. In laughter, other species phonate on the outbreath and inbreath, but humans only on the outbreath, and it seems that our cortical control of the larynx and diaphragm, along with the opposing relaxation on the inbreath, have submerged the tendency to phonate whenever air is passing through the larynx (Provine 1996).

The actual genes controlling the cellular developments involved in brain growth and maturation are often not specific to humans: for instance, one master switch implicated in the initial development of the neural tube is the gene *shroom*, so called because in mouse embryos with disrupted shroom genes the neural folds mu*shroom* outwards instead of forming a tube (Hildebrand and Soriano 1999, Hildebrand 2005). Typically, differences across species involve the timing and patterns of gene expression, not the existence of different groups of genes, or even forms of the same gene. In addition, different parts of the brain do show very different expression profiles. For instance, Khaitovich *et al.* (2004) compared 4,998 genes with a consistent signal in both humans and chimps. Of these, 2,014 genes differ in their level of expression between chimps and humans across the brain as a whole; around 1,400 in humans and 1,200 in chimps differ between the cerebellum and cortical tissues; but only four (well below the level of significance) show a consistent difference in expression level between Broca's area in the left hemisphere and the corresponding right-hemispheric region in humans. In fact, there seem to be no genes which are specifically under- or overexpressed in the human Broca's area relative to their level elsewhere in the prefrontal cortex. So, although there are plenty of candidates for genes involved in brain and/or associated behavioural differences between humans and chimps, these form diffuse patterns across the whole brain, rather than being localised in any obvious, or classically language-linked, region. Since we spent much of Chapter 6 arguing for a rather diffusionist picture of language in the brain in any case, this is not a great disappointment; but if you were committed to going out to look for The Language Module, you might not be best pleased.

This diffuse and overlapping organisation of at least some aspects of the brain might help us with an outstanding question about brain size. The human brain, as we know, is very considerably larger than our body size would predict; and yet we have suggested above that size isn't everything, and that wiring might count for a lot more. Why, then, evolve such a massively outsized brain, rather than a neater and more efficiently connected one? It might be developmentally and evolutionarily simplest to expand particular functionality in part of a diffusely organised brain by expanding the whole brain (Finlay, Darlington and Nicastro 2001). Bigger brains could be achieved by two strategies: either keep the expansion phase going longer, or reduce the number of cells dying. In the first case, altering the expression levels or timing of genes such as beta-catenin (which is part of the Wnt signalling pathway involved in the cycle from cell growth to division) could extend the timing of the initial neurogenesis and hence increase the number of cell divisions. On the other hand, during the proliferative phase of neuronal development as many as 50 per cent of the derived cells undergo apoptosis, or programmed cell death (Ryan and Salvesen 2003). Apoptosis is under the control of many extrinsic factors, and differentiating neurons effectively have to be persuaded chemically not to commit suicide. Altering the activity of some of the internal proteins involved in apoptosis, for example the caspases (particularly caspase-3 and caspase-9), has been shown to reduce the number of proliferating neuronal progenitor cells dying, resulting in dramatically increased cortical neurons and cortical folding in mice (Rakic 2000). Alterations in either of these genetic pathways could be expected to have dramatic effects on the overall size of the developing brain (an increase of 17 cell divisions of the precursor cells is predicted to give rise to a 131,000-fold increase in the final number of neurons, or roughly the difference in cortical size between shrews and humans; Finlay and Darlington 1995). However, this focus on general trends and cross-species patterns should not blind us to the variability that exists between individuals of the same species. Indeed, the chance nature of successful connectivities, and the contribution of external factors during individual development, would predict relatively extensive variation at this level.

7.4.2   Learning and post-natal development

Learning is the process by which signals from the environment are stored within the brain for later recall or use. As we have seen in previous chapters, discussions of language often polarise between views that foreground the contribution of learning and environmental input and those focusing on the contribution of innate or genetically specified structures or systems. As it turns out, though, learning itself is mediated by a diverse and complex suite of genes and gene products.

Different forms of learning have to be recognised, and each involves different gene products and neural circuits: their outcomes in terms of the memory types

Table 7.1. *Types of memory and storage areas*

Type of memory	Where stored
short-term working memory	pre-frontal tissue
long-term explicit memory of events (= episodic memory) and facts (= semantic memory)	throughout the cortex, via activity of the hippocampal formation
implicit memory (skills and habits)	striatum, motor cortex, cerebellum
emotional memory	retrieved and stored via the amygdala
conditioned reflexes	cerebellum

generally recognised are listed in Table 7.1 (see Markowitsch 2000). However, learning can often involve activation of more than one of these relatively independent storage types.

Learning begins with environmental input, but processing and storing it involves the production of vast numbers of genetically programmed synapses. Repeated stimulation of these synapses results in outgrowths of the receiving dendrites into two types of synaptic spur, small and transient ones present during learning, and larger and more persistent ones thought to mark the position of long-term memory traces. Although the precise relationship between learning, memory, intelligence and these synaptic formations is still not well understood, it is clear that a proliferation of synaptic spurs in a particular area will increase the thickness of the cortex; and Shaw *et al.* (2006) demonstrate a link between intelligence and dynamic aspects of cortical thickness during post-natal development in a cohort of 307 individuals between the ages of 6 and 20. In linguistic terms, associated theories of neural recruitment account for why the critical period in humans for language is rather extensive and variable, and also why it is harder to learn a second language once the first one has been acquired successfully or partially, although there might be a common-sense counter-argument that once the first one has been learned it should be easier to learn the second. Newport (1990) and Kuhl (2004) propose that language acquisition recruits particular neuronal networks in the brain, reinforcing those synaptic connections. However, other, non-reinforced networks will be lost, leading to an apparent inhibition of subsequent learning of different patterns. This provides a neurophysiological account of the critical period: some potential pathways are enhanced by language acquisition, while others regress, and if none are advanced by a particular period then all regress and no acquisition is possible.

Our understanding of the processes involved in learning is also advancing as a result of molecular genetic investigations of worms and fruit flies (Levinson and El-Husseini 2005, Yeh *et al.* 2005). The activation of a neuron depends on the balance of input from inhibitory and excitatory synapses. Each synapse is the meeting point of an axon and a dendrite. At excitatory synapses, the axon releases glutamate, which binds to receptors in the dendrite, increasing the like-lihood of the neuron attached to that dendrite firing. At inhibitory synapses, the axon releases a derivative of glutamate, gamma-amino butyric acid (GABA), with the opposite effect, making the neuron less likely to fire. Both glutamate and GABA are examples of neurotransmitters, so called because they cross the synapse. Each neuron 'decides' whether to fire or not on the basis of its 'monitor-ing' of incoming neurotransmitters, and the balance of these is controlled by gen-etic factors and environmental stimulation. Initial stimulation appears to result in small, transient spurs, which seem to be implicated in short-term memory.

Studies in animals indicate that long-term memory takes several hours to be established via more stable synapses, suggesting that major structural changes in cellular function, including gene expression, are involved. Two major groups of protein types are implicated in forming stable synapses in addition to the neu-rotransmitters: post-synaptic scaffolding proteins, and cell adhesion molecules. Scaffolding proteins control the formation of the synaptic spurs, and are them-selves the product of a massive number of genes (Kim and Sheng 2004). Long-term learning follows the stimulation of a dendrite and involves signals from specific brain areas, including the hippocampal formation, since damage to the hippocampus and related structures can lead to loss of new long-term memory while retaining old, already-established memory. The molecular details are as yet unclear, but the so-called NMDA receptor complex (after an artificial compound N-methyl-D-aspartic acid, which is similar to glutamate, but activates only this class of glutamate receptor) appears to be the glutamate receptor responsible both for the physical changes in the dendritic spines and the potentiation of long-term memory. Since this structure includes more than seventy-five different gene products, this implies the activation of many underlying genes as part of the process of memory storage (Husi and Grant 2001). Each of these genes could be subject to the mechanisms of mutation, selection and drift discussed above, pro-viding a vast range of potential inputs to specialisation for language or any other initially learned behaviour. Instead of a division between learning and innateness for language, we might want to think in terms of a flexible continuum. However, learning is just one of the prerequisites for language, and in the next section we ask whether language breakdown can help us identify other genetically mediated aspects of language.

7.4.3   'Language genes'

Above we have often talked of genes 'for' particular traits, even though we have also emphasised that since genes operate in vast interacting net-works affected by both each other and the environment, the concept of 'the gene for …' is at best oversimplified and at worst fundamentally flawed. Dawkins (1986: 297) gives the example of a cake recipe: if we follow every word of the recipe then the cake comes out perfect at the end, but we can't then reverse-engineer the cake by saying the slightly spongy part in the centre is due to the first word in the recipe, or the browner outside crumbs derive specifically from the second word; and it would be equally meaningless to remove any ingredient, like flour, and say that that was the gene 'for collapsing cakes' even though that might be the phenotypic effect. Through the previous chapters we have established that human language reflects a highly complex set of underlying structures and systems, which in turn are likely to be underpinned by a range of different, though potentially interacting genetic systems; we must therefore be particularly careful when we come to talk about genes 'for language'.

If we are going to go gene hunting for the inherited underpinnings of human language, the first problem is figuring out where to look. Underlying the ostensibly piffling 1–2 per cent genetic divergence between the coding regions of the human and chimp genomes are around 30–60 million individual changes, of which 15–30 million will be specific to the human line. Many of these will turn out to be completely neutral with respect to any phenotype of interest; but it increasingly appears that control of expression level and timing are as important in determining species differences as structural differences in genes and proteins. For instance, Enard *et al.* (2002a) show that humans and chimps are closest in expression pattern for genes related to the blood and liver, but that humans are highly divergent from all other primates when it comes to gene activity levels and expression patterns for aspects of the brain. Given the attention we paid to the brain in the previous chapter, this might not be overly surprising, but even narrowing down the locus of cross-species differences to this extent leaves a huge amount of genetic ground to cover as we sift for genes relevant to language.

One possible way forward, which again we might anticipate from discussions in previous chapters, is the traditional genetic examination of individuals with inherited language deficits. Finding implicated genes may help us identify major underlying networks of genes that participate in the control of structures involved in language, hence suggesting groups of genes as candidates for further study.

7.4.3.1   Genetic conditions affecting language and the brain

Developmental disorders of language and speech have clearly been shown to have an underlying genetic component in many cases (Bishop 2001). This means there will be relevant genetic variation in populations, and we can try to use this in two main ways to find out about the biological systems underlying language. One approach is to identify individuals with a consistent syndrome including inherited language deficits, examine the inheritance of the syndrome in affected families, then use genetic mapping techniques to localise the gene or genes underlying the phenotypic change: Rice, Warren and Betz (2005: 8) describe this as the top-down approach. The second, bottom-up, approach is to identify individuals who have mutations or disruptions of a particular gene or genes and characterise any differences between that group and normal controls either in acquiring or using speech and language. In other words, we can start either with the known phenotypic effects, or with the known genetic variation, and work towards the other.

Most genetic conditions that affect language development are also associated with other symptoms, either physical or neurological. Until recently, each condition has typically been studied in its own right, and there has been little attempt to characterise the similarities and dissimilarities between the linguistic consequences of these different syndromes, or to relate any differences in the cognitive profiles of affected individuals to their underlying causes. Nonetheless, the need for such comparative work is increasingly recognised by, for instance, Rice and Warren (2004), and Rice, Warren and Betz (2005). Comparison of the different conditions and their effects should ultimately lead to an understanding of how normal language development has been disrupted, and how these disruptions are correlated with or separate from other neurological deficits, therefore providing a greater understanding of how language is acquired under normal conditions (Bates 2004). We might then hope to clarify, for example, the relative importance of normal oral-motor skills and auditory memory in acquiring language, since different conditions might affect one of these systems or the other.

Currently, relevant disorders can be grouped into two major sets. For one group, the underlying genetic defect is clearly characterised but the relationship between this and the language pathology is poorly understood, as in, for example, fragile-X-linked mental retardation (OMIM 309550; Abbeduto and Hagerman 1997), Rett syndrome (OMIM 312750), and Down syndrome (OMIM 190685; Chapman and Hesketh 2001). In such cases, when the gene(s) underlying the condition have been identified, the specific mutation(s) have to be characterised in particular patients using molecular techniques before those individuals can be included in studies on language or cognitive deficits; they cannot be included solely because they appear to have the relevant phenotype. On the other hand, there are many developmental cases where the language deficit is clearly characterised but the underlying genetic contribution appears complex, as in autism (International Molecular Genetic Study of Autism Consortium 2001, Newbury and Monaco 2010). In these cases, where the relevant gene(s) have not been clearly or exhaustively identified, carefully agreed criteria must be used to determine that the observed phenotype 'counts' as a case of that specific disorder, as in the use of the Autism Diagnostic Interview – Revised (Lord, Rutter and LeCouteur 1994) and the Autism Diagnostic Observation Schedule (Lord *et al.* 2000).

Although progress is being made towards identifying the genetic pathology underlying these conditions, things are seldom simple, even when they initially appear to be. A case in point is Williams Syndrome or William-Beuren syndrome (OMIM 194050; Bellugi *et al.* 2000; and for a comparative approach, Rice, Warren and Betz 2005). This is a dominant condition, apparently resulting from a straightforward micro-deletion of around 1.6 Mb from chromosome 7q11.23. However, this deleted region is now known to contain over 20 protein-coding genes (Ewart *et al.* 1993), so that the notion of hunting 'the gene' for Williams Syndrome is clearly oversimplified. Moreover, the symptoms of the condition can include heart problems (supravalvular aortic stenosis (SVAS)), multiple peripheral pulmonary arterial stenoses, 'elfin face', short stature, characteristic dental malformation, infantile hypercalcemia, and specific neural defects. Intelligence is moderately reduced in around 75 per cent of individuals, with a mean IQ of 58, and a range of 20 to 106. There is typically slow initial growth and early cessation of physical and brain growth. Any or all of these phenotypic effects could reflect the loss of one gene from the deleted region, or each might conceivably map onto a separate gene or genes.

As we saw in Chapter 2 above, language skills, although not wholly normal, are relatively unaffected in Williams patients, and this apparent independence of language skills relative to other neurological processes was initially taken to support a modularity of these linguistic functions and their independence from other brain functions (Paterson et al. 1999). However, there is debate as to exactly how normal the language development and other neurological processes are in these children, and it appears that the cases with relatively normal IQ are also those with relatively normal language (for a discussion see Meyer-Lindenberg, Mervis and Berman 2006). Rice, Warren and Betz (2005: 18–19) suggest that Williams children typically have delayed development of both vocabulary and morphosyntax, though development in these areas then proceeds at a normal pace; nonetheless, adults with Williams Syndrome also appear to make errors in certain grammatical constructions, which might suggest that the end point of linguistic development remains somewhat deviant. In addition, (Somerville et al. 2005) describe an individual with a duplication of the Williams region (so, one-and-a-half times as much of this region as normal, and three times as much as in a Williams individual with the usual deletion); this is associated with severe developmental delay in language production. It would be difficult to preserve the claim that genetic mutations in this area have no effect on language, if duplicating the very same region were consistently to correlate with serious language delay or deficit.

In an attempt to clarify the associations between specific genes and specific phenotypic effects, Frangiskakis et al. (1996) studied a large cohort of individuals with Williams Syndrome who showed characteristically poor visuo-spatial constructive cognition (that is, they had problems in mentally visualising an object as a series of component parts and drawing a replica from those parts). Two of these families had a partial Williams Syndrome phenotype; affected members had most of the specific Williams Syndrome cognitive profile and vascular disease, but lacked other Williams Syndrome features. Unusual, smaller than normal 7q11.23 deletions accompanied the phenotype in both families. DNA sequence analyses of the region affected by the smallest deletion (83.6 kb) revealed that the Elastin (ELN) gene and the LIM domain kinase1 (LIMK1) gene had been deleted. The latter is strongly expressed in the brain, and Frangiskakis et al. (1996) therefore suggest that it is implicated in the impaired visuo-spatial constructive cognition of Williams patients, while mutations affecting individual amino acids in the Elastin gene have independently been found to be inherited causes of vascular disease. However, several other genes from the common deletions are also expressed in brain tissue and have been suggested as possibly contributing to the full syndrome, and further studies on patients with only some aspects of the syndrome would be required to establish their influence, if any; but in this case, as in many others, we are dependent on the mutational 'experiments' nature carries out and on the presence of well-informed medical colleagues who might appreciate the significance of particular phenotypes for outstanding genetic and cognitive or linguistic questions.

If there were a single condition which we might seek to focus on in our quest for genes associated with language functions, it would surely be SLI, or Specific Language Impairment, which 'is defined as an unexplained failure to acquire normal language skills despite adequate intelligence and opportunity' (SLI Consortium 2004: 1225). Individuals with SLI develop delayed and/or abnormal language development relative to age- and IQ-matched controls; but SLI may in fact be several overlapping conditions rather than a single syndrome in any strict sense, since affected individuals may show impairments only in one area of the grammar, such as syntax, phonology, semantics or pragmatics, and only in production or perception. Furthermore, as Rice, Warren and Betz (2005: 11) note, 'A long-standing question is whether speech and language impairments are closely linked in children with SLI.' The main problem for some individuals might lie in speech production, but this in turn could influence the rate of language acquisition and the system achieved. There is a clear genetic component to SLI, as evidenced from a range of studies of twins and families; but again, 'family studies have failed to find any clear segregation between genotype and phenotype, and it is generally accepted that the genetics underlying SLI are complex, involving several genes that interact – both with each other and with the linguistic environment – to produce an overall susceptibility to the development of the disorder' (SLI Consortium 2004: 1225–6). This difficulty probably also reflects the emphasis of many studies on large groups of patients, which is likely to lead to the conflation of a range of related conditions, and to a lack of definitive results for any aspect or subtype of SLI in particular.

When researchers are faced with a consistent phenotype, or overlapping set of phenotypes, and do not yet know the gene(s) underlying these, they can employ two methods of gene hunting, using a candidate-gene approach or positional cloning. (The difference is becoming blurred, however, as a result of the Human Genome Project, which has provided a complete genetic map, allowing any kind of positional information to automatically identify a range of suggested candidate genes; most investigators would now therefore tend to use both approaches together.) In the case of Williams Syndrome, the candidate genes were the original 20+ genes in the deleted region; but for the neurological aspects of the Williams phenotype, the set of candidates has subsequently been narrowed down to the subset of those genes expressed in the embryological development of the brain or in the adult brain. If we concentrated on brain-expressed genes in attempting to identify candidate genes for SLI, however, we would still have the choice of around 10,000–20,000 genes, approximately half the human genome, all of which are switched on in the developing or adult brain: that is barely a meaningful restriction at all. Sometimes in neurological conditions, the sensitivity of a particular syndrome to drug therapy suggests candidate genes, as in attention-deficit/hyperactivity disorder (ADHD; OMIM 143465), a condition where symptoms in some individuals improve with methylphenidate treatment. This chemical is a stimulant of the dopaminergic signalling system in the brain, allowing researchers to investigate and discover mutations in genes involved in this pathway, most notably *DRD4* and *DAT1* (Hawi *et al.* 2005).

However, for most developmental disorders affecting language there are no biochemical signposts, so that researchers hunting for 'language genes' have to resort to positional cloning (Botstein and Risch 2003). The positional cloning approach, which used to be called reverse genetics, is the method of choice when you have a phenotype but no knowledge of the underlying genes or systems that might be involved. This approach is difficult for the majority of language disorders since it relies on the assumptions of a single affected gene, an easily characterised phenotype, and a clearly defined mode of inheritance – all of which are violated in conditions like SLI (Fisher, Lai and Monaco 2003).

The problem with SLI is not that there are no candidate genes but that there are too many. There have been two genome screens involving positional cloning; the first suggested candidates on chromosomes 16q and 19q (SLI Consortium 2002), while the second proposed candidates on chromosomes 2p and 13q (Bartlett *et al.* 2002). However, even with these inherent difficulties, it is possible to find intriguing results, and one, involving a single extended family, has turned out to be of considerable potential significance for work on evolutionary linguistics.

7.4.3.2   The *FOXP2* gene

In single large multigeneration families it can be possible to perform gene mapping using random selection of probes over the genome and looking for the co-inheritance of the phenotype and particular gene variants or markers. One such family, known as the KE family (whose pedigree is shown in Figure 7.4), clearly demonstrates both the promise and problems of this approach (see Fisher *et al.* 1998, Fisher 2005, Fisher and Scharff 2009).

A range of speech and language deficits has been reported in around fifteen members of the KE family, over three generations. There is some debate in the literature over the precise problems of the affected individuals (see Fisher 2005), but these are generally agreed to involve a central nervous system defect resulting in at least a developmental dyspraxia of the oral-facial system, leading to difficulty in controlling and coordinating mouth movements involved in speech articulation (Alcock *et al.* 2000, Watkins *et al.* 2002). Since affected individuals appear to be able to control simple facial and mouth movements outside speech, it seems that the deficit is specifically in the ordering of complex motor sequences, although whether this is a general problem made more obvious by the intense nature of such requirements in language or whether it is specific to language production is unclear. The affected individuals also exhibit a range of

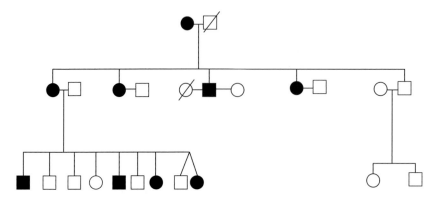

Figure 7.4 *Pedigree of the KE Family. Part of the KE family, redrawn from Lai et al. (2001: Figure 1). Only two of six sibships are shown in generation III. Circles represent females and squares males; affected individuals are shown as black. A clearly dominant inheritance pattern is seen, with each affected individual having one affected parent, and roughly 50% of an affected individual's children being affected.*

grammatical problems in production and interpretation of syntactic structures and in word-level inflection and derivation; for instance, they seem unable to use the regular English past tense inflections reliably and experience serious difficulty in generalising these regular patterns to novel, invented stems (Gopnik and Crago 1991, Gopnik 1999, Ullman and Gopnik 1999, Alcock *et al.* 2000).

The next step was to search for a gene or series of genes forming a pattern specific to the affected members of this family, and hence likely to be inherited from a common ancestor; in Rice, Warren and Betz's (2005) terms, this is a bottom-up strategy. Initially, of course, that gene could be anywhere, so there must first be a wholesale search of the genome; this DNA mapping identified a minimal region of chromosome 7 present in all affected individuals and containing around seventy genes (Fisher *et al.* 1998). Subsequent testing of a subset of likely candidate genes was significantly shortened by the identification of an unrelated child with developmental dyspraxia who had a chromosomal translocation involving chromosome 5 and this critical region of chromosome 7 (Lai *et al.* 2000). The breakpoint, where regions of the two chromosomes were interchanged, turned out to run through the middle of a previously uncharacterised gene; and screening of this gene in the KE family identified a base substitution present only in the affected individuals (Lai *et al.* 2001). Screening of a population of forty-nine dyspraxics also led to the identification of three other mutations in the same gene in individuals outside the KE family (Macdermot *et al.* 2005), and a further mutation, in another unrelated individual, is reported by Liegeois *et al.* 2001.

The next question is where this new gene fits into our typology of existing, known gene types. The gene sequence turned out to resemble a group of around forty others, each of which contains a so-called forkhead-box (FOX) domain. The FOX domain is a block of around 100 amino acids involved in recognising and binding to upstream control regions of DNA, and the FOX genes are suspected to be transcription factors which activate or inhibit the production of proteins, and are therefore involved in regulating many aspects of embryological development (Carlsson and Mahlapuu 2002). This specific gene, now named *FOXP2*, is highly conserved in evolutionary terms, being found in strongly similar forms across a whole range of species. For instance, *FOXP2* shows only one amino-acid change between mice and primates, which are separated by at least 130 million years (Enard *et al.* 2002b), indicating that 'the ancestral form of *FOXP2* was already vital for CNS development in the common ancestor of humans and rodents' (Fisher 2005: 119). In addition, the zebra finch version of *FOXP2* is 98.2 per cent identical with the human and 98.7 per cent identical with the mouse equivalents; this equates to 5–8 amino-acid changes since the last common mammal–bird ancestor around 320 million year ago (Scharff and White 2004: 334). Intriguingly, however, there are two further amino-acid changes in *FOXP2* just between humans and other primates, though the separation here is much more recent at approximately 5 myBP (Chen and Li 2001). We might then, continuing our genetic detective work, look for a reason behind this sudden and (relatively) rapid acceleration in the species-specific substitution rate of *FOXP2*. The obvious hypothesis would be that there has been positive selection for a new form of this gene in humans, and that this might well relate to some new or enhanced function: one immediate candidate for a significant behavioural difference between humans and other primates is, of course, language.

This hypothesis might initially seem to be encouraged by the finding that, in the brains of zebra finches (Scharff and White 2004: 336),

> *FoxP2* shows differential expression over development in the song nucleus Area X – a part of the special basal ganglia-like forebrain network required for vocal learning that non-learners do not possess. *FoxP2* expression in Area X stands out, slightly but consistently, from its expression in the surrounding striatum only during the time when young zebra finches learn to imitate song.

We should not, however, assume that we have found the (or even a) language gene. Even in zebra finches, *FOXP2* expression is not restricted to song-learning Area X, but is also found in non-vocal brain areas; and this is also true for a total of eleven bird species, some of which do not learn song vocalisations at all (Scharff and White 2004: 336). In mice and humans, *FOXP2* is also expressed in tissues outside the brain and is implicated in the development of the lungs, intestines and cardiovascular system (Shu *et al.* 2001). Within the brain in these same species, *FOXP2* expression patterns implicate it in controlling connectivity between particular brain substructures, suggesting an involvement in motor control (Fisher 2005). Although *FOXP2* cannot therefore be seen as uniquely involved with language in humans, it could reasonably be one of a range of genes whose original functions made them good candidates for co-option for the specific demands imposed by a developing capacity for language. The clear associations of *FOXP2* and its cross-species equivalents in brain development and vocal learning strengthen this hypothesis; as does the fact that fMRI in humans with *FOXP2* mutations has revealed abnormalities both in the cerebellum and in Broca's area (Fisher 2005, Liegeois *et al.* 2003). However, *FOXP2* is not involved in all cases of developmental language disorder by any means: Fisher (2005: 121) reports that the whole *FOXP2* coding region has been screened by several research groups without finding links to SLI, autism or developmental dyslexia. And *FOXP2* is only one of a substantial number of genes implicated in brain development which have been identified either through their association with particular disorders, or through general genomic screening: Clark *et al.* (2003) screened 7,645 matched gene sequences between chimpanzees, mice and humans and found hundreds of genes which appear to have been under active selection during human evolution. The part *FOXP2* might have played in this landscape of mutation, selection and interacting genes and effects can only become clear as we gradually understand more about the other actors in the drama.

7.5   Summary

We began this chapter by outlining the structure and composition of the gene, and the workings of its primary component, DNA. Genes have their effects through the construction of proteins, and while DNA stores information, proteins put its instructions into action by constructing and operating cells. However, genes also include regulatory regions which determine when, where and how much of the protein is produced. We saw that, though it is tempting to think about 'the gene for' a particular effect, this is virtually never realistic, particularly when we are discussing complex behaviours or deficits. In the second main section, we shifted focus from the individual to the population and considered the three main mechanisms of genetic differentiation, namely mutation, selection and drift. Variation between species begins at the level of variability within a population, so by understanding these forces, and measuring the diversity present within and between populations and species today, we can develop models allowing us to reconstruct the likely population histories for these groups, including factors like population size at the point of speciation and time since divergence. In the last section, we returned to the level of the individual, developing the idea of how genes build a phenotype, both during the initial period of embryogenesis and in the later and longer-term construction of the brain and of learned pathways within it. Throughout, as in the last chapter, we have used examples of genetically inherited disease and associated language breakdown or dysfunction to help us understand more about the genetic component underlying normally developing language. Here in particular, we saw that our knowledge is at a rather preliminary stage; some genes, like *FOXP2*, have been implicated in language functions, but the evidence is still partial and further investigation will undoubtedly uncover a whole range of further candidate genes.

The model we have proposed here is a complex one. Almost always, there is more than one gene per effect and typically more than one effect per gene; the environment, both internal in terms of other genes and external in terms of where the species and individual are, can alter the effects the genes have anyway; and the mechanisms of mutation, drift and selection interact in complex and sometimes unpredictable ways. This, however, is no bad thing: we have said from the start that language is extremely complex, and we should only be discouraged from looking for answers in the domain of genetics (or any other domain, for that matter) if we expect those answers to be simple. As Fisher (2005: 125) puts it, 'To my mind, the most exciting aspects of the *FOXP2* story are not the answers that have been provided (or hinted at) thus far, but rather the new questions that we are now able to ask, using this gene as a starting point.' As we move into the final set of chapters of the book, and focus increasingly closely on language itself rather than the underlying structures and systems of vocal tract, brain and genes which we need for it, we must take along those ideas of complexity and interaction. Our view, on the basis of the evidence in this and previous chapters, is that we are highly unlikely ever to find a single gene for anything complex or important; moreover, when there is a plausible contribution by a single gene to some complex behavioural or phenotypic effect, that gene often had a rather different function earlier in evolution, and has been co-opted rather than being designed from scratch. These ideas of multiplicity, graduality and non-optimal recycling turn out to put us firmly on one side of a pretty high fence when it comes to the evolution of language, as we shall see in the next chapter.

Further reading

This chapter has covered (admittedly not always in much depth) a whole range of different specialist areas, and we have found it best in this case to refer to overviews as well as research papers as we have gone along. We have also typically pointed out which of our recommendations are more general and introductory. There are, however, some other reviews and more popular books we can suggest here.

Although there are many available accounts of the structure and workings of genes in biology textbooks, they can be hard going for readers without a background in biology or chemistry; if you fall into that category, we recommend Jones (2000) as a much more approachable outline. For a detailed and highly readable discussion of the issues around sexual selection see Matt Ridley (1994), particularly Chapter 5; also relevant is Hamilton and Zuk (1982). Matt Ridley (2003) provides an extended argument for not seeing human behaviour as resulting either from nature or nurture, but for the stimulation of genetic programs by aspects of the environment. Gilbert (2003) is a very thorough, detailed and clear textbook account of developmental biology. Fisher (2005) and Fisher and Scharff (2009) provide an extremely clear overview of the *FOXP2* story so far, while

Marcus and Fisher (2003), in a slightly earlier review, give an accessible account of how transcription factors like *FOXP2* switch on other genes and hence influence development. Rice, Warren and Betz (2005) is an approachable overview of the linguistic phenotypes associated with a range of developmental disorders, and provides strong arguments for a more comparative approach.

Points for discussion

1.   Eldredge and Gould's idea of punctuated equilibrium has been widely derided as 'evolution by jerks' (a description sometimes attributed to Lewontin; note that Gould is said to have retorted that the alternative is 'evolution by creeps'). Read Eldredge and Gould (1972), and some opposing arguments, which might be Dawkins (1983: 101–9), or perhaps critiques by Daniel Dennett or John Maynard Smith. In your view, is this an alternative to Darwinian natural selection or not? Dixon (1997) adopts a punctuated equilibrium approach to historical linguistics, as an alternative to the conventional family-tree model; how well do you feel this works for language?

2.   The term 'drift' is used in historical linguistics: Sapir (1921: 150) famously comments that 'Language moves down time in a current of its own making. It has a drift.' How does this usage compare with the meaning of 'drift' in genetics? Could confusion arise from the different usages, and can you find evidence of it?

3.   In this chapter we have raised some problems encountered in population sampling in genetics. Are there similar difficulties and limitations associated with sampling in sociolinguistics, or historical linguistics? Do we find the same kinds of issues in attempting to compare the present day with the past? Have these problems been resolved in either linguistics or genetics to your satisfaction (or in anthropology, sociology, or any other field you might know or be able to find out about), and how could the different disciplines learn from each other?

4.   Find out as much as you can about mouse song and its links with *FOXP2* (you will have to use the internet). How was it discovered that mice sing; under what circumstances do they sing; and to what extent do you think the mouse is a good animal model for understanding the effects of genetic factors on brain and language development?

5.   *FOXP2* is one gene which has been linked with language, but more are being proposed all the time. Look for some literature on another 'language gene', and evaluate the methods and claims involved.

8 Big bang or cumulative creep? Saltation versus gradual, adaptive evolution

8.1 Overview

Since Chapter 4 was entitled 'Who, where and when?', we should really call the current chapter 'How and why?' In Chapters 5–7, we have built up a picture of the physical structures which enable language; but we have not yet got to grips with how and why those aspects of modern human vocal tracts, brains and genes developed in the first place. There are two central questions here, which will occupy us through this chapter, and continue to underlie some of the discussion in Chapter 9. First, did the underpinnings of language emerge as a single package, perhaps as the result of one monumental genetic mutation, the final step in a series of predisposing changes which took modern humans over an essential cognitive threshold, creating a new species in the process? Or are we dealing with a mosaic of interacting but essentially independent processes, following their own evolutionary trajectories but giving the impression in retrospect of working towards a common goal because they are all improving the fit of individuals to their environmental niche? Second, did the structures and systems we use linguistically emerge precisely to allow us to become linguistic animals, or are they by-products of different motivations and evolutionary paths, which fortuitously turn out to provide us with the capability for language as a kind of unexpected bonus? In short, this chapter will be concerned with whether the structures underlying language emerged punctually or gradually and whether they evolved specifically for language or language is a happy accident.

8.2 Saltation: language and the big bang

8.2.1 *FOXP2* versus Protocadherin

At the beginning of Chapter 11 of *The Language Instinct*, Pinker (1994: 332) sets out all the multifarious and wonderful things elephants can do with their trunks, from uprooting trees, through holding and drawing with pencils, to checking the ground for pits and obstacles, and smelling food as much as a mile away. 'Until now', Pinker continues (1994: 332–3),

> you have probably not given the uniqueness of the elephant's trunk a
> moment's thought. Certainly no biologist has made a fuss about it. But now

> imagine what might happen if some biologists were elephants. Obsessed
> with the unique place of the trunk in nature, they might ask how it could
> have evolved, given that no other organism has a trunk or anything like it.
> One school might try to think up ways to narrow the gap. They would first
> point out that the elephant and the hyrax share about 90% of their DNA …
> They might say … that the hyrax really does have a trunk, but somehow it
> has been overlooked … The opposite school, maintaining the uniqueness
> of the trunk, might insist that it appeared all at once, in the offspring of
> a particular trunkless elephant ancestor, the product of a single dramatic
> mutation. Or they might say that the trunk somehow arose as an automatic
> by-product of the elephant's having evolved a large head.

Pinker diagnoses very acutely, in the second part of this quotation, the tendency
of some theorists to see human language as absolutely unique, but also the consequent temptation to argue that it must have emerged through a virtually unique
set of circumstances. We don't want to be, in Robert Foley's words, just another
unique species, with our development regulated through the usual series of
mechanisms and chance events: we want language, our criterial property, to be
so amazing that it must have developed through a big bang. We want a fanfare of
trumpets. We want Language as The Speciation Event.

The problem linguists have always had with the family-tree model for groups
of languages derived from a single common ancestor is that it forces us to see
linguistic history in a particular, over-idealised way. If we see branches splitting, and two languages arising where there was once a single one, we are
tempted to reify the split, assuming that these two diverging languages and
their speakers never had any more to do with one another, and furthermore that
the split itself is a point in time (and perhaps even a datable point). Linguistic
reality is messier: trees are convenient fictions, because dialect continua are
far more common than the Pacific Island model of continuous split, settlement
and isolation. Continua, with borrowing and interaction and degrees of mutual
intelligibility, mean graduality rather than definite, once-and-for-all splits; they
mean that we must recognise intermediate and sometimes shared systems, and
indeed that variable behaviour may need to take centre stage in our thinking.
And yet there is something seductively simple about the tree, with its clean
splits and definiteness; almost seductive enough to make us forget how idealised it is.

We face the same problems when we talk about speciation. *Homo sapiens* is
one species; *Homo erectus* was another; and *Homo neanderthalensis* another
again. A family-tree model would show absolute splits: no interbreeding, no genetic exchange, and a gradual distancing in appearance and behaviour after the
bifurcation point. Seeing this as a point in the tree can lead some theorists to
envisage speciation as a punctual event, with a single, dramatic cause. If we (i.e.
modern humans) have language and nobody else has or had, so that language is
essentially definitional of modern humans, then we might expect to see in the

literature signs of a search for **the** genetic factor underlying language and its association with the split of *Homo sapiens* from our other relatives in the family tree. This search is, indeed, precisely what we find.

In the previous chapter, we surveyed evidence for the involvement of a large (and growing) number of genes in language acquisition, storage, breakdown and use, and in the evolution and regulation of the brain and other physical structures relevant to human speech and language. Putting this evidence together with the overview of human prehistory in Chapter 4, we see plentiful indications that these different structures, systems and behaviours, and therefore the evolutionary genetic modifications underlying them, date from a whole range of periods in the hominid and hominin lines. This has not, however, stopped a number of theories involving saltation, or big-bang, evolution being promoted, though we shall discuss only one example here.

Crow (2000, 2002a, b, 2005) sees himself as primarily involved in a debate over 'what genetic change accounts for the transition from one species to another and by what selective process it is retained' (2002a: 93). Since he also considers that 'The speciation of *Homo sapiens* and the origins of language are surely two sides of the same coin' (2005: 133), it is inevitable that he should search for a major genetically controlled distinction between humans and all other species which could validly be associated with the emergence of language, and therefore be seen as provoking the speciation. Crow cites the considerable asymmetry of the human brain, such that certain areas typically contain more tissue in the left hemisphere, and others in the right; this was noted at least as early as Broca (1877), and has been associated in earlier chapters with the clear lateralisation of modern human language now demonstrable through scanning techniques and in studies of language deficit. Alongside this asymmetry and lateralisation of functions goes handedness. Whereas all primates seem to show a preference for a particular hand, the distribution in populations is typically 50:50 in non-humans, while humans are roughly 90 per cent right-handed and 10 per cent left-handed or mixed across populations (McManus 2002: Ch. 9). Annett (1985) attributes this increase in right-hand dominance to a single (unidentified) gene; Crow identifies a possible candidate gene in the shape of ProtocadherinX/Y (*PCDHX* and *PCDHY*)

Crow (2002a: 98) argues that language develops at a different rate in boys and in girls, so that any candidate gene underlying our capacity for language might be expected to be sex linked; he further suggests a correlation between verbal ability and degree of lateralisation, and on this basis proposes that girls' brains may lateralise faster than boys'. His initial hypothesis was therefore that any candidate gene for lateralisation should be on the X chromosome, since XX girls would then have two copies, while XY boys would have only one. However, girls with Turner Syndrome, who have only a single X chromosome, have deficits in terms of tests for lateralisation as compared with boys, so that the Y chromosome must also be playing some role. Protocadherin becomes a key candidate for

Crow because it is expressed (that is, active) in the brain; moreover, through an evolutionary process of translocation, part of the X chromosome containing that gene, among others, has been copied onto the Y. Since many such translocated genes are down-regulated and hence do not maintain their previous expression levels, males with ProtocadherinX and a translocated copy of ProtocadherinY might therefore be expected to show a higher level of lateralisation than Turner Syndrome females (or males prior to the translocation) with only one copy of *PCDHX*, but a lower level than females with two copies of the fully expressed *PCDHX*. Crow then makes a conceptual leap, assuming that increased lateralisation can be assumed to correlate with greater verbal ability, and that women would prefer partners with two copies of the gene, as their language would be closer to that of the women themselves. ProtocadherinY would therefore be spread through sexual selection, since its bearers would be at a reproductive advantage.

Crow's connection of language and speciation arises from his conviction that 'language is an embarrassment for gradualist evolutionary theory' (2002a: 107), and that 'there is no evidence of a gradual accumulation of linguistic capabilities over a long period' (2002a: 94). Seeing language as essentially 'a barrier which no animal has ever crossed' (Max Mueller 1873, quoted in Crow 2005: 135) almost inevitably leads to an account in terms of saltation. However, Crow's hypotheses rely on a series of assumptions which can easily be called into question. First, Crow asserts that preferential handedness for a species is unique to modern humans; and second, he assumes a particular timescale for the developments he discusses, rejecting graduality on the basis that there would not have been time for protolanguage to develop into language between *Homo erectus* and *sapiens* (2002c: 10).

However, capuchin monkeys also show a preferential right-handed bias when manufacturing stone tools and using them for throwing, pounding and cutting (Westergaard and Suomi 1994, 1996a, b), so that directional handedness appears to have evolved independently at least twice. If species-level handedness is a consequence of the defining human mutation, then finding it in another species (and without language) is a problem; on the other hand, if directional handedness is a gradual, emergent result of specific patterns of tool use, we might anticipate finding it in more than one species of primate. There is evidence for lateralisation in the context of tool use relatively early in the hominin fossil record, from at least 1.4–1.9 myBP (Troth 1985). In fact, early Oldowan phase tools created by striking hand-held hammer stones against hand-held cores initially appear from 2.5 myBP, and are identified as tools (rather than as cobbles that have been smashed by river action) primarily because they occur in relatively high concentrations at particular sites. Since fossil hand bones identified as belonging to *Australopithecus afarensis* and *Australopithecus africanus* specimens both demonstrate specialised derived features that are adaptations for gripping, the potential for tool use, if not necessarily manufacture, must be pushed back to 3.2 myBP (see discussion and references in Panger *et al.* 2002).

Continuing with this problem of timescale, Crow argues that his 'speciation gene' causes a counterclockwise torque or twisting in the development of the brain, leading to added protrusion of the left posterior and right anterior parts of the skull and hence cerebral asymmetry; if so, we would expect (relatively) sudden changes in the fossil record co-incident with the spread of the newly speciated talking hominid. However, an asymmetrical pattern of this kind has been seen in at least 7/9 *H. neanderthalensis*, 11/12 *H. erectus* and 2/3 early *Homo* skulls (Holloway and Lacoste-Lareymondie 1982), suggesting that the change was much earlier than the *Homo erectus/Homo sapiens* split. Indeed, although evidence is less clear given the age and fragmentary nature of the fossils, there are indications that at least some Australopithecines showed similar asymmetries (Tobias 1995). Either all the species/individuals with this lateralised pattern had language, in some sense, or the advent of language is not directly traceable to that specific skull shape or any genetic mutation responsible for it. Furthermore, the translocation of sequences including *PCDHY* from the X to the Y chromosome appears to be datable to 3–5 myBP, soon after the human/ chimp divergence (see Tyler-Smith 2002 for a discussion); again, this does not fit well with the hypothesis that the crucial ingredients for the origin of language and the *H. sapiens* speciation event evolved after the *Homo erectus* stage. Crow (2005) acknowledges this problem of dating, and conveniently transfers the association with language and hence the potential speciation event to a later development involving inversion and deletion within the genetic elements transferred to the Y chromosome. He notes (2005: 147) that 'This change cannot at present be dated but must be later than the translocation', a logical consequence, though not necessarily particularly strong evidence in favour of his new hypothesis that the original translocation 'might be a candidate for whatever language-related innovation (the arbitrariness of an association between a sound and its "meanings" – maybe the advent of the "proto-lexicon")', while 'it is the latter change that is the candidate for the expansion of the lexicon and the introduction of syntax'.

Crow's argument is then compromised by issues of timescale; the apparent non-species-specificness of the human speciation event; and in its recent version, a retreat from one major saltation to several smaller sequential ones, a position perhaps less clearly distinguishable from the graduality Crow claims to reject. Crow (2005) even has to invoke a touch of genetic magic: if the trigger for the speciation event was not the original translocation itself, but a subsequent change on the Y chromosome, then surely males should have all the linguistic advantages, not the other way around. Crow suggests that 'Although it might seem as though a change on the Y chromosome could account for distinctive characteristics only in males, it is probable that by as-yet ill-understood mechanisms ... the presence of the gene on the Y influences expression of the gene from the X. Therefore a change may be expected in females as well as males' (2005: 146). We would feel a little more comfortable with this optimistic response if women actually had the Y chromosome with its mutated translocation in the first place.

Claims for saltation in the transition to language do not necessarily involve a particular time period or invoke the speciation of *Homo sapiens*. It is equally possible to propose that the (sudden) emergence of language was part of, or even explained, the (sudden) emergence of other human behaviours or cultural artefacts. In particular, there is common reference in the literature to a cultural explosion, the Upper Palaeolithic Revolution. Around 50–30 kyBP, we observe the apparently very swift development of different lithic traditions (or in other words, patterns of stone tools) throughout Eurasia, after a long period of stasis from around 250 kyBP. Along with these new stone tools goes the first clear evidence of art and symbolism, particularly in cave paintings: as Dunbar (2004: 30) puts it, 'Before this point, we have rather crude but functional tools; afterwards, there is a profusion of more delicately constructed implements (knifelike blades, borers, arrowheads) as well as items designed to serve functions other than mere food extraction; by 20,000 years ago, we are into awls and needles, brooches and Venus figurines.' The traditional story was that the older and relatively static suite of stone tools was associated with *Homo erectus* (though these did represent a step forward compared with the initial stone handaxes associated with *Homo habilis*); in Europe, the Lower Palaeolithic was taken to be the domain of the Neanderthals, with relatively little advance on the *H. erectus* technologies. The great leap forward of the Upper Palaeolithic Revolution is almost inevitably associated with the advent of *Homo sapiens*, and there is a tendency to associate that technological advance also with language.

The waters, however, are considerably muddier than this. There have been suggestions that some Neanderthal fossils are associated with deliberate burials, as with the Shanidar flower burial in Iraq, where a Neanderthal skeleton was found curled up in a foetal position, and surrounded by a great many pollen grains, mainly from plants associated with traditional medicine (Solecki 1975). This is controversial: on the one hand, the findings provoked considerable media interest (imagine the headlines – Neanderthals were not lumbering brutes, but the first flower children!). On the other, we find claims that the burial was not deliberate (there are other skeletons nearby which had clearly been involved in cave falls, and the accumulations of pollen might be hoards collected by jirds, *Meriones persicus*, the local burrowing rodents). Controversy apart, this is just one of a series of findings which link Neanderthals with behaviours we identify with *Homo sapiens* (see Trinkaus and Shipman 1993), perhaps bridging to some extent the great divide between the two species.

Furthermore, the previous interpretation of a sudden and radical change in behaviour in the 50–30 kyBP range is arguably an artefact of a Eurocentric approach. Recent intensive study of African lithic technologies suggests that individual elements of the cluster associated with the Upper Palaeolithic Revolution in Europe actually emerged individually and variably in different areas, arguably from 200 kyBP, and certainly from 90 kyBP onwards (McBrearty and Brooks 2000). Bearing in mind that the human brain probably reached its current size around 150,000 years ago, this encourages a gradualist and variationist view:

rather than envisaging the single explosion required by an assumption of 'full' language and symbolic control at a recent time, global evidence suggests cultural fits and starts in different populations with fine tuning throughout the last 200,000 years. For instance, Lewin (2005: 212) notes that recent excavations at the Blombos Cave in South Africa have revealed symbolic objects from 77 kyBP, while harpoon-like points and worked bone from Zaire have been dated to 80 kyBP, though these are not found in Europe until the Upper Palaeolithic. Although Dunbar suggests that, from a European perspective, 'the change is very sudden', he is equally careful to note that 'There is no one point in our history at which we can safely point and say: "Ah, and now we became human!" ... Perhaps we might be better advised to see the history of our species as one of increasing degrees of humanity which only finally came together as a unique suite a mere 50,000 years ago with the Upper Palaeolithic Revolution' (2004: 31). This variability is strongly suggestive of a long-term, gradual, cultural change, not an immediate, sudden, genetic one.

It seems, then, that the prevailing archaeological climate is shifting away from a revolutionary perspective towards ideas of gradual cultural innovation and exchange. We might anticipate a similar shift away from saltationist approaches to the evolution of language, with more emphasis on graduality and adaptation, and will explore these ideas in the sections below. This approach is much more in line with the emphasis on many genes with small effects which we have highlighted in previous chapters: biological evolution is typically slow and cumulative, not radical and sudden, and may acquire the appearance of directionality and uniformity in retrospect because one small thing has led to another. However, appearances can be deceptive, and although the outcome of a slow and additive series of steps can be indistinguishable after the fact from a macromutation causing an immediate and radical change, the latter is evolutionarily highly unlikely. The reasons for this are developed at length by Dawkins (1986), but are succinctly and memorably summarised by Pinker (1994: 361):

> natural selection is the only process that can steer a lineage of organisms, along the path in the astronomically vast space of possible bodies leading from a body with no eye to a body with a functioning eye. The alternatives to natural selection can, in contrast, only grope randomly ... When one organ develops, a bulge of tissue or some nook or cranny can come along for free ... But you can bet that such a cranny will not just happen to have a functioning lens and a diaphragm and a retina all perfectly arranged for seeing. It would be like the proverbial hurricane that blows through a junkyard and assembles a Boeing 747.

This view of many genes with small effects is reminiscent of our encounter in the previous chapter with *FOXP2*. The scientists investigating this gene have resisted the temptation to claim that this is **the** language gene, or **the** grammar gene: Fisher (2005: 119) notes that while 'Some may suggest that the validity of referring to *FOXP2* as a speech gene is simply a question of subjective semantics, and that the simplified view is important to make the research accessible to non-specialists … I have argued that the speech gene view is a damaging caricature of the available data.' There is a considerable difference between seeing *FOXP2* – or any other single factor – as one part of the jigsaw or as the whole picture, and again Fisher (2005: 124) counsels caution here: 'Rather than invoking selection of the modern human version of *FOXP2* as *the* major force that produced a population of chatty humans, it seems more reasonable to place evolution of this gene modestly within a framework of other events that contributed on our pathway toward linguistic prowess.' We commend this policy of biological modesty to all those working on the evolution of language.

8.2.2   Recursion: the keystone[1] of language?

In the previous section, we considered the possibility that a single macromutation in a single gene might have led to the sudden emergence of language, or to a change in the brain which then inevitably preconditioned language (and indeed the speciation of modern humans). Our conclusion is that there are far more likely to have been hundreds of genes which, through mutation and potentially selection in the human lineage, have contributed in their own small way to the gradual development of the human capacity for language. Some of these might improve production or perception; some might make the brain bigger and allow more scope for new skills to develop. Either way, a gradual approach is more in keeping with the usual forces of evolution, which tend to operate by changing and building on what is already available, rather than starting from nothing and trying for a quick-fix answer.

However, it is possible to take an essentially saltationist approach to language without identifying or hypothesising a single candidate gene: recall Pinker (1994: 333) on the elephant's trunk, which, it might be suggested, 'somehow arose as an automatic by-product of the elephant's having evolved a large head'. Perhaps evolution is proceeding on its usual slow and gradual way, when a particular, criterial property or structure emerges, with a completely transformative effect both on the affected species at that time, and on its future development. That property might or might not emerge suddenly and catastrophically, through saltation: either way, its influence is just as far reaching, since it causes some physical, behavioural or conceptual threshold to be crossed.

This is exactly the approach taken by Hauser, Chomsky and Fitch (2002, Fitch, Hauser and Chomsky 2005, and see also Fitch 2010), who identify a particular feature, namely recursion, as 'the only uniquely human component of the faculty of language' (2002: 1569). Strictly, they see recursion as the only ingredient in the faculty of language in the narrow sense (FLN), as opposed to the faculty of language in the broad sense (FLB), which includes systems and properties

---

[1] As readers will note from the discussion of spandrels below, it is traditional for writers on the evolution of language to include at least one architectural metaphor; this is our contribution.

either unique to humans but not specific to language, or not unique to humans at all. Rather than trying to identify a particular 'language organ' in the brain, an approach which has gone out of fashion following the discovery, mainly through the imaging techniques considered in Chapter 6, that brain activity associated with language is highly diffuse, Hauser, Chomsky and Fitch (henceforth HCF) prioritise a criterial quality of language in the mind.

There are two key aspects of HCF's proposal, namely the distinction between FLB and FLN, and the hypothesis that recursion is the only property of FLN. Before proceeding to consider these in turn, it is worth defining recursion, though this property will be familiar already to readers who have taken even elementary courses in syntax, or from discussions of the design features of human language. Recursion is carefully defined by Parker (2006: 3) as 'the embedding at the edge or in the centre of an action or object one of the same type. Further, nested recursion leads to long-distance dependencies and the need to keep track, or add to memory.' Embedding one linguistic structure into another of the same type produces increasingly complex sentences of the following kind (where it is perhaps worth noting in passing that these four increasingly complex examples really tell the story of this section, except that 'are concerned' and 'understand' in the last two cases are somewhat understated):

Only recursion really matters.
Hauser, Fitch and Chomsky say [*only recursion really matters*].
Pinker and Jackendoff are concerned that [*Hauser, Fitch and Chomsky say* [*only recursion really matters*]]
We understand that [*Pinker and Jackendoff are concerned that* [*Hauser, Fitch and Chomsky say* [*only recursion really matters*]]].

As Parker notes, these cases show that we can embed one structure, here a whole clause, into another structure of the same type; in these examples, all the embedding happens at the edge. It is clear that processing the longer and more heavily embedded cases is more taxing in a range of ways, and this potentially puts pressure on both language structures (which will be more readily processed and understood if they include signals to help listeners, in Parker's words, to 'keep track'), and language users (because we need considerably enhanced memory to be able to process the last example compared to the first).

If recursion is coterminous with FLN, what is in the FLB? According to HCF (2002: 1570), 'FLB includes an internal computational system (FLN ...), combined with at least two other organism-internal systems, which we call "sensory-motor" and "conceptual-intentional".' The discussion in HCF is somewhat vague on precisely what counts as part of the FLB and what does not (they leave open the option of other, currently unspecified systems as well as the sensory-motor and conceptual-intentional cases), but they clearly exclude 'other organism-internal systems that are necessary but not sufficient for language (e.g. memory, respiration, digestion, circulation, etc.)' (HCF 2002: 1571). HCF accept that some systems they locate in FLB might be part of FLN, though they propose that any such hypothesis would have to be tested against comparative evidence from other species: hence, the property of categorical perception was once thought to be uniquely human, but has now also been identified in chinchillas, macaques and some birds, meaning that it could not have evolved specifically for human language. It seems, then, that categorical perception and the like 'may be part of the language faculty and play an intimate role in language processing' (HCF 2002: 1572), but can only be part of FLB, not FLN, insofar as they are not unique to humans and/or did not evolve specifically for language. On this basis, HCF propose that the sensory-motor component of FLB might include properties like vocal imitation, discrimination of sound patterns, biomechanics of sound production, and constraints reflecting vocal tract anatomy; the conceptual-intentional system would include symbol use and non-linguistic conceptual representations, theory of mind, referential vocal signals, and intentional signalling.

HCF proceed from this distinction of the broad versus narrow senses of the faculty for language to a number of major evolutionary implications. These are predicated on the suggestion that FLN only includes recursion, the property by which language achieves discrete infinity, or the use of a finite number of elements (such as words or structures) to create a potentially infinite range of new utterances, each of potentially infinite length. In particular, HCF (2002: 1573) propose that 'If FLN is indeed this restricted, this hypothesis has the interesting effect of nullifying the argument from design, and thus rendering the status of FLN as an adaptation open to question.' Although they accept that aspects of FLB may well have evolved gradually from traits not originally involved in communication, they doubt 'whether a series of gradual modifications could lead eventually to the capacity of language for infinite generativity' (HCF 2002: 1573). Instead, they suggest that 'FLN may approximate a kind of "optimal solution" to the problem of linking the sensory-motor and conceptual-intentional systems' (2002: 1574), and that the details of FLN may not have been shaped directly by natural selection, but are instead determined by the constraints of FLB and of human organisms in general. Insofar as there is any contribution of natural selection to the evolution of FLN, it may reflect an initial development of recursion for non-linguistic reasons, perhaps 'to solve other computational problems such as navigation, number quantification, or social relationships' (HCF 2002: 1578), in which case the question is how and why this computational technique was extended in humans to language.

In any case, HCF clearly see recursion as absolutely vital to human language: indeed, if we take their position on FLN literally, any communicative system constructed without recursion would not qualify as 'human language'. This potentially gives recursion a special status, as the keystone of human language. In architectural terms, when an arch is constructed, it consists of a series of stones arranged from two sides in a curve; but these will each need to be supported separately until the topmost apex stone, or keystone, is put into place. At that point, the entire structure **becomes** an arch, with its own identity and structure, and is self-supporting. To extend this metaphor to language, all the elements of FLB may have evolved independently, but it is only when recursion develops that these are connected into a system, and human language as we know it emerges.

HCF (2002) is an ambitious paper, written by a well-known and interdisciplinary team of authors, and published in a very prominent scientific journal. It is not at all surprising that it has provoked a great deal of comment and debate, not all of which we can hope to summarise here. However, one high-profile reaction came from Pinker and Jackendoff (2005), who argue essentially that the approach of HCF (2002) is strongly coloured by Chomsky's theoretical commitment to the Minimalist Program, in which syntax and notably recursion are emphasised, and other aspects of language played down. This may cast doubt on HCF's attempts to draw more general conclusions on the evolutionary mechanisms involved in language.

Pinker and Jackendoff (2005) do agree that it may be useful to distinguish some version of the FLB from the FLN, and that the use of the biological comparative method to assess the species-uniqueness or otherwise of particular traits is vital and welcome. However, they raise a whole host of questions over the recursion-only hypothesis, and over HCF's conclusion that this challenges the view of language as adaptive. For instance, HCF seem to assume that it is straightforward to assess what aspects of language capacity are unique to language, while Pinker and Jackendoff (2005: 202) contend that 'The answers to this question will often not be dichotomous. The vocal tract, for example, is clearly not exclusively used for language, yet in the course of human evolution it may have been tuned to subserve language at the expense of other functions such as breathing and swallowing.' Likewise, while HCF see the human conceptual system as independent of language, Pinker and Jackendoff argue that certain concepts (including temporal expressions like 'week', aspects of kinship systems, and supernatural, sacred and scientific concepts) probably could not be learned without language. Similarly, Pinker and Jackendoff propose that, while some aspects of human audition are essentially continuous with hearing and perception in non-human primates, these may be overlaid by some innovations specific to language. Thus, individuals with pure word deafness can recognise environmental sounds but fail to understand words; and newborn human babies seem to prefer speech to non-speech sounds, and moreover seem to attend to speech using parts of the brain involved in language for adults. As we saw in the previous chapter, many speech and language deficits can be traced to particular genetic loci, none of which has yet been shown to affect only recursion. As Pinker and Jackendoff (2005: 218) point out, the recursion-only hypothesis 'predicts heritable impairments that completely or partially knock out recursion but leave people with abilities in speech perception and speech production comparable to those of chimpanzees. Our reading of the literature on language impairment is that this prediction is unlikely to be true.' In fact, these problems of gradience do not only affect the relationship between FLN and FLB, but also between what is in FLB and what is outside it. Take memory, for instance: this is clearly placed outside even FLB by HCF, but it would be impossible to process more complex cases of recursive syntax in the absence of a rather well-developed memory capacity. Does this mean that memory is actually part of FLB, or in fact logically part of FLN, or does it lie beyond both, but challenge the clear distinction among these three components?

Pinker and Jackendoff also note that HCF make little or no reference to phonology (as opposed to perception and articulation) or to words and word learning, for which there is no room in the FLN if it is restricted to syntactic recursion. Even in syntax, there seem to be properties which do not directly involve recursion, and yet are very plausibly specific to humans and to language: Pinker and Jackendoff (2005: 215–16) cite word order, agreement, case-marking, and the long-distance dependencies holding between a relative pronoun and a verb at the other end of the sentence. Pinker and Jackendoff (2005) and Parker (2006) both consider Everett's (2005) assertion that Pirahã lacks recursion, and expresses similar concepts using other structural options, mainly juxtaposition of clauses – yet 'Pirahã very clearly has phonology, morphology, syntax, and sentences, and is undoubtedly a human language, qualitatively different from anything found in animals' (Pinker and Jackendoff 2005: 216).

If it is possible to have recursion-free natural human languages, and if in addition there are other properties humans have which seem to be tied to our use and acquisition of language, and which other animals seem not to manifest, it seems fair to say that the recursion-only hypothesis is in trouble. The FLN seems certain to include more than recursion, and in fact may not include recursion at all. Pinker and Jackendoff ask what might then have motivated HCF to make the hypothesis in the first place, and suggest first that it emerges from a wish to dissociate language from adaptive accounts involving natural selection, since 'If language per se does not consist of very much, then not much had to evolve for us to get it' (2005: 219). Secondly, it may depend substantially on Chomsky's wish to defend his current theory of syntax, the Minimalist Program, which characterises language as an optimal, perfect and stripped-down system where a single operation, Merge, builds binary trees and links sound and meaning. The problem with this theory is that so much actual linguistic structure has to be disregarded or made peripheral for it to work: Pinker and Jackendoff (2005: 220) argue that the original version essentially discounted phonology, most of morphology, word order, lexical entries, processing and acquisition, making it strikingly parallel to what the FLN can accommodate and what is exiled to the FLB. This kind of emphasis of simplicity and elegance is characteristic of a number of formal linguistic theories (including Optimality Theory for phonology, which is not immune to such theoretically driven paring down by any means); but the counsel of perfection means disregarding the messy normality of surface language data. Our own view is that biology, and language as a partial manifestation of biology, is not perfect, because evolution and life exist in a climate of compromise. Our kitchen table often has piles of books on it, because the kitchen is a congenial place to work; it wasn't designed for piles of books (and to be frank, is slightly too high for a laptop, so that one of us sometimes has to sit on a cushion), but this doesn't stop us using the table for eating or entertaining. Biological systems are comfortable with a degree of mess; and they have to be, because it is a fact of life. It can be ignored (or put in cupboards), but that does not stop it being real. And theories that try to account for the biological or linguistic world by pretending the irregularities, variation and gradual accretions are not there are, to our minds, fatally compromised on the grounds of realism.

As one might expect, HCF have reacted to Pinker and Jackendoff's critique (Fitch, Hauser and Chomsky 2005, henceforth FHC). Much of FHC reiterates that there may well be aspects of FLB which were not pursued in the 2002 article for reasons of space (potentially including phonology, for instance), and that FLN may include more than recursion, or indeed may be empty; but surely the original article must be judged at least in part on the concrete proposals made there, not on the many alternative interpretations which *could* be made? FHC rightly note that questions (and answers) on function are complex: 'what is bat echolocation for?' does not lend itself to a single answer synchronically (finding prey, navigation, finding mates …), and is difficult to address diachronically in terms of why echolocation evolved in the first place, since behaviour of course does not fossilise. Likewise, they argue that language can be used, and useful, for communication without necessarily having developed originally to fulfil this role: in particular, 'That recursion is useful is obvious; this does not automatically make it an adaptation in the evolutionary sense' (Fitch, Hauser and Chomsky 2005: 186). Specifically, Hauser, Fitch and Chomsky in both papers set up inner speech as an equally plausible role for language: it can of course be used to talk to other people, but its key function lies in talking to ourselves, working through alternatives and considering options purely internally.

Jackendoff and Pinker (2005) respond in turn, initially challenging FHC's view that questions about adaptation need to focus on either current utility (what a trait is used for by a species now) or functional origins (what it was used for in the first organisms to have it). However, Jackendoff and Pinker argue for a third possibility, namely current adaptation, or

> what the trait was selected for in the species being considered … Though this is often the most biologically interesting question about a trait, FHC provide no place for it in their dichotomy. And it is only by omitting this third alternative that Chomsky can maintain that nothing distinguishes the use of language for communication from the use of hair styles for communication, and that nothing distinguishes the utility of language for communication from the utility of language for inner speech. (2005: 212–13)

Though FHC contend that adaptive function is hard to determine precisely, as in their example of the multiple uses of bat echolocation, Jackendoff and Pinker reply that this is not a problem if we adopt the right level of resolution: we need not be too worried about whether the main use of echolocation is to find food or to navigate, but could cover both by saying its function is 'something like sensing the location and motion of objects in the dark. Likewise it seems odd to say that, just because we do not know whether primate vision evolved for finding mates or finding food, we cannot say anything about the adaptive function of the visual system at all!' (Jackendoff and Pinker 2005: 213).

Secondly, Jackendoff and Pinker (2005: 214) return to the distinction between FLB and FLN, which they accept in principle, but interpret in a crucially different way from FHC. The key to this controversy is just what it means to be unique: is something only unique if it is and always has been completely distinct from anything else, or can it be described as unique 'in graded terms: that the trait has been modified in the course of human evolution to such a degree that it is different in significant aspects from its evolutionary precursors (presumably as a result of adaptation to a new function that the trait was selected to serve), though not necessarily different in every respect'? As Jackendoff and Pinker (2005: 214) continue:

> FHC often apply the FLN/FLB distinction in absolute terms, using any similarity between a language function and anything else (speech perception and generic audition, word learning and fact learning, speech acquisition and vocal imitation) to sequester the function in FLB. It is no surprise, then, that they can relegate all the evidence we adduce for linguistic adaptations into FLB … We instead interpreted FLN in graded terms. The point of the FLN/FLB distinction is to identify those features of language that recently evolved in the human lineage (and which thereby help to answer the key question of why we have language and other apes do not). Evolutionarily speaking, nothing springs forth without a precedent, so if FLN is interpreted as 'uniquely human' in the absolute sense, it is hard to imagine that it could contain anything, in which case it no longer defines a useful distinction.

One option, then, is that we find FLN to be empty; alternatively, however, 'the current state of play suggests expansion of FLN. Conceptually, the FLB/FLN distinction makes sense; empirically, HCF's division is in the wrong place' (Parker 2006: 6). This view still leaves space for a distinction between FLN (which would contain systems showing significant adaptations for language, though they may be built on precursors that exist elsewhere), and FLB, with systems certainly used for language, like motor control, essentially unchanged from their other previous and current uses (and again, what about memory, for instance – where would that fit in?). Recursion, however, might or might not be one of those adapted aspects of FLN. Again, Jackendoff and Pinker challenge its inclusion by proposing that recursion can also be seen in human visual cognition and music, where we also find hierarchical arrangements of units which can be repeated and extended with no theoretical limit, producing non-linguistic examples of discrete infinity.

Much of the rest of Jackendoff and Pinker's argument depends on the distinction between two theoretical views of syntax. Whereas Chomsky's approach focuses on a regular syntactic core with stored lexical items combined by rule (or by Merge in the Minimalist version), Jackendoff and Pinker favour a view based on constructions, attempting to integrate the many idioms and irregularities which cannot comfortably be reconciled with Minimalism. Their claims for the evolutionary superiority of their constructional approach, however, depend on an assumption of the relative priority of development of particular aspects of modern human language. In particular, they claim that a constructional view 'makes it natural to conceive of syntax as having evolved subsequent to two other important aspects of language: the symbolic use of utterances ... and the evolution of phonological structure as a way of digitising words for reliability and massive expansion of the vocabulary' (Jackendoff and Pinker 2005: 223). This part of the argument belongs in the next chapter, when we turn to the concept of protolanguage and its transition to modern human language, and therefore cannot be evaluated fully here.

For the moment, however, it is worth summarising the key objections raised by Pinker and Jackendoff/Jackendoff and Pinker, who challenge the HFC/FHC approach as too closely tied to a series of unrealistic dichotomies. Just as a formalist, Minimalist approach to grammar exiles a range of semi-regular and messy (but perfectly usable and interpretable) constructions to the linguistic periphery, so the FLN/FLB dichotomy, where interpreted narrowly, may impose an overly strict division on what is actually a gradient between highly adapted human capacities and those barely modified at all from their precursors. Absolute dichotomies (whether they involve core and periphery, human and non-human, or FLN and FLB) inhabit the clear-cut but essentially fictional universe of the binary-splitting family tree which we have encountered (and worried about) before. In just the same way, such dichotomies seek to impose a black-and-white perspective on a comfortable but somewhat irregular world full of shades of grey. However, if we are to argue, with Jackendoff and Pinker, for a world of relative shades rather than absolute colours, we must also show that a reasonable and testable evolutionary account can be given of the development of language in this kind of gradient framework, and it is to this challenge that we turn in the next section.

8.3   A gradual, adaptive view of the evolution of language

8.3.1   Escaping from 'just-so' stories: language as beneficial

Adaptive arguments are difficult precisely because they are too easy. Think of those pictures that sometimes appear in magazines or on television programmes, where the reader or viewer is challenged to identify some unlikely-looking implement by saying what it is for. Humans can very straightforwardly impose the most bizarre interpretations, which they rapidly become convinced are the only possible answers: it's got lots of little holes in it, so it must be a cheese grater; and it's that shape because maybe cheese came in different shapes at the time when it was made; and so on. This is a kind of reverse-engineering 'just-so' story, like the fables that start from the observation that elephants have trunks and conclude that they got them because a curious young elephant stuck its nose in the river and had it stretched beyond the point of no return by a crocodile, or that bears have no tails because an ancestral bear was tricked by a fox into putting its tail into a frozen river to fish and had it cryogenically removed.

The same problems potentially arise when complex systems are described in evolutionary biology as being good for a particular function, and therefore as having evolved in order to fulfil that same function. The previous section has shown that it is important to distinguish between what a particular trait does now, and what it might have been selected for; but certainly adaptation is a recognised and perfectly legitimate evolutionary mechanism, and Dawkins (1986), among others, has discussed at length how complexity and apparent design can arise from an initial mutation and subsequent selection, giving rise to a gradual accumulation of beneficial properties (Dennett 1995). The problem is the potential unfalsifiability of such adaptive accounts, which are in danger of being satisfying on the bedtime story level but intellectually rather flimsy. As Hauser and Fitch (2003: 158) put it, 'the construction of historical narratives of language evolution is too unconstrained by the available data to be profitable at present ... At best, this practice provides a constrained source of new hypotheses to be tested; at worst it degenerates into fanciful storytelling.'

Two clarifications are necessary before we go further in this discussion. First, Jackendoff and Pinker accept without question that using the biological comparative method to investigate similarities between animal systems and human language is essential. But while this method is necessary to progress in evolutionary linguistics, it is arguably not sufficient, especially if some aspects of the human language faculty do turn out to be specific to humans and to language, in which case they will need to be investigated in their own right. Comparative studies can test possible hypotheses on the functional source of a particular innovation, but they must also be supplemented when we turn to the consideration of how that trait is now used in a modern species, and a properly developed adaptive account might well show us how and why that trait left its original moorings in an ancestral species and shifted to its current function. Second, it is possible to recognise the poor quality of some adaptive 'explanations' without ruling out the existence of good ones: Jackendoff and Pinker (2005: 213) rightly challenge the argument that 'Adaptive explanations can be done badly, so no one should ever attempt to do them well.' We shall bear this in mind as we consider possible adaptive accounts below. Such accounts ask how and why an initial, random change might become cumulative, or an initial small beginning might develop into something much more complex. Returning to Chapter 2, can we reasonably regard language, or aspects of language, as 'designed', and therefore as the product of adaptive complexity? What factors might be active in the selection of language-friendly adaptations, and what evidence can we gather for their likely operation?

Pinker and Jackendoff (2005) note that HCF provide three arguments against the involvement of natural selection in the evolution of the language faculty, namely that language is not designed for communication and may fulfil communicative functions rather poorly; that language provides a perfect mapping between sound and meaning, which means it is unlike biological systems in general and needs a different explanation; and that FLN developed for cognitive purposes other than language and cannot therefore be said to have been under selection for language. As Pinker and Jackendoff (2005: 223) observe, 'these hypotheses challenge a more conventional evolutionary vision of language, according to which the language faculty evolved gradually in response to the adaptive value of more precise and efficient communication in a knowledge-using, socially interdependent lifestyle'.

To some extent, the three objections to be found in the papers by Hauser, Fitch and Chomsky seem to sit rather uneasily together in any case. How can the perfection of the second point be reconciled with the inefficiency of the first, or the denial of natural selection in the third point combine with HCF's claim that recursion may well have been selected for some non-linguistic function like navigation or number? These apparent contradictions do resolve to an extent when we recall that 'language' is being used in several different senses, the key one for HCF/FHC being the internal computational system of the language faculty defined narrowly, FLN. There remain, however, a number of possible objections to these claims.

First, if FLN initially evolved for reasons other than language, such as navigation, this can be explored through the biological comparative method (by considering other navigational systems in other species), and this will certainly tell us more about the remote ancestral past of aspects of the language faculty. But invoking selection for other functions still means invoking selection. Let us say that FLN evolved piecemeal for reasons other than language, but then once evolved took on its role as the linguistic keystone, unifying a whole range of other disparate systems into the enabling structures for human language. This effectively buys us a saltationist perspective on language without an unlikely genetic macromutation: we have normal processes of gradual evolution, but language itself is special because it arises spontaneously from the outcome of these processes in another domain. However, Chomsky *et al.* cannot have it both ways. Either natural selection and adaptation were involved in the evolution of FLN, or they weren't; and if they were, whatever the motivation at the time they were going on, the resulting system is in no sense **biologically** special, and might subsequently have been subject to further selection for linguistic purposes.

Likewise, if language does not fulfil communicative functions optimally, that does not necessarily mean aspects of language were not selected for communicative reasons. Chomsky's view of language as being primarily for inner speech and the expression of thought fits well with his own emphasis on the purely internal, computational aspect of language; but this is not the only conception that matters, and evolutionary linguists might equally be interested in universals of external language structure and use, and in what those might tell us about underlying systems of mind and brain. It is certainly true that language is used very substantially for communication – especially if we allow 'communication' to cover not just cases of direct transfer of factual information from one person to another, but also the conveying of social information (I'm listening to what you are saying; I'm interested in you; neither of us comes from round here; I don't trust you). On the other hand, many biologically evolved systems work tolerably well without doing their job perfectly, precisely because a gradual, adaptive model of complexity has to contend with evolutionary leftovers and build on less than ideal foundations. So, human teeth are terrific, but they do not always fit very well into a *Homo sapiens* shaped mouth, and impacted wisdom teeth, for instance, before dental surgery and antibiotics, could be very nasty indeed. One of the most frequently cited cases of adaptive complexity is the human eye; but in at least one respect, it is 'imperfect'. The light-sensitive cells in the retina point towards the back of the eye, and not towards incoming light, so that light has to pass through the cell body (causing some 'scatter' and a consequent reduction in sharpness of focus). Furthermore, the axons for these cells have to take a circuitous route to the optic nerve because they are on the wrong side of their cell bodies, and have to pass through the retina; this creates a blind spot, though that is to some extent compensated for in the brain if we have good binocular vision. In this respect, the octopus eye is closer to 'perfection', since here the light-sensitive cells point toward the source of light. In no way, however, do these facts threaten the very secure hypothesis that eyes, in whatever species, are primarily for seeing, and that the forces of evolution will have worked differently in different environments (depending on the species, the starting point, and the other systems present) to produce an outcome that provides vision of some kind. Since many organs and systems involve multiple functions, there is often an additional layer of compromise to take into account, as with the larynx, for instance, which functions in mammals both to protect the airway and to produce sound (Hauser and Fitch 2003).

Where does this leave us with Chomsky's assertion that language is a perfect mapping between sound and meaning? Recall that perfection is only claimed for FLN, and hence for recursion. The interesting challenge posed by this proposal of perfection is that it necessarily goes against any idea of gradual adaptive evolution. 'One might ask what the relevance of the possible "perfection" of language is to its evolution. The idea seems to be that nothing less than a perfect system would be in the least bit usable, so if the current language faculty is perfect, one could not explain its evolution in terms of incremental modification of earlier designs' (Pinker and Jackendoff 2005: 227). There are two responses here. First, we have already seen that the linguistic framework in which Chomsky is operating leads many aspects of language to be sidelined as peripheral; that does not stop them being part of human languages, but they are certainly not perfect. Similarly, many aspects of the FLB, however that is ultimately defined in detail, will have evolved through gradual accumulations of beneficial mutations and have clear analogues or homologues in other species, meaning we can track incremental developments. The second issue here is about the utility of such partial systems, and here we turn to one of the longest-running debates in evolutionary theory.

The argument about perfection, and the evolutionary one-offness of FLN, goes along with a challenge to gradual adaptive evolution based on the uselessness of an imperfect, partial system. This has been a recurring theme in Chomsky's own writing: for instance, he has suggested that 'In the case of such systems as languages or wings it is not easy even to imagine a course of selection that might have given rise to them. A rudimentary wing, for example, is not "useful" for motion but is more of an impediment. Why then should the organ develop in the early stages of evolution?' (Chomsky 1988: 167). Refutations of such arguments can be constructed for partial wings (which might be very useful for balance) and, as we shall see in the next chapter, can also be developed for linguistic communicative systems lacking much of the structural complexity of human languages today. However, most discussion in the literature involves the utility or otherwise of part of an eye, or of a whole eye with poor vision, and it is perhaps most clearly articulated by Richard Dawkins, who points out that (1995: 89):

> You can play tennis with quite blurry vision, because a tennis ball is quite a large object, whose position and movement can be seen even if it is out of focus. Dragonflies' eyes, though poor by our standards, are good by insect standards, and dragonflies can hawk for insects on the wing, a task about as difficult as hitting a tennis ball. Much poorer eyes could be used for the task of avoiding crashing into a wall or walking over the edge of a cliff or into a river. Eyes that are even poorer could tell when a shadow, which might be a cloud but could also portend a predator, looms overhead. And eyes that are still poorer could serve to tell the difference between night and day, which is useful for, among other things, synchronizing breeding seasons and knowing when to go to sleep.

The important thing is that there is a continuum from really fabulous vision through not very good vision to no vision at all, not just a black-and-white, absolutist distinction between eye = vision and no eye = no vision. By way of further evidence, Dawkins (1986: 81) notes both that 'In a primitive world where some creatures had no eyes at all and others had lensless eyes, the ones with lensless eyes would have all sorts of advantages', and that when you lose your glasses, you don't feel you have to just close your eyes until they turn up again.

Of course, the eye is a particularly good example, because we know from current variation among species that there are many kinds of eye, some much more complex than others, but all used for (sometimes rudimentary) vision. We also know from application of the biological comparative method that 'the eye' is a misnomer: eye-like structures have evolved at least forty times, independently, in different species and groups (see again Dawkins 1995, Nilsson and Pelger 1994). Language could, by the same token, be seen as a rather poor example, because on Chomsky's reasoning we are unable to find similar but simpler systems in our own or other species; moreover, we need to be able to show, to refute the single-step evolution part of the argument on perfection, that whatever underlies language could have evolved from some demonstrable starting point.

Before exploring this further, it is worth noting also that the normal course of gradual evolution, resulting in the accumulation of complexity, will in itself be constrained by just that starting point and by the environment in which the process takes place. Consequently, finding some aspects of the eventual system which are apparently maladaptive (like the backwards human eye discussed above), or at least not perfect, does not contradict an overall assumption of utility and adaptation. It is certainly adaptive for bones to be hard, and calcium is hard, so using calcium is a good adaptive solution; but calcium is also white, which is not adaptive for bones (though not especially problematic either). We can put forward and defend an adaptive view of bones, which is not derailed at all by the observation that whiteness is contingent, a 'just-so happens' part of the story (see McMahon 2000: 145). An overall system can be adaptive and useful, but to assume this means every contributory element must also be selected for and maximally functional amounts to 'hyper-selectionism', which Darwin rejected (Newmeyer 1998: 309).

If individual components of an adaptive whole can be neutral, or even maladaptive, or not selected for at all, then we need to recognise both non-adaptive traits and non-adaptive evolutionary mechanisms by which these can come into being. The second point is unproblematic in principle, as we have already discussed one such mechanism in Chapter 7 above, in the form of drift, and we should equally include, as Pinker and Bloom (1990: 4) observe, the direct influence of the environment and accidents of history. In practice, Gould and Lewontin (1979) suggest that many adaptationists pay lip service to these other mechanisms but tend to sideline them in favour of adaptive accounts, and this is clearly to be guarded against: the various alternatives must be considered and the arguments weighed up in each case. As for non-adaptive traits, the most famous example here also comes from Gould and Lewontin (1979).

The dome of St. Mark's church in Venice is mounted on rounded arches, and between each pair of these to the sides, and the roof above, we find a triangular space, narrower at the bottom and broader at the top. In architectural terminology, these spaces are known as spandrels, and in St. Mark's (and many other ecclesiastical buildings designed on similar principles), they are occupied by mosaic pictures. 'Each spandrel contains a design admirably fitted into its tapering space. An evangelist sits in the upper part flanked by the heavenly cities. Below, a man representing one of the four Biblical rivers ... pours water from a pitcher into the narrowing space beneath his feet' (Gould and Lewontin 1979: 581–2).

There are two ways we can see these spandrels. One is to regard them as essentially functionless structures which arise accidentally from other independent developments: since nature abhors a vacuum, the spandrels are appropriated for another purpose. On the other hand, if we are too firmly entrenched in adaptationist mode, we are apt to see the spandrels as designed for the mosaics, as if the artists and architects had felt an overpowering need to site these very pictures somewhere in the church and had come up with the dome-plus-arch plan expressly to accommodate them. The narrow faculty of language, in Chomsky's terms, might well be a spandrel: a happy accident that arises from a conjunction of other forces and mechanisms (recursion as selected for navigation, say) and then turns out to be good for language, just as the spaces in St. Mark's turned out to be good for mosaics. However, returning to Chomsky's own objections above, the initial status of FLN as a spandrel does not mean that it could not be subject to natural selection subsequently: 'putting a dome on top of four arches gives you a spandrel, but it does not give you a mosaic depicting an evangelist and a man pouring water out of a pitcher ... To get the actual mosaic you need a designer. The designer corresponds to natural selection' (Pinker and Bloom 1990: 710). Again, we see the absolute dichotomy between natural selection on the one hand, and other forces like drift on the other, as artificial: these forces interact, and do so over long time-spans and series of environments.

Where, then, does this leave spandrels? 'The Gould and Lewontin argument could be interpreted as stressing that since the neo-Darwinian theory of evolution includes nonadaptationist processes it is bad scientific practice not to test them as alternatives to natural selection in any particular instance. However, they are often read as having outlined a radical new alternative to Darwin, in which natural selection is relegated to a minor role' (Pinker and Bloom 1990: 5). Pinker and Bloom go on to show that this second reading is not well motivated: forces like drift, models like punctuated equilibrium, and spandrels all turn out to be perfectly compatible with a neo-Darwinian approach to evolution. Whenever we find ongoing evolutionary processes, we are likely to find by-products, which might at the time they arise have no function. Alternatively, a structure which did at one point in evolutionary history have a function may have lost it, because of other developments in the organism of which it is part: so, humans no longer have tails, but we still have tail-bones, and here we find down-regulations to the extent that the remaining structures no longer interfere with other systems; but that is all. Where a leftover structure is essentially neutral, there is no reason for selection pressure to act further against it, and it will persist in that state. Nonetheless, whenever we see a structure without a function, we have a potential locus for future selection, just as church artists seized on the empty and coincidental spandrels to produce beautiful mosaics. Similarly, as we shall see in the next section, selection can step in and work on a structure which has lost its function, or add a further function, recruiting a pre-existing trait for a new purpose.

Even if FLN, or any other system used for modern human language, turned out to be a spandrel, then, this would not rule out its subsequent selection for linguistic purposes. However, to maintain the option of an adaptive account, we need to address two remaining issues: the usefulness (or the reverse) of 'half a language', and the benefits that might follow from developing language in the first place.

We can demonstrate that half an eye is still useful in two ways: there are other species which have poorer, less-complex eyes, but still show demonstrable advantages over creatures without vision; and there are individuals within a species with less acute vision who are advantaged by correcting or compensating for this in some way (one of us has a very positive view of contact lenses). Language is difficult because it is a combination of the physical and the behavioural, and behaviour does not fossilise, so that we cannot trace the evolving complexity of language through our ancestors. In the next chapter, we shall discuss in more detail theories about the development from elementary protolanguage to modern human language, and will consider evidence from 'living fossils' of various kinds, including child language and pidgin languages. For the moment, however, we clearly find a large number of species which communicate vocally (among other ways), and evolutionary advantages do seem to follow from these: vervet monkeys, as we have seen, have calls that appear to refer to different predators, and taking the appropriate avoiding action in response to these will frequently lead to the preservation of those vervet monkeys and their genes. Here, of course, we are begging the Chomskyan question: this argument is not relevant to human language if human language is not useful for communication but is primarily for inner speech. Yet there are many cases where language transparently is used for communication, and returning to the second eye-related example above, we find many cases of children being referred for speech and language therapy because society sees it as an advantage to them to develop their capacities as fully as possible. Parents typically raise concerns about their children's language, not because they cannot think or talk to themselves, but because they are experiencing difficulties in expressing themselves successfully and communicating with other people.

Again, there are two key problems with this kind of pretheoretical defence of the communicative utility of language. How do we deal with allegedly maladaptive elements, and how do we demonstrate 'utility' in its true biological sense? To some extent, it is to be expected that some aspects of language may be either maladaptive or simply less streamlined than others, because of the piecemeal operation of evolution in the construction of the many and varied systems underpinning it: maladaptivity is therefore potentially just as good an argument for a gradual, adaptive account as for a saltationist one. However, we must be cautious in interpreting 'utility': as we have already seen, Chomsky considers language as primarily serving non-communicative functions, so we cannot necessarily assume that improving or derailing communication is the issue here. In biological terms, if a trait is useful to an organism, it must mean, all other things being equal, that an individual with that trait has a greater likelihood of leaving viable offspring than another individual without it. Lightfoot (1991: 68–9) ridicules this view for one particular aspect of syntactic theory, suggesting that 'the Subjacency Condition has many virtues, but I am not sure that it could have increased the chances of having fruitful sex'. Working at this level of detail might not be helpful: for one thing, as theories of language change, their ingredients will change too, so that in Minimalism the idea of Subjacency (which refers to the permissible distance a constituent can move within a sentence or between two co-dependent constituents) is no longer captured by a free-standing principle. If the effects of Subjacency turned out to be useful in terms of communication, as Newmeyer

(1991: 15) suggests they may be, then this cuts no ice in Chomskyan terms at all because communication is essentially irrelevant. But even if we accept that communication is the essential purpose of language, we then have to proceed to show how this can benefit individuals and the species.

Biologists tend to struggle with the idea that communication should **not** automatically be seen as advantageous, and that more subtle and varied examples should not serve individuals better: as Ridley (1999: 105) suggests,

> It is easy to conceive how it was advantageous for our ancestors on the plains of Africa to share detailed and precise information with each other at a level of sophistication unavailable to other species. 'Go a short way up that valley and turn left by the tree in front of the pond and you will find the giraffe carcass we just killed. Avoid the brush on the right of the tree that is in fruit, because we saw a lion go in there.' Two sentences pregnant with survival value to the recipient; two tickets for success in the natural-selection lottery, yet wholly incomprehensible without a capacity for understanding grammar, and lots of it.

Similarly, Pinker and Bloom (1990: 9) argue for the benefits of recursion by noting that 'it makes a big difference whether a far-off region is reached by taking the trail that is in front of the large tree or the trail that the large tree is in front of. It makes a difference whether that region has animals that you can eat or animals that can eat you' (1990: 28). Regardless of how and whether individual aspects of language might ease parsing or in fact make it more difficult in certain circumstances, there is plentiful evidence that being able to provide and understand information might lead to greater chances of survival and thence, indirectly, to a greater chance of leaving viable offspring. The analogy with Dawkins's eye also goes through: any vocal communication might be better than none because it allows alarm calls to be given at night or when your hands are full; more language might be better than less because it increases the complexity of what can be communicated and hence the scope for appropriate action, social sharing, and passing on of accumulated knowledge. Moreover, for the individual, bravura rather than competent language use could conceivably be rather attractive to the opposite sex.

Accepting all this still does not tell us precisely what context language might initially have evolved in, and precisely what advantages it might have conferred. As Számadó and Szathmáry (2006: 555) note, there are plenty of selective options for language evolution to choose from: 'There are scenarios focusing on almost all of the possible uses of human language one can think of: mate choice, mating contract, pair bonding, parent–offspring communication, gossip, rituals, grooming, tool making and hunting.' This is not necessarily a good thing. Having plenty of candidates for the selective function of early language should not necessarily convince us that it is an example of adaptive complexity. We are potentially back in 'just-so' story territory again. Számadó and Szathmáry (2006) try to tackle this underlying problem by working out what we can assume about the evolution of language, and on what basis, so that conflicting theories can be rejected, though the strategies for this evaluation still need to be determined and tested. We turn to some of these in the next chapter, when we review social aspects of language evolution, and possible tests involving simulations.

8.3.2   Exaptation and mirror neurons

In the previous sections, we have reviewed some suggestions that language is beneficial, and that selection for language abilities may therefore have formed part of our species' evolutionary history. However, we have also argued that there are very strong reasons for seeing the evolutionary emergence of the capacity for language as a gradual process, rather than invoking saltation. In cases like language, where we see a highly complex, highly integrated set of systems, it can be tempting to assume that the only possible source is instantaneous, since we cannot reliably retrace the steps by which the current whole has developed (though we will return to some hypotheses on these steps in the next chapter). On the other hand, we also know that evolution has a propensity for reconfiguring existing structures, regardless of their original function or provenance, rather than creating something new out of nothing. Such reconfigurations are known in biology as exaptations (Gould and Vrba 1982), defined by Gould (2002: 1232) as 'features coopted for a current utility following an origin for a different function (or for no function at all)'. By considering some possible cases, we may be able to reconstruct some of the evolutionary scaffolding which allowed modern human language to be constructed in a gradual, adaptive way. In doing so, we may also be able to demonstrate partial continuity between mechanisms associated with language in humans and mechanisms of other kinds both in humans and in non-human primates.

One possible example of exaptation which has been in the evolutionary linguistics news lately involves mirror neurons (Rizzolatti and Arbib 1998, Arbib 2003, 2005a, b). Studies of monkey brains have revealed that area F5 of the frontal lobe is involved in the control of hand movements for grasping. Furthermore, a specific group of neurons in region F5 are activated not only when the monkey performs an action itself, but also when it sees another monkey (or a human) performing that same action – hence the term 'mirror neurons'. This mirror system may provide the key to learning strategies for physical actions through positive feedback. Infant monkeys learn to carry out certain actions like grasping by observing and practising; crucially, these will both activate the same neurons, and as the infant's own attempts become more like the adult's action, we can assume that the match becomes closer.

The immediate question is what all this has to do with human language. First, the human brain area homologous with monkey F5 is part of Broca's area, which we have already seen is associated with certain language functions; it has also 'been shown by brain imaging studies to be active when humans both execute and observe grasps' (Arbib 2005a: 113; see also Rizzolatti et al. 1996, Grafton et al. 1996). The literature on impairment and imaging reveals a broader range of connections between language and motor skills. For instance, human language use crucially involves planning and sequencing. Conway and Christiansen (2001) note that humans tend to learn complex sequences of motor skills by planning all the necessary sequence of actions in advance, whereas non-human primates show a restricted ability to plan beyond the next few actions; chimps are in a sense intermediate, with some ability to plan ahead. Broca's aphasics often have a degree of constructional apraxia, and frequently experience problems when they are asked to reconstruct abstract, hierarchically structured diagrams from memory (Grossman 1980), while deficits in motor skills are also common in SLI children (Bishop 2002). Neural imaging studies (Fox *et al.* 1988, Rogalsky *et al.* 2011) and corticostimulation studies (Calvin and Ojemann 1994) both indicate a high degree of overlap between areas of the brain, notably in and around the posterior portion of the traditional Broca's area, involved in processing oral-facial movements and in phoneme perception. It has been suggested that this co-localisation of circuits required for motor planning and language processing arose as a consequence of exaptation of language structures from a pre-existing brain system developed to plan and control complex motor sequences and in particular throwing actions (Calvin 1983, Calvin and Bickerton 2000). Throwing is one of those behaviours that humans learn to do quite easily but which other primates are not so adept at; it is possible that it may have required a considerable degree of brain power, along with the development of increased memory and sequential processing algorithms (of the sort which also turned out to be required for the development of syntax and the production of complex, hierarchically arranged linguistic utterances). Both speech and throwing require rapid alteration of precise muscle group activations in a complex linear sequence, with accurately timed transitions, in order to successfully achieve the behavioural target.

Arbib (2005a, b) then argues that there are two crucial shifts in evolutionary terms. The first is from a system where only simple actions can be copied, which Arbib assumes to have persisted from the common ancestor of chimps and humans, to the human-specific potential for imitation of complex, compound actions. Secondly, grasping and other learned manual actions became communicative as well as practical: once gestures are used as signs, they can be conventionalised, providing the basis for what Arbib characterises as '"protosign", a manual-based communication system that broke through the fixed repertoire of primate vocalizations to yield an open repertoire of communicative gestures' (2005a: 115). This conventionalisation seems to have required further neurological innovation, since monkey mirror neurons for grasping only fire when the monkey sees both the relevant hand movement and an object being grasped, whereas, of course, a conventional use of a gesture to 'mean' grasping might involve a pantomime use of the gesture without any actual contact with an object. Increasing conventionalisation, and increasing distance from the original, practical use of a sign, would lead to protosign, which 'provided the scaffolding for early protospeech after which both developed in an expanding spiral till protospeech became dominant for most people' (2005a: 110).

This relates to the arguments in Chapter 5 above on gesture as a possible pre-requisite for spoken language: Arbib takes the view that gestural communication need not, and probably could not, have developed to the level of a full language system before vocal communication began to emerge, but rather that elementary communicative gestures were the initial stage, and that speech and gesture then developed and became more complex together. However, Arbib (2005b: 38) suggests a long-standing link between physical movements and sound, as 'Some mirror neurons in the monkey are responsive to auditory input and there are oro-facial neurons in F5 that control movements that could well affect sounds emitted by the monkey.' This connection between sound and manual gesture might underlie the human innovation of Broca's area as a controlling region for speech and for the vocal apparatus. Such exaptations are in line with Arbib's (2005a: 120) general view that 'Most of the stages of our evolutionary story are not to be seen so much as replacing "old" capabilities of the ancestral brain with new ones, but rather, as extending those capabilities by embedding them in an enriched system.' This line of argument is in turn extended by Molnar-Szakacs and Overy (2006), who implicate the mirror neuron system in human emotional reactions to music: both emotion and the production of music are active physical processes (think of the facial expressions and gestures we instinctively produce in particular emotional states), and the mirror system means we can 'read' emotional responses back when we hear music.

Once we start looking for possible cases of exaptation, we find them everywhere: mirror neurons might be the current favourite, but they are not by any means the only example. For instance, take pitch, which is a fundamental characteristic of speech and music, but also a subjective aspect of the sound based on perception of a predicted fundamental frequency, even if that particular frequency is absent from the input (Moor 2003). Recording brain activity in marmosets presented with complex series of clicks indicates that there are specific units which fire in response to the pitch of the signal; furthermore, a 200-hertz neuron will also fire in response to a signal that contains 800, 1000 and 1200 hertz even in the absence of the cognate fundamental frequency of 200, because the higher frequencies are harmonically related to that fundamental (Bender and Wang 2005). These special 'pitch' receptors appear to be responding in a way that is consistent with our human perception of the abstract concept of pitch, which allows for the detection of a musical note as being, say, F sharp, regardless of the instrument it is played on. These pitch detectors are relatively few, located in a small area of the marmoset's auditory cortex and embedded with a much larger number of units that respond to low-frequency sound in the 125–2000 hertz range near the boundary of the primary auditory area A1. This corresponds to a general area of activity detected by functional MRI scanning during processing of pitch in humans, although this is mainly active only in the right hemisphere (Penagos, Melcher and Oxenham 2004). It is as yet unclear how the relevant information is extracted, or indeed how these units are connected to the primary input stream from the ears or to the higher brain centres, but it is likely that the particular pattern of timing and frequency of specific hair cells in the cochlea will only stimulate these 'pitch cells' in the presence of a significant component of a particular pitch. Whether these are hard wired or 'learned' networks is also currently unknown, leaving the same open question of the balance between innateness and environmental learning that we find for human language in general.

These ideas of adapting and retuning pre-existing components of the brain (and indeed other relevant systems) for linguistic purposes have been central to this chapter, and indeed are central to the book. Arbib (2005b: 22) summarises his argument on the interconnected development of protosign and protospeech by suggesting that these culminate in 'language readiness', by which he intends:

> those properties of the brain that provide the capacity to acquire and use language. Then the hypothesis offered here is that the 'language-ready brain' of the first *Homo sapiens* supported basic forms of gestural and vocal communication (protosign and protospeech) but not the rich syntax and compositional semantics and accompanying conceptual structures that underlie modern human languages.

What, then, was the protolanguage of these early, language-ready brains? How can protolanguage be defined, and how far can we go towards determining what it might have been like? Moreover, how did we, or rather our ancestors, get from protolanguage to the complex systems we know as language today? These questions will be addressed (though not necessarily fully answered) in Chapter 9.

8.4   Summary

In this chapter we have mainly argued that there are no single, simple answers in evolutionary linguistics (which makes perfect sense since there are very few single, simple questions either). We began with a return to the genetic evidence which was the focus of Chapter 7, and a comparison of the case of *FOXP2*, which is typically seen as one small part of a complex story, with Crow's proposal that genetic changes associated with protocadherin constitute the crucial, single factor in language evolution. In the next section, we proceeded to a more general consideration of the opposing models of saltation – a once-and-for-all, mighty leap, for instance through a genetic macromutation – and gradual, adaptive evolution in the case of language. Here we made particular reference to Chomsky, Hauser and Fitch's distinction between the faculty of language in the broad sense, involving aspects of human cognition which are not specifically dedicated to language, and the faculty of language in the narrow sense, which might be restricted only to the property of recursion. We reviewed a series of criticisms of the FLN, and specifically of the idea that recursion is its unique property, and instead argued for a broader view of language as intrinsically involved in communication, not relevant only for internal monologue and thought.

However, if we are to consider gradual, adaptive evolution of the capacity for language seriously, then we must also be able to demonstrate that contentful and intellectually robust adaptive scenarios can be constructed, without undue reliance on 'just-so' stories. In the next section, we therefore considered alternatives to gradualism, and specifically the idea that language might be a spandrel, or that aspects of language might be maladaptive. We concluded that neither of these possibilities is incompatible with a gradualist, adaptive approach: language need not be perfect in order to be fit for purpose, and the whole system might be strongly advantageous even if parts of it are imperfect or bear signs of their evolutionary origins outside language or even communication. This led naturally to a discussion of mechanisms whereby systems or structures originally developed in connection with one function might in time come to play a quite different role, and specifically to the key notion of exaptation. In particular, the exaptation of mechanisms for grasping, mediated through the mirror neuron system, which involves activation both while an action is carried out and while it is observed, may have formed the evolutionary basis for certain hierarchical, highly structured aspects of speech production and perception. This provides a possible scenario for the mutual reinforcement and development of gesture and spoken language, which Arbib (2005b: 22) suggests might have culminated in the 'language-ready brain' of early *Homo sapiens*.

Further reading

We are now at the stage where a good deal of primary and summary literature is necessarily being referenced as we go along, since we have moved into the central questions of evolutionary linguistics. It may now be helpful to start consulting relevant chapters from key overviews, and in particular we recommend Christiansen and Kirby (2003), Tallerman (2005) and Wray (2002b). It is worth looking out for other volumes of the Proceedings of the Evolang Conferences; these include Cangelosi, Smith and Smith (2006), Smith, Smith and Ferrer i Cancho (2008) and Smith *et al.* (2010). We will refer to more papers from these edited volumes in the next chapter.

In terms of more specific aspects of the discussion in this chapter, McManus (2002) is a useful popular introduction to issues of asymmetry and lateralisation; McBrearty and Brooks (2000) provides a salutary review of our tendency to see revolutions in human behaviour where none may exist; and Dunbar (2004) gives a very accessible introduction of what happened where in human evolution for anyone who might need a quick reminder at this stage. In terms of recursion and more general discussion of the FLN/FLB debate, the key papers are Hauser, Chomsky and Fitch (2002), and Fitch, Hauser and Chomsky (2005), contra Pinker and Jackendoff (2005) and Jackendoff and Pinker (2005); Everett (2005) contends that Pirahã does not have recursion, and Parker (2006) is a helpful overview of the whole recursion debate.

Classic statements of the case for gradual, adaptive evolution and for apparent design involving natural selection can be found in Dawkins (1986, 1995) and Dennett (1995). The key statement on spandrels is Gould and Lewontin (1979), and the key refutation of their anti-adaptationist arguments is Pinker and Bloom (1990), both still very well worth reading. Számadó and Szathmáry (2006) give an overview of the selectionist landscape for the evolution of language.

The essential outline of exaptation in general is provided by Gould and Vrba (1982) (and there is much more in Gould 2002), while the term is extended into historical (as opposed to evolutionary) linguistics by Lass (1990). Finally, the literature on mirror neurons is often highly specialised, but Arbib (2005b) gives a short introduction aimed at evolutionary linguists, while Arbib (2005a) includes a series of commentaries which give an indication of the contentious issues in the mirror system hypothesis. Molnar-Szakacs and Overy (2006) provide an intriguing suggestion of connections between gesture, emotion, language and music, mediated by mirror neurons.

One interesting by-product of the FLN/FLB debate has been a re-opening of the whole issue of language universals – what are they, do they exist, and do they matter? Evans and Levinson (2009) approach these central questions through an examination of the typological data on language diversity which have come to light in fieldwork and grammar-writing. The commentators on the target article address these questions from as broad a range of perspectives as one might wish for – from Pesetsky, who counsels 'against taking linguistic diversity at "face value"', to Tomasello, who bluntly announces that 'Universal grammar is dead' ('Universal grammar is, and has been for some time, a completely empty concept. Ask yourself: what exactly is in universal grammar? Oh, you don't know – but you are sure that the experts (generative linguists) do. Wrong; they don't. And not only that, they have no method for finding out' (2009: 470)). Nobody can accuse evolutionary linguists of failing to engage with the big questions!

Points for discussion

1. Gould (2002: 1231–3) argues for the use of the term *exaptation*, and against the term *preadaptation*. What is, or should be, the difference between these terms? Why might Gould be so negative about one and so positive about the other? Does Lass's (1990) extension of the concept of exaptation into historical linguistics cast any light on the matter?
2. Bryan Sykes (2006: 125ff, in part of a very readable and entertaining account of the genetic history of Britain and Ireland) outlines the 'Genghis Khan effect'. Sykes reports a particular Y chromosome in Asia, which is unusually widely distributed and extremely frequent (carried for instance by 8% of men in Mongolia). He suggests that this marks descent from Genghis Khan, who appears to have fathered a staggering number of sons as well as establishing

a system of patrilineal inheritance. How far and how demonstrably, if at all, might results of this kind in terms of frequency and distribution of genetic markers be linked to language as opposed to other factors? Is it possible to distinguish language as a motivating factor from other influences?

3. We suggested earlier in this chapter, in a rather off-hand and unsubstantiated way, that 'bravura, rather than competent language use could conceivably be rather attractive to the opposite sex'. How might you go about either defending or refuting this suggestion? Can you find any examples from the sociolinguistics literature of cases where a particular register of language might be used for display or 'showing off' purposes, or where there appear to be social advantages to be gained from using language in a particular way?

4. If you are interested in phonology you might know about, or might wish to find out about, the theory of Articulatory Phonology (see for instance Browman and Goldstein 1986, 1989, 1992). The concept of gestures, which is central to Articulatory Phonology, is based on research on reaching. How far is this analogous to the grounding of protosign and protospeech in motor activities such as grasping, which is central to the mirror system hypothesis?

9   From protolanguage to language

9.1   Overview

In the last chapter, we derived the notion of the 'language-ready brain', which Arbib (2005b: 22) saw as providing 'the capacity to acquire and use language', and supporting 'basic forms of gestural and vocal communication … but not … rich syntax and compositional semantics'. In this chapter, we shall ask how clearly we can define and describe the product of that language-ready brain, which we might call protolanguage. Furthermore, once protolanguage had developed, how did it subsequently become more complex, to give us the variation and structure characteristic of modern human languages? These questions of structural complexity connect to issues of motivation as well as mechanism. We have argued for the admissibility of arguments based on natural selection in the evolution of the capacity for language; but should this extend to accepting adaptive accounts of the subsequent increase in linguistic complexity? Invoking natural selection for genetic specification of the brain or physical systems controlling language in general is one thing, but claiming direct genetic control of individual constructions or concrete elements of a language would be something else again. On the other hand, if natural selection is not involved in the transition from protolanguage to language, then what does explain that development? Are the arguments about motivations for enhanced structure in language affected or shaped by the nature of protolanguage itself? And if so, how confident can we be in our assertions about protolanguage and hence about the mechanisms for increasing linguistic complexity? Even if we were to conclude that learning rather than evolution is the major factor shaping the development of modern languages, which would therefore fall within the domain of cultural rather than biological evolution, is there any capacity for feedback from those learning mechanisms to the genetic level, or are these completely insulated from one another? Indeed, is it reasonable to distinguish between biological and cultural evolution at all?

A word of warning: this is a very long chapter. We have kept it as such because we see the argumentation as continuous, and feel that making a division between the sections on protolanguage to language and those on increasing complexity and biological versus cultural evolution would be essentially artificial. Worse, arranging a chapter division at any point here would appear to commit us to supporting elements of discontinuity in the evolutionary picture, where in fact we see none. In actual use, however, readers (whether as individuals or in class groups) might be well advised to treat this material as sufficient for two sessions rather than strictly comparable to one of the earlier chapters. Of course, that's evolution all over: we've 'designed' one thing, but environmental conditions will probably make it divide into two in actual practice.

9.2   Why protolanguage?

9.2.1   Natural selection and the language-ready brain

We have taken the view throughout this book that language is a mixture of the physical (since it crucially involves the brain, and for spoken language, the vocal tract and the respiratory system, for instance), the genetic (since these physical structures are built and configured under genetic control, and their particular form is part of our genetic blueprint as humans), and the behavioural (which helps us understand how and why the apparatus is used, for instance, in different social contexts). In the last chapter, we considered evidence for invoking biological evolution through adaptive natural selection in the case of the physical and genetic underpinnings of human language, and concluded that these arguments can be strong and convincing – even though sometimes there remains a danger that the constructer of strictly adaptive, evolutionary reasoning might slip over into the role of spinner of 'just-so' yarns.

Much of this adaptive reasoning involves an assumption that language was beneficial to our ancestors (and arguably is still beneficial to humans today) because of its communicative function – though note that this need not be its only function for the argument to go through. As Pinker and Bloom (1990: 9) summarise, 'communication of knowledge and internal states is useful to creatures who have a lot to say and are on speaking terms'. While this is an apparently simple assertion, it merits some unpacking. The having a lot to say part presumably implies a certain size and capacity of brain which allows analysis of both external and internal context; and we have already considered the evolution of the brain in some detail. However, being on speaking terms suggests a social order of some kind, where individuals have regular contact with others, and perhaps live in groups – and this takes us straight into the interaction of genetic, physical and behavioural (including social) systems in the development and use of language which is the essence of this chapter. It also raises the question of whether natural selection, perhaps motivated similarly by communicative functionality, is driving (biological) evolution in these later stages when language becomes more complex. In other words, if language confers advantages, then is more language, or 'better' language (whatever that means), necessarily more advantageous, and for the same reasons? These questions, and their implications for the possible division of biological from cultural evolution, will be the topic of later sections in this chapter.

For the moment, if we do assume that some early precursor of language had communicative benefits, we can use the well-known trajectory of natural selection to derive the properties of key language-relevant physical and neurological systems from pre-existing systems and structures which did something else (and often still do, in addition to language). This kind of early communication might have been primarily gestural or vocal, and indeed other primates tend to do a bit of both; modern non-human primates may have more nuanced gestures, but then modern humans have evolved a different vocal tract. Of course, this example shows clearly that characterising the development of protolanguage, or elementary human language, as a shift from nothing to something is over-simplistic. We know that other primates, and other non-primates for that matter, have communicative signalling systems, some of which, like the vervet monkeys' apparent signs for different types of predators, are moving towards satisfying certain design features of human language. This means that early hominins were not starting from absolute zero in terms of language development. Nonetheless, we do need to account for the shift towards protolanguage and specifically towards the productive association of a signal with a meaning via symbolisation techniques; and this kind of development usually means co-opting behaviour that previously had another use.

Throughout this chapter we will be considering both computer simulations involving artificial 'agents', and much more recent, and still pioneering, language games with human participants, which try to model or capture aspects of the evolution of language. However, both types of experiment typically focus on increasing complexity from a pre-existing, if minimal, system, which means they belong more naturally in the next section, when we turn to the (possibly long and complex) transition from protolanguage to language. Nonetheless, there is one recent example of research specifically seeking to understand 'the deeper problem of how audiences even know that signals are signals in the first place' (Scott-Phillips, Kirby and Ritchie 2009); or as Scott-Phillips *et al.* more elegantly put it, the question of how signalhood is signalled.

As Scott-Phillips *et al.* note, language in its modern forms (and arguably also in its protolanguage state) relies absolutely on a capacity for symbolic communication (see also Deacon 1997), which has been investigated extensively through methods ranging from computational modelling and robotics to game theory. Until now, this work has centred on inferring the meaning of a signal, rather than on identifying it as a signal in the first place. This is even true of the initial, ground-breaking work of Galantucci (2005), one of the first studies involving human participants rather than computer simulations; here, two individuals have to coordinate their behaviour, but they are supplied with a communication channel to use in order to do so. Scott-Phillips *et al.*, however, focus on how a signal is identified as such (and hence on the detection of communicative intention), and moreover on how this can happen in a novel medium.

Scott-Phillips *et al.* use an embodied communication computer game for two players, each represented by a stick man in a box divided into four equal squares;

players can move their stick men from square to square using the arrow keys. Each square is coloured red, blue, green, or yellow; and each player can see the other's divided box and stick man, but crucially cannot see the other's colours. The object of the game is for both players to land on the same colour at the end of the turn, which scores a point; players were paid for their time, but there was also an additional payment to the members of the highest-scoring pair overall. The game was played by twelve pairs of adult participants; the members of the pairs could not see one another or speak to one another. In each case there was a short initial training period, with an opportunity to ask questions, then a 40-minute period for play, during which teams played an average of 206.9 rounds. Success was calculated for each pair as 'their highest number of points scored in succession. This criterion means that the players cannot succeed through the sheer quantity of games played; they must instead find a way to communicate reliably and hence coordinate their behaviour with each other' (Scott-Phillips *et al.* 2009: 228).

Achieving all of this was difficult: for the twelve pairs who played the game, the scores were 83, 66, 54, 49, 39, 17, 14, 7, 5, 4, 3, 3, with the last five of these scores being at chance level only. Members of the failing groups either could not work out how to communicate the relevant information at all, or figured out a communication system that would work if both players knew it, but could not find a way of sharing it. On the other hand, five of the seven successful pairs developed a 'default colour strategy' (2009: 231), which has the following steps:

1.   Each player develops a strategy of moving to the same colour on each turn.
2.   Noting the other's behaviour, they eventually converge on the same default colour.
3.   This works until someone lacks the default colour, in which case the strategy either falls over, or …
4.   One of the players develops an idiosyncratic behaviour, maybe moving all the way around the box, which can be interpreted as 'No default colour!'
5.   This idiosyncratic behaviour may then become the signal for a specific one of the other colours, so that if the default colour is red (signalled by moving straight there as efficiently as possible), moving all the way round the box rather inefficiently then signals blue.
6.   Finally, signals for the other two colours are developed, so that for instance oscillating right to left might mean green, while up and down might mean yellow.

The two pairs who achieved success without the default colour strategy began to use unexpected behaviour (like right to left oscillation for green, say) right from the start, indicating that the key here is precisely that quality of unexpectedness, which helps signal signalhood as the intention.

Of course, there is a potential issue with Scott-Phillips' game which extends to any experimental work involving modern humans, namely the fact that the brains of these participants are not just language-ready, but language-equipped; they are all speakers of at least one modern language and have been through the process of language acquisition. However, the game does require players to shift from communicating in one modality to another where they will not usually apply their language capacity. It is therefore strikingly apposite to the development of protolanguage, which is highly likely to have involved the 'recycling' of cognitive and physical systems already in place for other purposes. Indeed, Scott-Phillips *et al.* (2009) assume that these communicative intentions could not emerge from nothing: the creation of a symbolic communication system first requires certain cognitive abilities to generate and store symbolic representations. We see here again the relationship between innate capacities and learned, or indeed created, symbols, which jointly contribute to language, and arguably also to protolanguage.

9.2.2 Motivating protolanguage

Modern human languages are both highly complex and highly structured; how, then, do we get from the product of the language-ready brain to here? The trajectory we argue for will clearly depend on the nature of that product: any journey depends not only on the destination, but also on the starting point, so the transition from protolanguage to modern human language can only be specified if we have cogent arguments about what protolanguage was like. Even more fundamentally, we first need to justify the assumption that the language-ready brain initially produced something different from and less complex than today's languages.

Taking Scott-Phillips' experiment as an analogy, the initial output was a set of symbols which allow players to signal colours; and this is undoubtedly helpful, both in real-world contexts and within the game, where being able to signal colours allows pairs of participants to win points. However, the system is limited to a small number of symbols for simple concepts like 'green' or 'red'; it is not possible within the confines of the game and the communication system to invoke the semantic or syntactic complexity of utterances like 'that kind of greeny-yellow on the flowers on the flouncy dress you wore to Emma's party'. In the simpler case, we find single 'words', no syntactic structure, and far fewer nuances in meaning; and it is tempting to suggest that something like that simple system is where human language started out, and the more complex one is where we have ended up, and to try to chart the shift from one to the other. However, there are many unanswered questions here. Bickerton (1990) first developed 'The notion that the earliest stages of language evolution involved a largely if not entirely structureless protolanguage' (Bickerton 2007: 515); but can we equate that protolanguage stage with anything found in the communication systems of other primates, in which case we have a continuity between animal signalling systems and human language, or is even protolanguage unique to our species? Did protolanguage become language through a gradual series of transitions, or in a single great leap forward? Were the motivations for protolanguage and for the transition to language of the same kind, and if so were those primarily or solely social, biological, cognitive, or a mixture of all of these?

It might seem logical to propose an initial, less-complex stage purely because there is a considerable distance between the communication systems of our closest primate relatives and modern human languages. However, this cannot be our only argument for protolanguage, as it begs a central question, at least for those of the discontinuity persuasion: why would the gap between modern human language and primate communication systems need to be filled if one cannot have led to the other? Bickerton (2007: 512), for instance, is fully convinced that 'nothing resembling human language could have developed from prior animal call systems'. Even if we accept this discontinuity argument (and readers will have observed that we are closer to the continuity side), there is another pressing reason to consider a protolanguage stage or series of stages. Unfortunately, we do not have fossilised language output from our hominin ancestors; and where we have physical fossil remains, those could have produced many different kinds of early language. However, we do have '*living fossils* … types of communication used by modern humans that are close to, but do not share all the features of, fully-modern language' (Kirby 2009: 674). Bickerton (1995) proposes three subtypes of living linguistic fossil, namely pidgins, the language of children under the age of two, and the output of language-trained apes, which Bickerton sees as based on 'pidgin signed English' (2007: 516). All of these simple systems have words which can be combined into longer strings, but structure is minimal, with little or no inflectional morphology, for instance, and seriously limited stylistic or socio-linguistic variability. As Kirby (2009: 675) observes, the missing ingredients are 'recursive embedding leading to indefinitely long sentences; prepositional structure based on a verb and arguments which are optional only if their meaning is recoverable; grammatical elements (such as agreement markers, conjunctions, case-endings etc.) that do not directly correspond to aspects of the meaning of a sentence, but rather have purely structural roles'. Such 'living fossil' systems are useful, and lend themselves to communication and interaction; but they do not have the flexibility and complex structure we would expect from an adult human first language today. They do, however, motivate a notion of protolanguage, and give us one initial picture of what that might have been like.

9.3   The nature of protolanguage

The discussion above has led us to Bickerton's (2007: 515) conception of 'a synthetic, pidgin-like protolanguage, differing from fully developed modern language in its vocabulary size, its lack of syntax and its lack of a modern phonology, but in no other significant respects'. This equates to what we might see as a common-sense account of language evolution: things started simple, and got (either gradually or suddenly) more complex. This synthetic account proposes that each protolanguage symbol, or signal, corresponded to a single, atomic meaning, much like a modern noun or verb; the later development of grammatical complexity allows these symbols to be combined in more flexible and complex ways, and to be modified either morphologically or phonologically to convey additional layers of grammatical or social meaning. However, there is a logical alternative, the holistic (or analytic) account, which argues that individual symbols in protolanguage corresponded to whole propositions, much like a full sentence in a modern language. From this earlier stage, where a symbol described a situation in the world in full, speakers gradually learned to segment out smaller units which each expressed a subpart of that total situational meaning, hence giving rise to individual morphemes. As Andrew Smith (2008: 99) neatly puts it, 'the synthetic account emphasises composing word-like units into sentences, while the analytic [= holistic; AMSM and RM] account stresses breaking sentence-like units apart into words'.[1]

The synthetic account of protolanguage is generally recognised as the 'standard model', and has a lot going for it in terms of its proximity to the 'living fossils' we encountered earlier. Pidgins, children under two and signing apes typically have single words or signs, and can combine these into rather elementary juxtapositions, often composed of noun and verb. There is often variation, though generally without stylistic, sociolinguistic or semantic significance, in the order of combination; and as we have already established, there tend not to be grammatical markers and may not be evidence of other grammatical categories. Heine and Kuteva (2007: 59) 'describe grammatical evolution in terms of a set of "layers"', the first two of which are nouns and verbs – or strictly speaking, noun-like units and verb-like units, since at an early evolutionary stage we must accept that these would lack many of the properties which we would regard as criterial for identifying and distinguishing nouns and verbs in modern human languages. For example, nouns are frequently defined as taking demonstratives, or as having markers for number, case and gender; and clearly, if we hypothesise a protolanguage stage without grammatical marking or any categories except nouns and verbs, those definitions would make no sense at all. Instead, Heine and Kuteva (2007: 59–60) define nouns and verbs as 'evolutionary primitives in that they are not derived productively from any other morphological or syntactic categories while they themselves commonly develop into other categories'. Heine and Kuteva's first layer consists of nouns only, and would correspond to a stage of one-word utterances, while the second layer also introduces verbs, and allows for mono-clausal propositions.

---

[1] The terminology here, it must be said, is a horror story. *Synthetic* and *holistic* are straightforward enough, and we will use these to refer to the two opposing accounts. However, *analytic* is ambiguous in the literature, and can be used for either model (see Wray 2002a: 113, fn.1, and Tallerman 2007: 2, fn.1, for alternative approaches).

This suggestion that noun-like and verb-like units are fundamental, and that they correspond to elementary concepts about objects or entities and actions or changes in state, is captured in Fitch's (2010: 401) description of 'a "lexical" protolanguage, with a large learned lexicon of meaningful words, but no complex syntax'. The suggestion is that these proto-nouns and proto-verbs could be juxtaposed into elementary linguistic structures, while still pre-dating syntax proper; and as Fitch (2010: 401) notes, 'This "syntax-final" model of language evolution is shared by a diverse group of scholars who disagree about almost everything else (e.g. Lieberman 1984, Bickerton 1990, Givón 1995, Jackendoff 2002).' To this group we can add Maggie Tallerman (2007), who argues fervently against the holistic view of protolanguage, as we shall see below. Arguments for the synthetic model draw, for example, from the psychological literature, with Andrew Smith (2008: 105–6) suggesting that 'There is much evidence that humans conceptualise objects and actions most readily at a basic level of categorisation (Rosch *et al.* 1976)', and that 'Basic level categories ... are cognitively more salient, maximally informative in distinguishing objects and actions, and more easily and quickly learnt by children (Taylor 1995).' Indeed, there are strong suggestions that this stage of elementary concept formation would have pre-dated modern humans, and hence should be regarded as a property of the language-ready brain: Hurford (2007: 331) develops this view in detail for our primate relatives, building up 'a picture of apes with quite a lot going on their heads. They are host to mental representations ... with the essential basic structure of human propositions, namely predicate-argument structure.' Apes, Hurford argues, can understand and attend to the global properties of a scene or situation on the one hand, and the local properties of up to about four participants in that scene. However, he is at pains to stress that 'the semantic predicate-argument structure attributed to animals is emphatically not to be confused with any linguistic structure, such as the Subject-Predicate structure of sentences'. It may correspond to, and in part lead to, those subsequent Heine-and-Kuteva second-layer linguistic structures, but it is not the same, essentially because 'The pre-linguistic mental representations so far discussed are all activated in individual heads and find no outlet (as yet in the story) in public communication of any kind ... Apes have rich mental lives, but keep their pictures of the world to themselves' (Hurford 2007: 332).

We therefore have two evolutionary developments to account for if we are to adopt the syntax-last, synthetic view of protolanguage, namely the shift from internal concepts to external linguistic expression, and the subsequent emergence of morphosyntax itself. Heine and Kuteva (2007) focus on the latter, stressing the ways in which an elementary lexicon could have developed into much more complex and modern systems through grammaticalisation (and other well-understood mechanisms of ongoing language change like analogy); and we return to the question of how complexity might have arisen, and whether this involved gradual progression through a series of stages or a more abrupt 'leap' (Heine and Kuteva 2007: 10) in 9.4 below. The striking omission from descriptions and defences of synthetic protolanguage is instead the former issue, the appearance of words. If you can't account for where those come from, after all, you are in a bad place trying to motivate a lexical protolanguage.

The question of where words come from in protolanguage can also be decomposed into two sub-issues: first, where would you get the physical stuff to express concepts, and second, why would you want it? The question of motivation is the topic of 9.5 below, and there we will be dealing with a complex set of issues around the evolution of hominin social structures. Turning to how internal concepts can be expressed externally if there is good cause to try, we return to an earlier discussion of whether language was initially based on sounds or gestures or both, and again reach the conclusion that this is immaterial: if gestures came first, we still need to account for their later translation, as it were, into sound. Bickerton (1990) explicitly associates his lexical protolanguage with *Homo erectus*, on the basis of brain size, migration out of Africa, and evidence of relatively sophisticated tool use. As Fitch (2010: 403) notes, *Homo erectus* may not have had a fully modern vocal tract, but we can certainly imagine vocal capability, or a mixture of vocal and signed utterances, which would have allowed for the limited needs of a noun and verb lexical protolanguage; and indeed, Studdert-Kennedy (1998, 2005) suggests that the need to accommodate an increasing vocabulary and hence to make more differentiable sounds may have motivated the evolution of the modern vocal tract. Still, however, we have intriguingly little indication of a line of development to that initial externalisation of concepts: as Wray (1998: 47) observes, 'it is not clear where … referential words would have come from'. Tallerman (2007: 18) rightly notes that the central question is about nouns and verbs, since we have mechanisms like grammaticalisation to help us account for other word classes; but her subsequent claim that 'Nouns and verbs more or less invent themselves, in the sense that the protoconcepts must be in existence before hominids split from the (chimpanzee) genus *Pan*', does not really address the question satisfactorily. Likewise, Fitch (2010: 403) seeks to downplay this question: 'Once the complexity of syntax is acknowledged, the specific form of early word-meaning pairs seems the least of our theoretical worries.' There is nonetheless a worrying gap here, which we need to acknowledge: if the development of syntax rests on having a lexical protolanguage to work with, we do need a hypothesis about where the lexical ingredients for the subsequent recipe came from.

On the contrary, a plausible hypothesis on the origin of its building blocks is one of the strengths of the holistic account of protolanguage. Holistic protolanguage is most closely associated with the work of Alison Wray (1998, 2000, 2002a, 2008) and Michael Arbib (2002, 2003, 2005a, b), who trace its elements more or less directly to elements of non-human primate communication – though it should be noted that Arbib inclines towards a gestural protolanguage, while Wray talks more about primate calls and speech. In contrast to the synthetic model, where the key units are noun-like and verb-like elements which label individual concepts, the holistic model argues that each element of protolanguage would have been linguistically unanalysable and referred to a whole situation. In holistic protolanguage, each unit would have been a longer, indivisible and arbitrary string of sounds which together constituted a whole proposition. Just as a non-human primate call might mean 'I want to play with you', or 'Watch out, there's a predator!', or 'Give me that food!', so a protolinguistic string would correspond in full to a warning, a greeting, or a request, for instance. Some of these ideas go back much further than Arbib and Wray, however, with Jespersen (1922: 422) also proposing the derivation of individual morpheme-sized chunks from earlier, unanalysable elements: 'What in the later stages of languages is analysed or dissolved, in the earlier stages was unanalyzable or indissoluble; "entangled" or "complicated" would therefore be better renderings of our impression of the first state of things.' Jespersen's summary (1922: 429), with emphasis in the original, could still apply to many accounts of holistic protolanguage: 'THE EVOLUTION OF LANGUAGE SHOWS A PROGRESSIVE TENDENCY FROM INSEPARABLE IRREGULAR CONGLOMERATIONS TO FREELY AND REGULARLY COMBINABLE SHORT ELEMENTS'.

The holistic model has some similarities to the synthetic one, at least in the types of arguments which are marshalled in its defence. In both cases, there are references to living linguistic fossils, albeit different ones; and both rest on developing a plausible transition for the shift from protolanguage to modern human language. While the synthetic model is motivated by early child language, pidgins, and the output of language-trained apes, with a transition to modern human language based on grammaticalisation, the holistic model emphasises formulaic language, and derives structure and creativity through a process of fractionation or segmentation.

Turning first to living fossils, Wray's work draws attention to a common but previously under-researched characteristic of language, namely the existence of formulaic utterances. Many theories of language, and specifically of syntax, stress the importance of flexible units which can be identified in one utterance but then used in many more: so, when a child hears 'I saw the cat', she might subsequently use 'the cat' in a whole range of other contexts, creating new utterances which she may neither have heard nor uttered before, but which are grammatically and semantically well formed and interpretable. However, Wray points out that there are other constructions which, although they may seem to be composed of individual words in the conventional way, cannot be changed internally, but operate as whole, inalienable chunks. Some formulaic utterances are manipulative, where the speaker aims to make the hearer react in a particular physical or emotional way (*watch where you're going*; *I'm sorry*; *would you be so kind as to VP?*); others are idioms (*to kick the bucket*; *to cry wolf*; *to make a song and dance about it*), or just very common, regularly used constructions (*how do you do*; *fancy seeing you here*). Even though some of these constructions could be analysed, the indications are that they are stored, retrieved and used as prefabricated strings referring to a whole situation, or aiming at a particular result; they are processed swiftly, and treated as if they are single items rather than composite units. We use a remarkable number of these holistic utterances in everyday conversation, and Wray argues that they speed up retrieval, production and processing time precisely because they do not require analysis. Wray suggests that these formulaic utterances, especially the manipulative ones, correspond at least in kind to the holistic calls of chimpanzees, for instance, and hence that 'the holistic strategy for expressing manipulative messages in phonetic form may be considerably more ancient than the analytic strategy ... What we have inherited is not the forms themselves, but the strategy of using holistic linguistic material to achieve these key interactive functions' (2002a: 115).

We then face the opposite problem from the one we encountered earlier in discussing synthetic protolanguage: whereas the holistic model can deal with the emergence of the initial units through assumed continuity with primate vocal calls, we still have to motivate the development of individual, substitutable words in modern human language, since formulaic utterances like *How can I help you?* today clearly do consist of words (even if they behave somewhat differently and less flexibly than in non-formulaic contexts). The process Wray and Arbib propose here is one of fractionation, whereby accidental overlaps of sound segments across utterances which share an element of meaning are reinterpreted as signalling that specific meaning. So, the whole and arbitrary strings *tebima* and *mutapi* (Wray 2002a: 119) might have meant 'give that to her' and 'give that to me' respectively, and there is no scope here for fractionation; but if it so happens that the arbitrary string *kumapi* meant 'share this with her', then a protolanguage user may have observed that the sub-string *ma* occurs in both *tebima* and *kumapi*. Since these both also share the meaning 'her', in due course *ma* might have been analysed out from the arbitrary wholes, to become a subunit or proto-word meaning 'her' which could then be used in any utterance to convey that meaning.

If this capacity to derive analysable, separable units is a bonus, there are other aspects of holistic protolanguage which have met with severe criticism. Bickerton (2003) and Tallerman (2007) consider a whole range of difficulties for the holistic model, and while some of these are countered by Kenny Smith (2006), there remain some challenging problems. Bickerton's main argument is that it would simply not be possible in practice for a community to agree on the sentence to which a particular holophrase, or holistic, unanalysable vocal string, corresponded; and without such agreement, there is no way of segmenting out a subunit with a regular connection of sound and meaning. For example (Derek Bickerton's blog 2006), a formulaic warning about a sabretooth tiger could be understood as 'A saber-tooth tiger is approaching!', or 'Look out, danger on the ground!', or 'Quick, guys, run up the nearest tree!' All of these might, in manipulative terms, elicit the right behaviour and lead to the preservation of the individuals, tribe and even species; but it is hard to imagine how a single vocal element with a single meaning could be excised and used elsewhere if speakers in the same community did not converge on the 'right' correspondence of form and meaning. Bickerton also throws down the challenge of explaining how negatives and interrogatives could be generated in a holistic protolanguage (a challenge which we extend to you in the points for discussion at the end of the chapter).

There are also inherent problems with the mechanism of fractionation. Tallerman (2007) argues that Wray and Arbib tacitly take a fairly complex phonology, and the existence of discrete or at least separable phonetic segments, for granted, even though there is a strong case for phonology as a later development (see Studdert-Kennedy 1998, 2005). But how are our protolanguage users to figure out which sounds are 'the same', and which subunits within a holophrase can therefore be regarded as 'the same'? What sort of variation between sounds would be an obstacle to fractionation? Tallerman (2007: 9) takes Wray's (1998: 55) example of *pademe* 'give her the apple', and *mapatu* 'give me the nuts', from which an overlapping sequence *pa* could be segmented out and come to mean 'give'. What, though, if the *pa* in *pademe* has an aspirated voiceless stop, and the *pa* in *mapatu* starts with a voiced stop, standing as it does between two vowels? On what basis would the protolanguage speaker regard these as 'the same'? The fundamental difficulty here is that learning phonology in a modern language means in large part learning which sounds count as 'the same' and which do not; and that does not follow solely from the acoustic signal, but from which sounds signal a difference in meaning and hence belong to different phonemes, or what the limit of tolerance is to how differently instantiations of 'the same' sound can behave in different contexts. Yet we can only define phonemes on the basis of morphemes, which rely on an agreed mapping of sound to meaning; and that is precisely what the initial stages of holistic protolanguage lack. Without a phonological system, we can't do sound-to-meaning mappings; with one, holistic protolanguage isn't holistic.

There is clearly a challenge here to segmentation; but there is a further challenge to learnability. Andrew Smith (2008: 103) is careful to concede that 'word meanings are far from unambiguously evident', so that problems of mapping are to be anticipated to some extent even in the synthetic model. However, he stresses that the really important issue in learning or understanding language is the ability to infer meaning from context, so that hearers can reconstruct what they think the speaker intended. Smith argues that the meanings we reconstruct most easily from context are those we learn most readily; and those in turn are the basic categories, corresponding to nouns and verbs again. If we oppose simple categories like 'spear' or 'eat' to a holophrase with a complex meaning like 'take your spear and go round the other side of that animal and we will have a better chance of being able to kill it' (Arbib 2005a: 118–19), Smith suggests that 'there is an obvious mismatch between the latter meanings and the very simple meanings most happily learnt by humans, based on basic-level categories' (2008: 106). Modern speakers of modern languages can cope with complex meanings because we have the linguistic structure to give us extra cues; but

> [a] unitary holophrase contains no structure by definition: context and general pragmatics provide the only evidence for semantic reconstruction, and it is implausible in such circumstances that any meaning could be reconstructed

> to such a degree of specificity, complexity and intricacy. On the contrary, it
> seems reasonable that, without linguistic cues, the more complex and elabor-
> ate the semantic representation, the *less* likely the meaning can be faithfully
> reconstructed. (Smith 2008: 106)

Hence, if an utterance is unstructured, its meaning must be pretty simple; and
that is not what is being proposed for holistic protolanguage. Even if Fitch (2010:
500) does argue that some of the more complex cases, like Arbib's above, 'were
given tongue in cheek', there is a long attachment to the idea of complex, holistic protolanguage meanings, with Jespersen (1922: 425) already proposing that
'Primitive linguistic units must have been much more complicated in point of
meaning, as well as much longer in point of sound, than those with which we
are most familiar.' The consequent problem of meaning recovery supplements
Tallerman's (2007: 16) criticism that holophrases, being arbitrary phonetic
strings, would have been extremely hard to learn; and the smaller the inventory
of agreed holophrases, the less opportunity for a substantial number of overlapping units interpretable as sharing the same meaning.

Much of the attraction of holistic protolanguage lies in the continuity Wray
claims to demonstrate between primate calls and protolanguage on the one hand,
and protolanguage and formulaic utterances in modern languages on the other.
Yet Tallerman (2007: 15) points out one fundamental mismatch between modern
formulaic utterances and the holophrases envisaged for protolanguage: while the
latter must by their nature be open to segmentation and reanalysis,

> [m]odern formulae seem to survive exactly *because* they are unanalysed,
> even (sometimes) unanalysable (in terms of the speaker's grammar). Why,
> if modern speakers don't spend any time analysing holistic strings, should
> early hominids have done so? Why aren't we constantly breaking them
> down into their component parts, and thereby eliminating them from the
> language? Instead, what actually happens is that formulae are hugely resist-
> ant to change – they are one of the few aspects of language that persists,
> often for centuries, when lexicon, syntax, morphology and phonology have
> all changed … The question is, then, why were holistic strings not equally
> resistant to decomposition in protolanguage?

Turning this on its head, it seems that for holistic protolanguage, we can get some
approximation to the initial units (the holophrases) as an analogue to primate
calls; but getting to modern human language involves a rather counterintuitive
process for constructions of the formulaic kind. In the case of synthetic protolanguage, we can get from a simple lexical system with noun-like and verb-like
atomic elements to modern language complexity largely through processes like
grammaticalisation, which are well known from attested language histories; but
we have trouble with the fundamental units. We might have a preference one way
or the other, but nobody wins outright.

This might encourage us to consider whether this is, in fact, a knock-out con-
test. Is it the case that protolanguage must have been either wholly synthetic or

wholly holistic? Several recent accounts recognise that both the development of complexity through mechanisms like grammaticalisation, and the recognition, analysis and extension of common elements into general patterns, are highly likely to have had their place in the evolution of language. Kenny Smith (2006: 321) suggests that we need not choose between holistic and synthetic proto-language: 'the two theories ... seek to explain different aspects of linguistic structure and seem to be compatible, at least potentially', though he proposes a temporal difference, with the holistic model coming first, leading to the develop-ment of nouns and verbs, and hence feeding into the evolution of syntax through grammaticalisation. Smith is not the only one to propose that protolanguage is not temporally unitary but encompasses more than one stage; Jackendoff argues for several (nine partially ordered steps in 1999, and eleven in 2002: 238), though none of these maps straightforwardly onto a holistic or a synthetic stage. Even this view might be too regimented, however; Andrew Smith (2008: 110) con-cludes that 'Protolanguage probably contained units with varying degrees of semantic complexity, and its complexification into modern language was a grad-ual process of increasing complexity arising from discourse, through the same processes of re-analysis and analogy which underpin contemporary language change.' What would have mattered on this account is whether meanings were reconstructable from context or not; and there may well have been some con-texts (involving social formulae for greeting or parting, to take just one example) where holophrases would have been easily interpretable enough. Some criticisms of the holistic approach seem to us extremely convincing where applied to holis-tic protolanguage as the whole system, but it may be worth considering what we gain by suggesting that both synthetic and holistic elements, with the consequent mechanisms for deriving complexity associated with them, may have formed part of the initial system, as indeed analysable and formulaic units still do in modern languages.

9.4   From protolanguage to language

9.4.1   Arguments for abruptness

Protolanguage, then, may itself have been developed through various stages and over a long period. We can make certain deductions about its char-acteristics and level of complexity if we assume it resembled 'living linguistic fossils' like pidgins, the language of young children, or the signs of language-trained apes. Views differ on whether its shape can be traced back to elements of primate calls and/or gestures, or whether there is a clearer break with those earl-ier systems; but even a discontinuist like Bickerton (1990: 128) accepts that 'the level of representational systems achieved by some social mammals amounts to a stage of readiness, if not for language, at least for some intermediate system such as protolanguage'. However, as Bickerton (1990: 128) further observes, 'the gulf between protolanguage and language remains an enormous one'; and our next task is to consider how the former could have become the latter. Along the way, we can ask whether our conclusions in turn reflect any greater light back on the nature of protolanguage itself, as we have already seen that opinions vary considerably (for instance, as between the holistic and synthetic camps).

The discussion in the previous section has been couched, if not very explicitly, in a gradualist framework; but there is one high-profile alternative, proposed by Bickerton himself. If, as Jackendoff (2002: 235) puts it, Bickerton's protolanguage is 'modern language minus syntax', then the question is when and how syntax developed; and at least in 1990, Bickerton argues that this could have happened very quickly and non-gradually indeed. Again, this argument is based partly on linguistic fossils: in his Language Bioprogram Hypothesis (Bickerton 1984, for discussion see McMahon 1994: 10.5, Aitchison 1983, 1987, 1989a, Thomason and Kaufman 1988, McWhorter, 1997, 2005), he suggests that protolanguage still exists in modern humans. For the most part, it is not relevant, being overwritten by the modern languages we hear and learn as babies; but if for whatever reason our access to language data is interrupted, then we can fall back on our interior blueprint for protolanguage, which will at least keep us going in the basics we require for successful communication. The most famous case where the bioprogram has been invoked is in the swift transition within some communities from pidgin to creole, as discussed in Chapter 2.5 above for Nicaraguan Sign Language; there is now also the further intensively studied case of Al-Sayyid Bedouin Sign Language (Sandler *et al.* 2005). Just as the rapid grammatical expansion from pidgin to creole can in some cases happen within a generation, so Bickerton argues that the evolutionary transition from protolanguage to language in our species may have been catastrophic, involving the co-option of existing neural structures previously used to negotiate primate social life (see 9.5 below) for the hierarchical organisation of syntax. Initially, this 'catastrophic' model of the emergence of syntax may seem too close for comfort to the saltationist view, with big-bang results purported to arise from single mutations, which we challenged in Chapter 8.2. In fact, however, co-option of existing structures for new functions or into new modalities is a classic evolutionary move, equally compatible with a more gradualist approach (which Bickerton seems to be moving towards in Calvin and Bickerton 2000).

It is nonetheless true that Bickerton strongly prefers terminology which suggests a catastrophic or abrupt origin for syntax, and hence stresses the singularity and species-specific nature of syntax as the unique component of modern human languages. This certainly lends itself to combination with claims of a genuinely saltationist kind, and with a range of conceptions of what might constitute syntax or its absolutely criterial feature or property. For example, we might invoke the idea of the Faculty of Language Narrowly defined, or FLN (Hauser, Chomsky and Fitch 2002, Fitch, Hauser and Chomsky 2005 and see chapter 8.2.2), and hence identify syntax with the capacity for recursion. In any case, we seem to have reached a position of stalemate over this single-step enabling of syntax, as to some extent we did over the holistic versus synthetic models above. On the one hand, Bickerton (1998: 357) argues that 'The burden of proof … lies squarely upon the shoulders of those who would claim the emergence of syntax as a gradual process or a series of several events, rather than a single, catastrophic process'. On the other hand, Fitch (2010: 407) suggests the opposite: 'While Bickerton's pidgin/creole transition example provides a reasonable argument that sudden change *could* have happened, it by no means demonstrates that it *did* happen. Far more support is required to make this argument compelling.' The key question is where we are to find such support for either side of the debate; we seem to have progressed as far as we can through rhetoric alone.

One way of approaching the issue is to ask, for any given view of protolanguage, whether 'we can find a plausible evolutionary trajectory that will take us from this protolanguage to full human language' (Kirby 2009: 674). Tracing possible evolutionary courses in this way might take us beyond conjecture to some evidence, albeit indirect, in favour of one conception of protolanguage over others. However, by their very nature, developments in evolutionary time are likely to be lengthy and complex; even the swift expansion of homesigns into Nicaraguan Sign Language took place over years in real time, and is consequently not something we can rerun, or that we are especially likely to be in the right place to observe in action. Kirby and his colleagues have approached this problem by making their observations in the laboratory, where interactions involve artificial agents, and where evolution and learning can be greatly speeded up, and a range of other environmental parameters carefully controlled.

9.4.2   Computer simulation and the emergence of language

Many of the initial computational simulations in this area have been developed by linguists in the Language Evolution and Computation research unit at the University of Edinburgh (www.lel.ed.ac.uk/lec/). Initially, these simulations tended to involve artificial agents, and hence involved the transfer of techniques from research on artificial life (or ALife) to questions on the evolution of language. Kirby (2002) provides an overview, suggesting that we need to approach explanations for syntax, for example, through such simulations (2002: 210):

> because linguistics does not have a way of tackling the complexity of the interaction of the actual processes underlying the origins and dynamics of language. Just as ALife since its conception has made sense of the array of facts about biological systems that are known by studying in miniature the processes that give rise to these systems – so too I hope it can help provide an explanatory underpinning for linguistics.

Often, such work involves these artificial agents developing an ability to communicate in an increasingly structured way, in a context where the experimenter controls all the starting conditions for the agents, and all the external parameters they are subsequently exposed to. In these experiments, computer-generated populations of agents with 'language-ready brains' are allowed to communicate for several generations. The creation of a language system, if any develops, is then recorded to establish the resultant characteristics of that system, and the steps through which it arose. Since each simulation can be performed over many thousands of generations and can be rerun with single and or multiple changes to the initial state of the agents, proponents of this approach argue that we can gain an understanding of which characteristics of the internal and external environment have been important in establishing the nature of the emergent languages as we observe them.

Kirby (1999, 2000), for example, modelled an initial community of agents which might want to talk about 100 possible situations. Each situation involves two of five proper nouns, and one of five verbs, so we have composite meanings like 'Mary hit John' or 'Andrew sees Anna'. They also have access to five syllables, which they can combine in any way, and repeat any number of times, giving a potentially infinite set of outputs. Combining each of the 100 meanings with a single output would provide a recipe for a simple, holistic protolanguage, if we assume that one output will mean all of 'Andrew sees Anna', without a separable word for 'Andrew', for instance. Agents enter the simulation with no linkage between their meaning space and the possible outputs, and have to learn how to speak by querying a random selection of those agents already present in their community and remembering what their outputs for queried meanings were. Outputs for the same and overlapping meanings already stored in the grammar are compared at the time of 'hearing', and a learning algorithm simplifies the stored grammar to minimise its size. Initially, there is little or no communication, since querying a neighbour with no output stored for a particular meaning will result in no response, and therefore no learning. However, the simulation allows random new outputs, as well as occasional errors where a stored meaning is linked with a new output, so from time to time a novel link between output and meaning is produced by chance and stored by the speaker and learner. Hence, over time each agent will develop its own small corpus of output–meaning pairs, allowing it to communicate about just a few of the hundred possible meanings; but there is still no population-wide pattern of shared meaning–output pairings.

These small, individual grammars, with relatively little overlap between agents, persist for a lengthy period in the simulations. After many generations, however, a second stage is entered, characterised by a rapid increase in the proportion of the hundred meanings that can be expressed in interaction, because a substantial number of agents agree on the appropriate expression for a meaning. Grammars remain unstable during this phase, with the number of output–meaning pairs shared by the whole population see-sawing up and down, and the output used to express a particular meaning shifting between generations. However, this is followed relatively quickly by a third stage in which an effectively stable shared system is created. Now, the number of meanings that can be talked about collaboratively rises to the maximum 100, but there is a trend towards synthetic rather than holistic storage and combination of smaller, lexical, elements, so that the size of the average grammar drops to a stable, efficient and low value. The possibly somewhat surprising result of these simulations is that starting from a phonologically rich and semantically complex holistic protolanguage, the simulated populations always converge, after hundreds of generations, on stable, language-like, synthetic syntactic systems.

Such simulations, to be sure, have their critics. Bickerton (2007: 522), for instance, challenges the realism of Kirby's (2000) work, agreeing that 'If you go on doing this for long enough, apparently, populations of simulated individuals will converge on the same string for the same meaning', but asking 'what is the likelihood that actual hominids randomly produced invented strings of symbols for indefinite periods of time? When nobody could figure out what they meant until a stable structure had developed?' While Bickerton suggests that there is too little initial information for agents to work on, making the timescale for language emergence unrealistically long, Tallerman (2007) and Studdert-Kennedy and Goldstein (2003) criticise Kirby for including too much information in the initial settings of his agents' behaviour. As Studdert-Kennedy and Goldstein (2003: 238) put it, 'His initial utterances are semantically holistic, but consist "physically" of discrete symbols ... Thus, a necessary condition of compositional syntax (discrete phonetic units) is included in the initial conditions: compositionality can only emerge because "holistic" utterances readily fractionate along the fault lines of their discrete components.'

So, while some authors suggest particular simulations start from too simple a foundation, others critique the same studies for including too much structure in the initial settings and effectively biasing the results. The practitioners of simulations are well aware of these difficulties and have been exploring ways of including more realistic assumptions to investigate stages of language evolution that are otherwise unavailable for examination. As a result, they argue that certain aspects of the transition from protolanguage to modern language emerge as priorities for future study (for an overview, see Kirby (2002) or Steels (2003)).

One area that has been particularly clearly illuminated in recent years is the importance of cultural transmission for the development of an emerging language system. Kirby, Dowman and Griffiths (2007) characterise language as a balance between innateness and culture; and although it has long been recognised that language results from the interplay of three factors, namely biological evolution, individual learning, and cultural transmission, they argue that the influence of learning and cultural transmission has often been underplayed in the study of language evolution, while biological evolution has been over-emphasised. The key question here is how we are to explain similarities across languages, which are not randomly and endlessly variable; instead, they manifest various commonalities or linguistic universals (Croft (1990), Hawkins (1988)). A proportion of these will result from general cognitive or physical constraints, like the nature of memory; but others appear to be structural and specific to language alone. A

Chomskyan approach would hold that these universals reflect innate constraints arising from the biogenetic structure of our language faculty; and if we accept this, it follows that the main explanatory factor for the development of human language must be biological evolution acting on these structures. Kirby, Dowman and Griffiths, however, while focusing on attempting to make their simulations more realistic, have established that many of these universals could have arisen as a consequence of cultural transmission.

Kirby, Dowman and Griffiths start with the observation that 'The linguistic behavior a learner is exposed to as input is itself the output of learning by other individuals. Similarly, the language the learner acquires will ultimately produce data for a later generation of learners' (2007: 5241). Kirby and colleagues refer to this sequence as iterated learning, generically 'a process in which an individual acquires a behavior by observing a similar behavior in another individual who acquired it in the same way' (Kirby, Cornish and Smith 2008: 10681). This models Chomsky's (1986) distinction between I-language (the internalised form of a language based on the neural connections within an individual's brain, or more abstractly their grammar) and E-language (the externalised form of a language manifested in utterances). In simulations involving this Iterated Learning Model, what the learner learns is completely dependent on the language of the previous generation spoken in the presence of the learner, and a language is conceptualised as a collection of pairings between meanings and classes, where a class is a set of possible ways of expressing a signal (such as past tense).

During learning, only a random subset of possible pairings is presented to the learner, reflecting real-world limitations on the conversations a child might hear; and the size of this subset imposes a 'bottleneck' on the proportion of cultural artefacts (words, traditions, behaviours) that are actually transmitted successfully across generations. One positive influence on learning is regularity in the structure of the language, which results in a higher predictability of pairings between meanings and classes; and in the simulations, 'it is the number of training examples, the cultural bottleneck, that determines how systematic languages become' (Kirby, Dowman and Griffiths 2007: 5243), irrespective of the degree of preference for regularity imposed on the learning 'rules' of the agents. In other words, the tighter the bottleneck and the fewer the learning opportunities (within reason, of course, as there has to be some exposure to the system if it is to be learned at all), the more regular the system becomes over time. It follows that regularity in language structure can be seen as an adaptation of the language, not the learner, which makes language more learnable under adverse circumstances. Here, the predictions of the Iterated Learning Model map onto an observable historical linguistic difference in the tendency of common and less common items towards regularisation, since the circumstances for learning might be considerably less adverse for items which are encountered frequently. Kirby, Dowman and Griffiths (2007: 5243) nuance their proposal on the emergence of regularity, suggesting that:

> Regularity is adaptive for infrequently expressed meanings because it maxi-
> mizes the chance of being understood by another individual with different
> learning experience to you. It is less relevant for frequently expressed mean-
> ings because there is a greater chance that two individuals will have previ-
> ously been exposed to the same form. In fact, irregularity might be preferred
> for these meanings if, for example, it enables the use of a shorter and there-
> fore more economical form.

This matches well-known historical links between frequency and irregularity:
persistently irregular items tend to be those that occur most frequently, and also
often those learned earlier by children (note that the verb 'to be' is commonly
irregular). Likewise, there is a general inverse relationship between frequency
of use of lexical items and size or length. For further discussion, see McMahon
(1994), Dąbrowska (2004) and Bybee (2007).

Returning to the essential findings of simulations under the Iterated Learning
Model, the main point is that the initial agents did not require a preset tendency
to develop regularity or rule-based grammars, only a weak bias in favour of mini-
mising the complexity of their stored grammar. The appearance of a strong linguistic universal, in the shape of structural regularity, actually results from this
weak initial bias combined with the existence of a cultural bottleneck, rather than
requiring an innate, biological predisposition for particular structures. Different
runs across different simulations generate the appearance of common design features; but far from needing to posit that these are hard wired into the agents, they
develop within the language systems themselves as a result of constraints on
the cross-generational learning context. Extrapolating from artificial agents to
humans, this means we must be careful to focus not only on how our ancestors
may have evolved to be able to learn language, but also on how languages may
have evolved to maximise their own learnability in the environment created by
those language-ready minds. This is not to rule out evolution of brain structures
after the protolanguage stage, but it does shift the emphasis: as Deacon (1997:
122) puts it, 'the brain has co-evolved with respect to language, but languages
have done most of the adapting'.

Many aspects of language structure have now been examined using a simulation approach involving iterative learning: these include the emergence of word-
order universals (Kirby 1999), the regularity–irregularity distinction (Kirby
2001), recursive syntax (Kirby 2002), and compositionality (Smith, Brighton and
Kirby 2003). However, as long as simulations of this kind are being conducted
in silico, with artificial agents, they are subject to the criticism of lack of realism
outlined above: a question mark will always remain over how similar the agents'
responses might be to those of real people (or early hominids) when presented
with a set of stimuli in a specific context. Recently, investigators interested in
the importance of cultural influences on language development have therefore
moved their experiments into a laboratory setting where human subjects, rather
than artificial agents, are involved in the learning and transmission of 'alien'
languages.

9.4.3   From artificial agents to alien languages

In the previous section, we saw that in simulation studies, artificial agents can develop syntactic 'languages' with many signs of linguistic universals, starting from minimal initial learning states, mainly through the action of cultural transmission. However, a common criticism of these studies is that the learning algorithms used by the agents are very far removed from the complexity of behaviour shown by modern humans and that the minds of speakers of protolanguage(s) were probably more like those of modern humans than the computer agents. As Derek Bickerton pointed out on his website:

> Of course in computer simulations you can just go on repeating your competing analyses through hundreds or thousands of rounds until the thing finally gets ironed out. Luckily, 'agents' in computer simulations can never whack one another upside the head. But if you persist in using for 'bear' the morpheme that I have determined means 'red', that's exactly what I'm going to do to you. At best I'll treat you as an idiot and have nothing more to do with you – that's if you haven't first come to a similar conclusion about me ... People just will not put up with the kind of mindless tomfoolery that agents in simulations perform as a matter of course.

One way round these problems is to replicate the types of situations modelled in the simulations, but replace the artificial agents with real people. This experimental approach has one obvious limitation: we cannot use subjects who have not already learned at least one language. However, we can get them to work with a system they don't know, or in a different modality, to try to get at some of the underlying cognitive mechanisms. In effect, experimenters create artificially controlled conditions for the initial development or transmission of novel 'alien languages' – then we see what, if anything, happens.

In one of the first experiments of this kind, Galantucci (2005) examined the establishment of novel signalling systems between pairs of participants playing a computer game. Once successful communicative systems had developed, Galantucci increased the complexity of the tasks required of the players, or the computer environment in which they were performed (or both); in turn, this led to modifications in the systems. Players sat at two linked, but physically separated computer screens and moved their cyber-selves through a computer environment consisting of a series of rooms. In addition, they each had a moving digitising pad connected to their partner on which they could 'write' signals; but due to the movement of the pad, neither letters nor iconic pictures could be sent. Success, both in the initial and subsequent games, required coordinated movement of both participants' agents in the computer environment, but they could only 'see' each other if they were in the same 'room', so success hinged on the pair developing a means of indicating their positions and intended movements to each other. Participants were paid for their time and told that success in the game would result in higher rates of payment. Despite these motivations, some of the human participants did indeed show rather Bickertonian responses, with at least one out of the initial ten pairs failing completely to develop communication, even after 160 attempts. One member of this pair, player B, used signals inconsistently, resulting in his partner becoming frustrated and ignoring any attempts at further communication, 'moving his agent into the closest reset square and … orienting the face of his agent in the direction opposite to Player B's agent' (2005: 761, fn. 9) – the computer equivalent of a good slap in the face.

On the other hand, most pairs did learn how to communicate in a rudimentary fashion in order to play the game; some learned very quickly, others more slowly, but most pairs innovated a signalling system based on identifying their position in the computer environment to their partner. While proving that humans are capable of developing communication through novel modalities, the design of the experiment left very little room for extension of these elementary communication systems beyond something equivalent to the protolanguage stage.

Recently, several studies have combined this laboratory-based approach with iterated learning protocols borrowed from computer simulations to further study the likely trajectory of language development. For example, Kirby, Cornish and Smith suggest that they have produced 'the first experimental validation for the idea that cultural transmission can lead to the appearance of design without a designer' (2008: 10681), confirming some of the more interesting results obtained from computer agents. The design of their experiment was rather akin to that of observing second-language acquisition in the laboratory, but with the interesting and novel twist that the system to be learned did not start out as a fully developed language; the best human comparison for this paradigm is therefore the experience of first-generation creole-acquiring children confronted with a very elementary pidgin as their input data.

Participants began with a training session, during which they were told they would be asked to observe and learn an 'alien language'. Initially, participants sat in front of a computer screen on which appeared a range of objects, each with a distinctive shape, colour and movement pattern. Each of these objects was labelled with a string of random syllables written on the computer screen in lower case letters; thus a green triangle moving smoothly across the screen might be associated with *tuge* and a bouncing, red square with *kihemiwi*. Five colours, three shapes and three movement patterns were used, giving a total of twenty-seven different stimuli. During training, only fourteen randomly selected stimulus-plus-label pairs, the 'seen' group, were presented to each participant in a randomised sequence. However, after the training sessions, the participants were given testing sessions during which they had to label a set of stimuli that included both those they had seen and the thirteen 'unseen' stimuli, with the string of syllables they thought the alien might use for each stimulus. In the initial round of play, the training 'labels' for all twenty-seven stimuli were generated randomly by the computer, but the responses of all first-round participants were then recorded, and each was used to provide training labels for a second-round participant, whose responses were then in turn used as the input for the next participant and so on for ten 'generations' of learning and production. The experimental design therefore constitutes a diffusion-chain study of iterated learning over ten generations with a learning bottleneck in each generation arising from the presentation of only half the output data set from the previous generation during training (remember that participants then had to use their own intuitions about the system to think of likely labels for the other half of the stimuli). None of the participants knew they were learning from the output from other participants, nor that their own output would be used as the learning input for later participants: all they were trying to do was to learn the language as best they could and reproduce it as accurately as possible without change or improvement (Kirby, Cornish and Smith 2008: 10682).

Extrapolating from the results obtained with artificial agents, Kirby, Cornish and Smith hypothesised that the languages would show cumulative adaptation over the course of the experiment, resulting in an increase in the learnability of each language, along with increased predictability of mapping between the visual symbols and their labels. This was indeed the case; overall error rates between learning input and testing output fell from around 80 per cent to less than 20 percent over the ten generations, while structural regularity increased. For instance, in one diffusion chain the originally unstructured language with twenty-seven distinct meaning/label pairings had stabilised by generation 8 so that all objects moving horizontally were labelled *tuge*; all objects moving in a spiral had become *poi*; and bouncing objects were labelled *tupim* if square, *tupin* if triangular and *miniku* if circular. However, this increased learnability, which follows from increased structure and predictability, is here gained at the cost of underspecification, since the evolved language can no longer describe every stimulus unambiguously – *tuge*, for example, does not tell us anything about the colour of the object. Each language, then, faces simultaneous but competing selection pressures to be both maximally learnable and maximally expressive; and to model this, Kirby, Cornish and Smith performed a second set of experiments in which they filtered the training data so that 'If any strings were assigned to more than 1 meaning, all but 1 of those meanings (chosen at random) was removed from the training data' (2008: 10684). This process prevented the development of gross underspecification as a strategy for maximising learnability. Nonetheless, learnability **did** increase during the experiment, leading to systems like the example shown in Figure 9.1.

The authors conclude that 'the culturally evolving language has adapted in a way that ensures its successful transmission from generation to generation, despite the existence of a bottleneck on transmission imposed by the incomplete exposure of each participant to the language' (2008: 10685). Remembering that at no time during the experiments did any of the participants believe they were altering the alien language, we have a situation here in which 'Cumulative cultural adaptation without intentional design' (2008: 10684) has resulted in a system with increased learnability, increased regularity and structure, but without a loss in expressivity.

	Green	Blue	Red	
	n-ere-ki	i-ere-ki	renana	□
moves forward	n-ehe-ki	i-aho-ki	r-ene-ki	○
	n-eke-ki	i-ahke-ki	r-ahe-ki	Δ
	n-ere-plo	i-ane-plo	r-e-plo	□
bounces	n-eho-plo	i-aho-plo	r-eho-plo	○
	n-eki-plo	i-aki-plo	r-aho-plo	Δ
	n-e-pilu	i-ane-pilu	r-e-pilu	□
spirals	n-eho-pilu	i-aho-pilu	r-aho-pilu	○
	n-eki-pilu	i-aki-pilu	r-aho-pilu	Δ

Figure 9.1 *Generation 9 of an evolved language (after Kirby, Cornish and Smith 2008: 10684)*

These experiments, although convincing, indicate only part of what might have gone on during the early stages of the development of languages. Crucially, they concentrate on participants sharing information in pairs and agreeing on what counts as the same in the communicative act; yet speakers also diverge. Each of the individual diffusion chains in these experiments reached a different end point; there were similarities in structure, but differences in detail. Sociolinguistics and the study of language change in progress both point to the importance of differentiation between speakers and the emergence of novel ways of speaking for the development of historical language change. Moreover, differences between the speech of individuals in modern human languages are essential in allowing them to draw important social and geographical conclusions about each other. The importance of social factors for the development of early language structure is only now beginning to be investigated experimentally. Steels and co-workers have examined the importance of groups in the rapid development of a shared lexicon via the interactions of collections of simulated agents in the form of robots – the 'talking heads' experiments (Steels 2003). However, as we have argued above, it is only in experiments utilising human agents that the importance of social motivations can be explored.

Although there are as yet very few studies in this area, the experiments of Roberts (2008, 2010) provide an excellent illustration of the point. Roberts' experiments examine the behaviour of participants in a competitive team game, who can only communicate via a made-up 'alien' language. In the initial pilot experiments, players were allocated to teams, then pairs of individuals were chosen to play the game. Playing pairs could be either from the same team or opponents and the rules of the game meant that the players had to negotiate and decide whether to share particular resources (meat, grain, fish, water and fruit) by giving some of their resources to their playing partner. Resources exchanged during the negotiation phase carried twice the value for the receiver that they had had for the giver, and the aim of the game was to become the team with the highest overall resource allocation at the end. So the most successful individuals were likely to be the team that shared as many resources as possible but only when their playing partner belonged to the same team. The game was played on computers; the players could not see each other and had no access to natural language clues as to their playing partner's affiliation – so again, the only cues to identity involved modification of the alien language.

Roberts reports that successful groups quickly adopted one or more of five different strategies for identification of allies: 'secret handshakes', or shared greetings; mimicry; identifying salient idiosyncratic features of another's language use; alteration in the pattern of language use; and changing the language to increase expressivity (Roberts 2008: 180). One particularly successful team demonstrated all five approaches and made innovations such as repeating the name for each resource in turn with a frequency that matched the level of that resource the individual already possessed, simultaneously increasing the direct information-carrying capacity of their alien language and acting as a marker for group membership. Significant levels of innovation were demonstrated in this game, with over 120 novel forms being used by round 3; but there was no evidence that any player altered a form deliberately to be different (Roberts 2008: 182). The changes that did occur could all be interpreted as stemming from initial errors in language use that were copied and established as consistent aspects of the alien dialect that could be used to mark group membership. In subsequent experiments (Roberts 2010), socially relevant factors were shown to affect the creation and rate of divergence of the alien dialects, with the presence of intergroup competition and a high frequency of individual interactions being particularly salient factors in rapid divergence and complexification.

While such studies are still in their infancy, they serve to confirm that many of the conclusions about the origin, spread and increase in variability and linguistic complexity inferred from experiments with artificial agents can be replicated using human agents. However, while this addresses some concerns over simulated agents not replicating human-like behaviours, there are still at least two remaining issues in viewing these studies as representative of the early stages of the transition from protolanguage to modern language. The first of these, which we have mentioned already and probably just have to accept as an intrinsic concern over the use of any modern human informants, is that they all already have language: the mythical 'blank slate' remains stubbornly mythical. It is possible to take some evasive action, such as aiming to provide a system which participants are unlikely to recognise as anything like a language they know – and experimenters are well advised to rule out people (like trainee linguists) who are used to thinking about language and externalising their knowledge in ways that other speakers aren't – but in essence there is no way of getting round this. It is true, however, that in attempting to model the development of protolanguage to language, we may not need to be too concerned about participants having a capacity for language, because that would need to be present in some form, genetically and in terms of brain structures, in order for post-protolanguage development to have taken place anyway.

The other possible criticism is that thus far most of these experiments have involved manipulation of graphical representations or written text, rather than anything closely analogous to speech, or a plausible precursor to modern language. We might argue over whether protolanguage was vocal or gestural, but there are few passionate defenders of an orthographic protolanguage. At best, these experiments therefore deliver an analogical model of what might have happened; though having said that, good analogies can be helpful in understanding the fundamentals of a case, and may well constitute an improvement over purely speculative hypotheses building. For example, with reference to the nature of the linguistic signal, Galantucci (2005) found that when his participants were faced with an increase in the complexity of the task and online environment in his second and third experiments, they modified their initial communication systems in different ways to accommodate the increase in informational content required. Although individual pairs used very different signs and signalling systems (representing different mental states and outputs), all shared three characteristics: each sign needed to be perceptually distinct, produced by simple motor sequences, and tolerant of individual variation in production each time they were made. It seems likely that these characteristics would need to be present in any human signalling system, regardless of medium. For the understanding of vocal communication, such underlying design constraints may well have been the initiating factor for the development of phonology via natural selection acting on the vocal apparatus and neural circuits for speech production and perception. Larger signals are more distinct, but require more effort to create; small signals may have only a limited information-carrying capacity. Hence simple design constraints predict concurrent selection pressure for segmentation of larger utterances and the neurological apparatus for hierarchical creation and analysis of concatenations of shorter distinct segments (Liberman, Cooper, Shankweiler and Studdert-Kennedy 1967). At the same time, selection would also be active to increase the distinctiveness of individual segments making up the concatenated signal, or at least the auditory cues identifying each segment, resulting in a vocal tract capable of increasing the range of distinct sounds produced.

Whatever the modality, these studies indicate that under reasonably realistic conditions, variability and errors in production will create potentially structured variation that can result in subsequent changes in language use after reinterpretation by learners. We have in fact moved from language evolution into the realms of language change, and can start to feed in our knowledge of factors that generate and influence language change to model ever greater complexity, like grammaticalisation (Heine and Kuteva 2002, 2007, Tomasello 2003). In the next sections, we will start to ask what might have motivated such a progression and turn to an examination of the relative importance of biological and cultural mechanisms in the development of modern languages.

9.5   Motivations for increasing complexity

In tracing possible evolutionary histories for modern human language, we must be concerned with both motivations and mechanisms. To some extent, we have already committed ourselves to a view of the motivation for the emergence of protolanguage based on the essentially adaptive nature of communication, and the exaptation of pre-existing physical and cognitive systems as scaffolding for emerging communication systems. However, the motivations for the subsequent development of more complex and highly structured language systems need not be the same. If we can agree that language evolved in part to facilitate the transmission of information (noting and accepting that this is a rather non-Chomskyan interpretation, since his preference rests on the capacity of language to embody inner thought), then we can perhaps rephrase the issue of motivation in terms of what information was ripe to be communicated, and hence what pressures for change and development were acting on early people and early languages.

One traditional view has been that language evolved to facilitate communication during the development of hunting behaviour in early hominids. In this scenario, more complex language is seen as increasing the likelihood of success for the hunting party, and hence the survival of the group. Similarly, technical information exchange and mental planning during tool use and tool manufacture have been adduced as the reason for early language development and complexification (see discussions in Gibson and Ingold 1993). However, each of these scenarios seems somewhat suspect as a primary adaptive rationale. Technical communication of this sort never seems to represent more than a minor component of day-to-day human interactions, and these activities are often performed in relative silence; moreover, it is hard to imagine that hunting would be more successful with the hunters shouting instructions at one another.

An alternative set of hypotheses suggest that the primary motivation for vocal communication development is its social functions. In these scenarios, talking serves to transfer social information from the speaker to the hearer, or to mediate social relationships (where the actual information transferred can be pretty vacuous, but being in the right context and the right frame of mind to chat to the right people is everything. Think Facebook). Social interaction may indirectly provide information about an individual's fitness as a mate (the Scheherazade Effect: Miller 1999, 2001), or, more directly, might involve negotiating details of food sharing and cooperative group behaviour (Deacon 1997) or reassurance and care giving to infants (Falk 2004).

Robin Dunbar has observed that in social primates the average number of individuals in a social group is positively correlated with neocortical volume (1993, 1997). Being in a cohesive group is undeniably beneficial for animals faced with hostile environments (Clutton-Brock 2002) and, in the initial stages of hominid evolution, our ancestors were relatively small compared with the marauders around them, and vulnerable to predation. The advantages of groups are many, but one aspect of this is that a group can perform tasks that as individuals would be difficult or impossible, such as the defensive action against predator big cats observed in wild groups of common chimpanzees. However, such behaviours required cooperative or altruistic interactions that may have a cost for the individual, and such behaviour is often seen as problematic from an evolutionary standpoint since it requires the spread of genes which may on first appearances be detrimental to individuals carrying them. Primate group cohesion today is dependent on a hierarchy of dominance, and on coalitions that are established and maintained via a process of social grooming. However, short of an industrial revolution providing grooming machines, you can only groom one conspecific at a time; and the time spent grooming hence increases as group size increases, and can become a major activity for even moderately large groups, working against the benefits that would otherwise follow from cooperating. Dunbar uses the fossil evidence to suggest that hominid group size may well have increased to the point where the time required for physical grooming became a limiting factor to further group expansion (which we must assume was itself an adaptation beneficial to individual and gene survival). From this point, further group expansion and the consequent social bonding required to maintain larger groups relied on a more economical and less pair-centred alternative to grooming; and this, in Dunbar's terms, was verbal communication. It is perfectly feasible to chat (or in Dunbar's terms, to gossip) with more than one conspecific at a time, combining social facilitation with (some) information exchange, and accessing the benefits of belonging to a larger group without spending too long on group maintenance. Even with language, such groups are necessarily finite, partly because of the demands on memory of storing and accessing sufficient social information about fellow group members to make them feel appropriately special. There has been much discussion of 'Dunbar's number', the hypothesis that even for modern, socially mobile and technologically enabled humans, the real and meaningful circle of friends and acquaintances will rarely exceed the 150 predicted from social psychological studies of primate neocortex size.

The discussion of groups and their benefits above suggests that the development of language may well be grounded partly in cooperative behaviour, and in its capacity to access and facilitate such cooperation. The origins and maintenance of altruistic behaviour are reviewed in West, Griffin and Gardner (2007). Basically, two types of benefit can establish altruistic genes and cooperative behaviour: direct benefit to an individual, or indirect benefit through the survival of related individuals who share the same genes – this is the concept of inclusive fitness (Hamilton 1964). Of course, mutually beneficial behaviours can double the benefit through direct and indirect benefits to the individuals performing the behaviour. Since a new allele for a costly behaviour is only likely to spread through a population if the behaviour preferentially benefits individuals carrying that allele, there is a strong reason for mechanisms to evolve that will focus such behaviours on individuals who are more likely to share common genetic variants, or relatives. Hence, this is often referred to as kin selection; and in turn it is important for individuals who are likely to show such behaviours to be able to demonstrate kin recognition. One way this could be achieved is for such an allele to have three separate effects: firstly, creating a phenotypic marker to signal those individuals who have it; secondly, an ability in carriers to recognise this marker; and lastly, a tendency to perform altruistically towards others with the marker. Individual examples of such alleles are extremely rare, for, as pointed out by Dawkins (1976) when discussing this so-called 'green beard effect', they are extremely unstable genetic systems that can be easily 'invaded' by cheats who carry the phenotype of the green beard but not the tendency to help others with green beards. They also rely on variation in a genetic marker which selection will quickly remove from the population by fixing the marker allele if it is success-ful (so, once everyone has a green beard, you can't tell which are your relatives by their green-beardedness, and some other phenotypic social signal had better come along instead).

Much more commonly, altruistic behaviour is found in nature associated with a learned marker for discriminating kin, associated with a relatively constrained pattern of geographical dispersal of individuals from their place of birth. A good example is found in long-tailed tits, where adults who fail to find a mate in a particular year will spend that breeding season helping to feed the offspring of a mated pair in which at least one of the pair has song characteristics shared with the helper. Long-tailed tits learn song characteristics as infants in the nest, so similar song patterns act as an indicator of common early raising environ-ment and are good predictors of genetic similarity (Sharp *et al.* 2005). Nettle and Dunbar (1997) have argued that the evolution of linguistic variation in the form of dialects and languages is a particularly good marker for group cohesion, and in simulations is a stable evolutionary strategy which is resistant to invasion by even highly mobile 'cheats' or 'free-riders', who can otherwise move from group to group taking benefit without the costs of reciprocating. As Roberts (2010) argues, linguistic variation can provide a range of signals of group membership which are both relatively plastic yet sufficiently costly to operate as signals of group membership. Moreover, these are robust within a generation, but relatively malleable over longer spans of time.

Another evolutionary stable form of cooperative behaviour that can evolve in group-living animals is reciprocal altruism (Trivers 1971) or more colloquially 'tit for tat' behaviour. In this form of altruism an individual, the perpetrator of an altruistic act, will help another member of the group irrespective of relatedness, unless that group member has previously failed to provide help for the perpetra-tor. This strategy relies on being able to accurately identify and remember indi-viduals in a group and how they have acted previously, and by its very nature is predicted to be involved with the development of long-term memory and recall in hominids developing increasingly large and stable groupings. Episodic mem-ory, the ability to recall events that have happened in the past and the sequence of those events (see Tulving 2002), is clearly entangled with the development of such social calculations. It is difficult to establish to what extent other species utilise episodic memory but there is no reason to restrict this form of memory, which clearly requires the identification of agents, actions and goals, to our lin-guistic species alone (but see Hurford 2007: Ch. 3 for discussion). Yet these types of memories are not created only as a mental language stream, as we can see from cases like Helen Keller's; her language came late, and yet she retained episodic memory of her pre-linguistic life (Donald 2001: 240). Indeed, such episodic memory seems almost a prerequisite for the reciprocal altruism demonstrated by social primates such as baboons and chimps. It seems reasonable therefore that, at an early stage of hominid history, **co-**evolution of vocalisations (or individual gestures) with the neural structures required for reciprocal altruism and cooperative group structure may well have taken place. Language-generated variation encoded in verbal handshakes and idiolectal novelties will then have served as useful markers for individual identification, as we saw in Gareth Roberts' alien language games.

In effect, a positive feedback loop can be envisaged, in which social cooperation may have become an important survival strategy for some groups of protohumans, generating the need for group and/or individual recognition signals. This may in turn have led to larger, more stable groups with more complex interactions, which in turn increase the requirement for coordinated action, social cooperation and larger brain capacities with increased memory and mental processing. This spiral of feedback could have been sufficient motivation for the development of the language-ready brain, both to keep track of the memberships of a group and create and store the individual episodic histories of interactions within that group. The development of a memory type in which agents, actions and goals are already established can be seen as one pre-evolved mental template that could be exapted for the development of syntax, with its subjects, verbs and objects. As further syntactic complexification continued, this feedback spiral may still have contributed to the shift from protolanguage to modern languages. However, we shall argue in the next section that the development of syntax may have been preceded by, or more likely occurred concurrently with, fundamental developments in phonology which effectively scaffold the syntactic changes.

9.6   Phonology – still a mystery?

Typically there has been a focus on syntax as the differentiating factor between protolanguage and fully modern human languages; this is based on a known characteristic of the 'living fossils', where pidgins tend to have extremely simple word order and only main clauses, while creoles develop variability and subordination, for example. Much of the emphasis in modern linguistic theories has also been on syntax. However, in recent work on evolutionary linguistics we are starting to see emerging something that might almost amount to a consensus (itself somewhat unusual for this field) which actually relies on the importance of phonology. Contributions from a whole range of different directions seem to point to phonology as the key to understanding aspects of the transition from protolanguage to language, as well as providing a measure of continuity between the two; and it is not impossible that this perspective can resolve the tension between synthetic and holistic models of protolanguage for good measure (in a way we hinted at earlier). As we shall see, this is to do with the centrality of sound to the communicative act. This does not mean gestures could not have been a step on the journey, too, or that syntax is not relevant; though it might turn out (rather satisfyingly from the perspective of one of us, who is a phonologist) to be secondary. Most importantly, and highly relevantly in social terms, we must start not with speech, but with music.

Darwin (1871) recognised from his detailed comparative knowledge of human and other animal systems that human speech, although unique, shared characteristics with birdsong. Moreover, modern humans show highly charged emotional responses to 'music' understood in a wider sense (and taken to include singing, chanting, rhythmical activities such as marching and dance); and if such responses are evolutionarily deep rooted, this might make such activities strong contenders for the role of an original signalling system, generating group cohesion and the sense of 'sameness' that has been found to underlie success in cooperative games. As Mithen (2006: 25) expresses it:

> Music and language are universal features of human society. They can be manifest vocally, physically and in writing; they are hierarchical, combinatorial systems which involve expressive phrasing and are reliant on rules that provide recursion and generate an infinite number of expressions from a finite set of elements. Both communication systems involve gesture and body movement … Yet the differences are profound … Spoken language transmits information because it is constituted by symbols, which are given their full meaning by grammatical rules … linguistic utterances are compositional. On the other hand, musical phrases, gestures and body language are holistic: their 'meaning' derives from the whole phrase as a single entity. Spoken language is both referential and manipulative … Music … is principally manipulative because it induces emotional states and physical movement by entrainment.

Others have also noticed the similarities between language and music, but have often dismissed music as 'auditory cheesecake, an exquisite confection crafted to tickle the sensitive spots of at least six of our mental faculties' (Pinker 1999: 534). In other words, music is a parasitic system (or spandrel), perhaps relying on the neural developments that accompany the origin of language and the modern human mind, and originating alongside language or subsequently. Mithen (2006) argues persuasively against this interpretation in two main ways. Initially, he makes a strong case based on a range of evidence (including studies of language/musical savants, acquired and congenital amusia, and brain imaging) for the independence of many of the neural systems used in processing aspects of music and language, implying at least partially independent developmental and, by implication, evolutionary history. Secondly, he emphasises that aspects of musical development are apparent before language developments in the maturation of newborns, and indeed that much of the early vocal interaction between mother and child has a highly musical nature. Music, with its emotional and holistic characteristics, is also closer to the vocal communication systems found in other primates than is spoken language.

This has led Mithen to propose that our immediate ancestors *Homo heidelbergensis* utilised a communication system called 'Hmmmmm', an appropriately onomatopoeic shorthand for holistic, multi-modal, manipulative, musical, and mimetic. Indeed, this might have existed in some form all the way back to the *australopithecines* – though starting out as merely Hmmmm (lacking the final m of mimetic culture). This earlier Hmmmm includes aspects of vocal calling, gestural signalling, shared or altered emotional states and shared activity and attention, and would be compatible with an early stage of hominid neural development prior to the existence of full symbolic logic and 'theory of mind', in perhaps *Homo habilis*, or even late *australopithecines* (Donald 1991).

A musical protolanguage is an attractive proposition, since it effectively provides an intermediate stage which allows the gradual development of many of the aspects of the physical and neurological structures required by modern language. It also bypasses the problem of justifying the transition from manual signing to vocal calling as it already includes a combined modal signal at the protolanguage stage. The rapid brain expansion early in the *Homo* line can then be interpreted as at least in part associated with the development of something akin to dance and singing or rhythmic chanting from the earlier primate calls and gestures. These will have acted as a cohesive marker of shared group membership and cultural heritage and as a form of extensive social grooming associated perhaps with the development of tool use/manufacture, larger group sizes and the initial spread of 'culture' through copying (mimesis). In this model, music in its broadest sense is an older communication system that pre-dates modern languages and may have allowed and encouraged both the development of spoken language and the further development of the human mind.

Mithen sees the break-up of this holistic protolanguage stage as a relatively late event in evolution, and one which distinguishes our own species from the Neanderthals who shared Europe with us up until thirty-five thousand years ago. However, we would question this late date for the transition, as recent results from the Neanderthal sequencing project indicate a small but significant introgression of Neanderthal genes into the modern gene pool in non-sub-Saharan-African populations (Green *et al.* 2010). This is most easily explained by interbreeding between Neanderthals and the ancestors of non-African human populations at some point in the last one hundred thousand years. It is unlikely that such exchange of genetic material would occur without some form of communication between the participants, so we would favour an earlier (though perhaps still very gradual) transition from protolanguage to some form of spoken language.

While there might be continued discussion over the dating of this transition, there is a growing consensus about the key role played by the development of a complex synthetic phonology in place of the emotionally linked holophrastic signals of the early protolanguage stage (regardless of whether those were signed, danced, sung, hmmmmmed, or otherwise vocalised). We can imagine a relatively gradual change in the protolanguage over time associated with the modification of the muscle groups and actions involved in facial expression/ signing and vocal signalling systems initially inherited from the primate ances- tors. Originally mainly under emotional or physiological control, as the neuro- logical structures of the brain evolved, these systems will have been available for co-opting under conscious control for use in communication of information other than the current emotional and physical state of the signalling animal. At some point during this stage, individuals developed a shared interest in the exchange of propositional information: who did what to whom, where, when and why, all of which constitutes the types of information important in large social primate groups, and highly appropriate for Dunbar's gossipy replacement for grooming. We can also assume that this development may have reflected an increase in the mental processing capacity of our ancestors, while increasing interest in group manipulation and interactions would plausibly have motivated the expansion of memory systems and establishment of symbolic logic and a theory of mind.

All this is likely to have led to a requirement for an expanded lexicon, in order to communicate about these shared interests. At some point of lexical complex- ity the error rate for identifying and learning individual holistic signals will have become limiting, if each concept were to be coded for by a unique vocal/gestural product. We noted above for the alien language experiments that one outcome of the bottlenecks in iterative learning is an increase in regularity and a decompos- ition of the learned holistic signals, initially presented as a unique string for each stimulus, into a small number of learned sub-strings that could be combined itera- tively into a large number of more complex output signals. This ability to gener- ate combinatorial infinity from finite units ('discrete infinity'; Studdert-Kennedy 1998, 2000, Studdert-Kennedy and Goldstein 2003) underlies the development of syntax, but is also a vital first step in creating a phonology capable of com- bining the relatively small number of straightforwardly distinguishable phonetic gestures possible with the human vocal tract (probably even fewer in the initial stages of the transition) into the large number of uniquely identifiable 'words' stored in a modern lexicon – or, as Studdert-Kennedy (2005: 49) puts it, '[the] capacity to form an unbounded set of meaningful words by combining a finite set of meaningless discrete phonetic units'. Without such a lexicon, and hence without a particulate phonology in place, recursion and other aspects of syntax would be impossible.

What is the nature of these discrete phonological or phonetic units? Although the simplest answer might be to consider the consonants and vowels of speech as the 'principal level' for phonology, there are reasons to question these as the fundamental level for dissecting the speech stream. We have the capacity to prod- uce many hundreds of slightly different 'vowel' sounds and variation in normal speech is a fact of life, both from one speaker to another, and even within the same utterance from a single speaker; yet most languages split the vowel space into three to six contrastive regions, with only a few languages recognising as many as twelve (Maddieson 1984, Schwartz et al. 1997). Vowels are therefore perhaps more appropriately thought of as neural constructs, or phonemes, which do not have any specific reality outside of the brains of a group of hearers and speakers who agree on their phonetic 'reality' only as a result of shared developmental acquisition. Similarly, consonants, as phonemes, are also 'phonological categories' determined by 'different processes of phonological attunement (or accommodation) among speaker-hearers in different language communities' (Studdert-Kennedy 2005: 65; cf. Browman and Goldstein 2000). As such, these segments, which we are used to thinking of as the ultimate building blocks of phonology, might better be thought of as 'later' additions, requiring as they do the interaction of a language-ready brain with the norms of a speech community, and hence, arguably, an already linguistic external environment. What then would be the primary step in phonologising the originally holistic verbal/gestural signals?

Our first clue comes from theoretical phonologists, who have long recognised that their domain of study is in fact divisible into at least two separate subdomains, as expressed in Harris, Watson and Bates (1999: 493):

> It is now widely accepted in the theoretical literature that phonological representations combine two quite distinct organisational subsystems, PROSODY and MELODY. Prosodic structure (roughly equivalent to Jakobson's (1971) FRAMEWORK) comprises a hierarchy of domains which define relations between segments within phonological strings. The terminal nodes of this hierarchy are skeletal syllabic positions, which are gathered into syllabic constituents (onsets, nuclei and rhymes). At this level, representations code such relations as a segment's syllabic affiliation, its contribution (if any) to syllable quantity and its phonotactic adjacency to neighbouring segments. Syllabic constituents themselves are grouped into larger domains, including the foot and the prosodic word ... Melody (Jakobson's (1971) CONTENT) codes those characteristics of a segment's make-up that are manifested as phonetic quality, including such properties as labiality, palatality, occlusion, friction and voicing. These categories are assumed to be deployed on separate autosegmental tiers, in recognition of the fact that each is independently accessible by phonological processes.

While this separation of the two types of phonology seems intuitively appealing, the terminology is certainly not settled, with segmentals also, and somewhat confusingly, sometimes referred to as melody (see McMahon 2007: 162–4, for a discussion of these and other terms relating to 'the two phonologies').

In terms of the early transition from holistic protolanguage, several observations suggest that prosody is in fact the older and more universal of the 'two phonologies'. Variation in prosody does exist between languages and dialects, but prosodic features are learned very early in language acquisition (Bloom 1973, Snow 1994, Vihman 1996), with babies as young as 4 days discriminating between their parents' language and an unfamiliar one on the basis of prosodic patterns (Mehler *et al.* 1988), while by 8 months they are able to accurately mimic adult pitch patterns (Cruttenden 1986), well before segmental speech patterns are manifest. Indeed, it seems highly likely that exaggerated prosody in infant-directed speech (IDS) may be significantly involved in the acquisition of language, despite the overall pattern of prosody in IDS showing a clearly similar pattern across languages as diverse as English, Japanese, Xhosa and Chinese (Fernald *et al.* 1989, Fernald 1992). McMahon (2005a, 2007) has argued that the early acquisition of prosody reflects a substantial innate component for prosodic phonology, since the correct acquisition of prosody requires 'knowledge' of underlying structures that are underdetermined in the phonetic signal – the only aspect of phonology where the argument from poverty of the stimulus can reasonably be brought into play. Children can only learn their language's stress system, for instance, with the assistance of information, such as the position of syllable boundaries, which is not audible in the speech signal (Dresher and Kaye 1990). On the other hand, Carr (2000) argues that 'there are, if anything, more data available to the neonate than is strictly required for phonological acquisition' of segmental patterns. Hence the vowel and consonant inventory of a child's language can be easily acquired from the linguistic input alone, but prosodic features require an innate component of information. Similarly, phonological theories that are based on assumptions of underlying universal parameters or constraints, like Optimality Theory and Metrical Phonology, perform better when applied to prosodic aspects of language and tend to become problematically abstract when attempts are made to apply them to the segmental phonologies. Optimality Theory in particular is arguably better suited to representing the mental processes involved in analysis and production of prosodic behaviour; other frameworks, such as Lexical Phonology and Articulatory Phonology, are more often applied to segmentals (McMahon 2000, 2005a, 2007).

Thus, it is possible to describe and define prosodic phonologies using a limited set of options that are innate and universal across all human populations, implying an underlying genetic programming that is shared and therefore could pre-date the origin of all human groups. Indeed, in comparative studies of ourselves with other primates, aspects of prosody can be seen as conserved features shared with many forms of primate vocalisations. As an example, in human speech a gradual decrease or declination of fundamental frequency across an utterance, with a final rapid fall, acts as a marker of the end of an utterance and is so widespread as to be considered a universal feature of all human languages. A similar pattern is also seen in vervet monkeys; and indeed, youngsters perform declination so early and so convincingly that they have to learn to maintain pitch in order to keep their turn; initially, they are constantly interrupted by adults (Hauser and Fowler 1992). Similarly, rising intonation, which in some surveys has been found to be associated with questions in more than 85 per cent of languages (Ultan 1978), has clear similarities to the 'inquiring pant hoot' call used by travelling male chimpanzees who apparently wish to establish whether other chimpanzees are within earshot (Goodall 1986).

Prosody is also more closely related to emotion than other aspects of speech. In particular, although intonation has a linguistic, phonological organisation, 'sometimes against our will, it signals or helps signal information about our sex, our age, and our emotional state, as part of a parallel communicative channel that can be interpreted by listeners (even some non-human ones) who do not understand the linguistic message' (Ladd 1996: 1). These paralinguistic messages 'are non-propositional and difficult to paraphrase precisely, and yet in many circumstances they communicate powerfully and effectively' (Ladd 1996: 33), exactly fitting the form of communication posited for Hmmmm above. Variation in prosody is particularly often observed in interactional contexts involving greetings, farewells and formulaic language used in social settings (Cruttenden 1986). Prosody therefore bridges the gap between primate calls and modern speech, and is an important component of those human interactions that we have argued elsewhere to have been essential components of the early function of verbal communication. Evolution does not tend to 'dispose' of prior systems when a novel mechanism comes along: rather, we find throughout the animal world new systems built on top of older systems, often without complete replacement, and accompanied by some modification of the old. We can thus conjecture that prosodic phonology today may be the remainder of the initial phonological system that evolved to permit the segmentation of holistic signals into smaller units, and that it retains the same role, although perhaps substantially modified and overlaid, in modern speech.

Once a prosodic phonology had evolved as the vocal component of the protolanguage, it opened up the possibility of developing the neural and physical apparatus required to further divide the stream of speech into component parts and hence create the possibility of segmental phonology. At the same time, this could have provided the motivation for further evolution of the vocal tract to produce an increased number of maximally distinct vocal outputs. One testable prediction of this scenario is that we might expect separate neural regions to be involved in prosodic and segmental phonological analysis. While prosody might share regions of analysis with those controlling primate calls, segmentals might be processed elsewhere in the brain, assuming that the two types of phonology evolved in sequence (although we might also suspect that many subsystems would also be shared; there is probably a level where a sound is a sound is a sound). These hypotheses seem to be borne out, at least to some extent. As we saw in Chapter 6, there are at least two parallel processing neural pathways after the initial primary processing of speech. Linguistic prosody appears to be analysed and processed in the right hemisphere on a timescale of 150–300 ms (Poeppel *et al.* 2008), compatible with the almost universal syllabic duration found in the world's languages (Greenberg 2005); but parallel processing, on a timescale of 20–80 ms which is commensurate with (sub)segmental analysis of vocal gestures, can also be observed, with subsequent higher-order processing being predominantly in the left hemisphere (Poeppel *et al.* 2008).

Interestingly, music shares the right-hemispheric processing pathway, and in 'musically naive' listeners a left-ear preference (compatible with a right-hemispheric processing) can be observed in the recognition of melodies (Bever and

Chiarello 1974). Trained musicians, on the other hand, show a right-ear dominance in melody recognition, and Bever and Chiarello interpret this difference as a result of the fact that 'musically experienced listeners have learned to perceive a melody as an articulated set of relations among components' (1974: 538), while untrained listeners simply hear the tune as a holistic whole. Perhaps here we are seeing the operation of a 'shared syntactic integration resource' (Patel 2003, 2010): emotional and holistic responses to music mediated through right-hemisphere processing are being supplemented in trained musicians by the use of the hierarchical processing architecture, perhaps shared with language, in the left hemisphere. Competition during development for a shared neural resource could explain the observations that babies are born with a high sensitivity to the absolute value of tonal variation, in effect absolute pitch, but lose this in most cases as they acquire language (Saffran and Griepentrog 2001). This loss appears to be associated with the development of a categorical phonology that permits the child to, for example, assign a phonemic identity to a vowel in the speech stream regardless of whether it is spoken by a large man or a small woman (Kuhl 1979, 1983), abstracting away from differences of pitch. At seven months, infants have problems recognising words in isolation that they have heard in continuous speech when the sex of the speaker at test is different from that of the original exposure, but by ten-and-a-half months this is generally no longer a problem (Houston and Jusczyk 2000). So the ability to determine the **relative** pitch contours of a speaker's output rather than the **absolute** pitch of each element appears to facilitate segmentation, and as a result the absolute pitch assessment is downgraded.

In a similar time frame to the acquisition of speech, 'perfect pitch', the musical ability to recognise and **name** a given note in isolation, shows a critical period, with less than 4 per cent of musicians who began training between 9 and 10 years of age reporting having perfect pitch as adults, as opposed to 40 per cent of those whose first training began before they were 4 years old (Baharloo *et al.* 1998). There is apparently genetic variation underlying this skill, since it runs in families and shows a higher frequency in Asian populations than in Europe, for example. However, as we have indicated above, early and continued exposure to musical experience also appears to favour the development of perfect pitch, so that familial exposure or other environmental effects cannot easily be ruled out. The rarity of perfect pitch appears then to result from a common process of 'unlearning' an innate neural ability during early post-natal development, to facilitate the development of a segmental phonology, rather than reflecting any lack of the appropriate brain and acoustic architecture in all but a few individuals. This is consistent with the observation that perfect pitch is more frequent in speakers of tone languages and musicians who are exposed to music early (Deutsch *et al.* 2005), both groups that have conflicting influences acting during the relevant period of brain development.

This model predicts that the phonological system responsible for prosodic analysis is indeed an innate universal human capacity that has been largely shaped by genetic changes before the last common ancestor of all modern peoples. Segmental phonologies and syntactic structures, on the other hand, while possibly resulting from innate capacities, appear to be much more variable and group specific, perhaps indicating the importance of cultural effects during the learning of an individual's language system in determining their detailed structure. In the next section we will suggest some of the biological factors that might have motivated the evolution of speech and language and ask to what extent we can determine whether or not the human language faculty is still under evolutionary selection, or whether in fact cultural effects are now the dominant factors in determining the form of modern language systems.

9.7   Biological and cultural evolution

Is it reasonable to ask, or feasible to expect a sensible answer to, the question of when and to what extent language 'evolution' has shifted from biological adaptation to cultural artefact? Back in Chapter 1, we noted that it is essential to discriminate between the evolution of the structures and capacity for language learning, and the evolution of the linguistic systems (behaviours) that we can observe in use today. Evolution in the sense of changes in allele frequencies is still happening within the human species, and the effects of recent positive natural selection have been detected by examining the results of the human HapMap project (International HapMap consortium 2005, Pardis *et al.* 2005). In particular, these studies have identified many genetic regions that have been selected in specific human populations but are not yet at fixation within those populations. Hence, we can assume these regions have been recently, or are currently, under positive natural selection, and to some extent that these selection pressures have resulted from human cultural change (Laland, Odling-Smee and Myles 2010). The selected regions include genes involved in resistance to diseases such as malaria and Lassa virus; genes involved in the metabolism of food stuffs, particularly sugars such as mannose, sucrose and lactose; genes involved in fertility and reproduction, skin pigmentation, hair development and skeletal morphology; and genes regulating fat deposition and metabolism of environmental chemicals (Sabeti *et al.* 2006, 2007, Voight *et al.* 2006). Although these studies do detect some evidence of selection on genetic variation affecting neural development, it is nowhere near as significant an effect as that seen in comparison of the human and chimpanzee genomes (Evans *et al.* 2005, Mekel-Bobrov *et al.* 2005).

It would seem, then, that these neural genes have certainly been selected in our ancestors since we split from the chimpanzee line, but that most of the change pre-dates the origin of our own species and cannot therefore explain variation between human languages. There is, however, one specific counterexample to this generalisation, which involves tone languages. These are not uniformly distributed across the globe, being common in sub-Saharan Africa and Southeast Asia, while non-tone languages are more frequent in Europe, the rest of Asia, and Australia. Bob Ladd and Dan Dediu (2007) have noted that this geographical distribution of tone languages is intriguingly similar to that of the ancestral alleles of two genes, *ASPM* (Abnormal Spindle-like, Microcephaly-associated; see the genetic database OMIM entry *605481) and *Microcephalin* (*MCPH1*, OMIM entry *607117). Both these genes have variants which can have extremely detrimental effects, causing microcephaly, a congenital disease where skull size and brain are far smaller than normal, typically leading to severe, though variable, developmental delay.

Dediu and Ladd compared the distribution, across 49 populations, of 28 typological linguistic features and 981 genetic variants, including one from each of *ASPM* and *Microcephalin*, that had been observed to differ in frequency from one human population to another. The linguistic features included phonemically contrastive tone versus non-tone use, but also whether codas are allowed in syllables (meaning closed syllables like *hate* would be well formed as well as open, codaless ones like *hay*); whether onset clusters are allowed; whether cases are marked using affixes; the dominant orders of subject and verb, and genitive and noun; and whether there is a passive construction. Overall they detected no relationship between the genetic and linguistic variants apart from a strong correlation between the *ASPM* and *Microcephalin* variants and linguistic tone.

It is difficult to interpret exactly what this sort of correlation means. The genetic variants are estimated to have arisen 6,000 and 37,000 years ago, respectively, and so must have spread rapidly through human populations, possibly because they confer some kind of advantage, and therefore are under active selection pressure. However, we must not, as Dediu and Ladd make clear at their further information site (www.lel.ed.ac.uk/~bob/tonegenessummary.html) assume that this adaptive advantage specifically involves language, or that their results should lead us to make any value judgements about tone versus non-tone languages, or that humans cannot learn languages of a particular type unless they have particular genes. In fact, we do not know for sure **what** the selective advantage of these genetic variants is, though perhaps a minor cognitive bias is a possibility since we know they are connected with the growth and development of the embryonic brain through the link of other variants of the same genes with microcephaly. On the other hand, positive selection of the genetic variants may have been for completely different reasons in these populations, since, for instance, Ponting (2006) suggests that *ASPM* is implicated in sperm mobility. Indeed, the apparent correlation might in the end turn out to be accidental or spurious (as Mark Liberman points out in his Language Log posting on the Dediu and Ladd paper, 'even in a distribution created by completely random effects, without any meaningful connections among the variables studied, **something** has to be way out in the tail' – http://itre.cis.upenn.edu/~myl/languagelog/archives/004554.html). What Dediu and Ladd actually show is that the distributions of two genetic variants and typological features co-vary, and that further experiments are needed to fully assess connections between the two, if any. However, we might speculate that the two variants that have spread for unrelated reasons have slightly altered some aspect of their carriers' brains, in effect altering the environment

for language learning. As a result, groups of languages have altered their structure somewhat to adapt to the 'altered environment' by becoming more 'learnable', taking us right back to the Iterated Learning Model and discussions of the
adaptability of languages through cultural evolution with which we began this
chapter. The change involved certainly would not prevent any child from learning any language, whether that language uses tone contrastively or not. Indeed,
the difference may be only a very mild and almost impossible to measure bias
in probability of language acquisition, but over several generations of language
transmission, this small bias in neurological structure, combined with a learning
bottleneck, may have reduced the frequency of contrastive tone-using languages
in populations with the new genetic variants.

Animals and plants are generally fairly well adapted to the biological niche
(the position within an ecosystem that is occupied by a particular species) that
they evolved to exploit. The concept of a biological niche is at least as old as
Darwinian Evolution (though the original use of the term is usually credited to
Grinnell 1917) and has been subject to many different descriptions and interpretations over the years (for a history and overview of the niche concept see
Polechová and Storch 2008). An organism's niche consists of not only the physical environment that the individual finds itself in, but also the plants and animals, including members of its own species, that it interacts with throughout its
life. But also, as Darwin himself noted for his earthworms in 1881, the action
of an organism can alter the nature of the niche to which that individual and its
descendants are exposed. Hence the consequence of evolution in the past can
alter the niche within which an organism exists today (Odling-Smee, Laland and
Feldman 2003, Kendal, Tehrani and Odling-Smee 2011).

We cannot begin to understand the origin and development of language as a
biological characteristic of humans without reference to the part our ancestors
played in shaping their environment and how this was influenced by and in turn
influenced the evolution of a language-capable brain and the subsequent development of languages.

Tooby and DeVore (1987) integrate the many behavioural and cognitive features unique to humans (including language, long childhood, reliance on tools,
variable diets, variable cultures and complex social interactions), to form their
'Cognitive Niche' hypothesis. They assume that human ancestors developed a
response to the dangers and challenges represented by the evolved structures of
other animals and plants (teeth, size, speed, poisons and spines, for example)
that involved the evolution of unique **behavioural** responses centred around
cause-and-effect reasoning. Effectively, they postulate that our ancestors began
to evolve behavioural flexibility and the ability to out-think evolution by creating
cultural responses that could change faster than the challenging species could
evolve to combat those responses. Such behavioural flexibility is not without
potential costs, as outlined, for example, by Mayley (1996). Firstly, a response
that requires learning may require time and experience for an animal to perform
it correctly, time during which the individual's fitness may be less than optimal,

possibly necessitating extended periods of parental care which could restrict the reproductive output of that genotype. Secondly, periods of learning may put the learner at direct risk from poorly performed behaviours, such as jumping out of trees onto their heads rather than their feet, or indirectly by making them vulnerable to predation during the acts of learning. Thirdly, the neural anatomy required for flexible behaviour may incur more direct 'costs' in energy terms to develop the necessary brain structures in comparison to those required for simple innate responses (we use around 25% of our resting metabolism to maintain our brains as opposed to 8–10% for other primates, Leonard and Robertson 1992). Finally, the act of having to explore your environment and learn the correct responses compromises the time available for foraging, mate identification and predator avoidance, potentially reducing the fitness of such individuals directly. Here we have a common tension in considering evolution: behavioural flexibility may be adaptive in an uncertain environment, but if the environment is relatively stable from generation to generation, natural selection would be predicted to favour the development of innate responses. Perhaps this is one reason why language is unique to our own species – a simple accident of variable environmental conditions associated with an already social primate.

Some of these costs are of course minimised by 'sharing' them in groups, as protection from predators or care of young learners may both be easier with many pairs of eyes. Hence, once this course of evolutionary response had started, larger hominin group sizes (with consequent increase in the repertoire of possible behavioural responses) became favoured and further advanced the survival and reproductive chances of individuals in those groups, resulting in increased reliance on behavioural flexibility. It might be tempting to see this as a Lamarckian argument suggesting that behavioural responses developed in one generation somehow directly improved the genetic legacy in the next. This is not the case; it is rather an example of a phenomenon called genetic assimilation, often referred to as the Baldwin Effect after one of the three Victorian researchers, Baldwin, Lloyd-Morgan and Osborne, who discussed it together in 1896 (see Richards 1987 for a detailed history of the Baldwin Effect, and Baldwin 1896). The potential importance of 'genetic assimilation' in language evolution has been emphasised by several authors (e.g. Waddington 1975, Pinker and Bloom 1990, Briscoe 1997, 2002a, b); but the term can be quite misleading, as it has no single definition, and has been invoked by different authors to explain completely different outcomes of the interaction between behavioural and genetic variation, without adequate reference to the conditions assumed in their formulations.

Genetic assimilation will occur when three factors are present. Firstly, there must be a character that can vary during an individual's lifetime either by development, maturation or learning; secondly, the degree of variability of this character from individual to individual needs to be partially under genetic control; and thirdly, an environment must exist which favours the survival or reproductive success of individuals who develop this character to the greatest extent, or at the greatest speed, during their lifetime. If this environmental pressure persists for many generations, alleles will develop within a population that favour a more rapid acquisition of the character. One extreme end point of this process, for example, is individuals born with a particular behavioural response as a genetically programmed instinct rather than the result of any learning at all, a process sometimes referred to as developmental canalisation (Waddington 1942).

The problem with the terms genetic assimilation and the Baldwin Effect is that both are cover terms for the interaction of behaviour, development, genetic programming and environmental influences, and as such they can be applied to explain distinct and apparently contradictory aspects of evolution. In one guise, as in Waddington's developmental canalisation, the Baldwin Effect results in a previously flexible developmental program becoming increasingly inflexible, eventually creating an innate behaviour in place of a learned response. This outcome of the Baldwinian argument has been used to motivate the gradual creation of an innate and detailed domain-specific language acquisition device (LAD) as a consequence of the co-evolution of language and the neural structures required (Pinker and Bloom 1990, Briscoe 1997, 2000). Here, flexibility in early communication behaviour (behavioural plasticity) allows individuals to 'experiment' with more complex and structured forms of communication, creating linguistic communities at the protolanguage stage who agree by learning 'how to do' aspects of phonology or syntax. Individuals within these communities who then learn these 'rules' more quickly because of underlying genetic variability may then be 'more fit' genetically, and leave more offspring. Language evolution then proceeds as an ever-increasing complexification of the innate mechanism of language acquisition in concert with ever increasing complexity in the language systems created. In this form the Baldwin Effect looks attractive to evolutionary linguists as it bypasses one of the main stumbling blocks to the gradual evolutionary development of a LAD devoted to communication that underpins Chomsky's 1972 statement that 'It is perfectly safe to attribute this development to "natural selection", so long as we realize that there is no substance to this assertion' (Chomsky 1972: 97). In other words, it allows for an environment of learned communicative behaviour to be created before the genes for innate language structures arise by mutation, and hence when these mutations do arise their effect can be expressed within this communicative environment and they can subsequently be acted upon by natural selection.

The Baldwin Effect has been widely utilised in computer simulations to demonstrate that virtually any aspect of a putative language acquisition device **could** have evolved; but we need to be careful not to fall into the trap of overusing it to justify the inclusion of anything, or indeed everything, that we might wish to put into a LAD. Canalisation, with a consequent reduction in behavioural flexibility, as used by Waddington, is not a necessary genetic consequence of the Baldwin Effect. Indeed, since increased speed of learning can be achieved in many different ways genetically, depending on the relative costs/benefits of behavioural learning in a variable environment and the degree of correlation between the individual effects of genes and the behaviours concerned, the Baldwin Effect can operate by creating **greater** behavioural plasticity, rather than by the creation of instinctive and invariant responses. Improved language learning could have been facilitated by improved domain-specific learning modules or equally well by genetically controlled statistical learning strategies acting during development (for a discussion of the nativist–empiricist debate, see Spencer *et al.* 2009).

As we said above, the evolutionary landscape had to remain relatively stable from generation to generation in order to favour genetic hard wiring of instinctive responses, and, as Deacon (1997: 329) has pointed out, 'The relative slowness of evolutionary genetic change compared to language change guarantees that only the most invariant and general features of language will persist long enough to contribute any significant consistent effect on long-term brain evolution.' Deacon sees the evolution of general cognitive abilities and mental flexibility as the underlying evolved structures created by genetic assimilation in an ancestral lineage increasingly dominated by flexible behaviour, with language as one of the learned behaviours that result. He proposes that there is no need to posit genetic encoding of every language structure per se; rather, he suggests that brain mechanisms involved in behavioural flexibility, such as symbolic logic, recursion and different types of memory, are the focus of evolution. In his extreme formulation, evolution has created a species with a particularly flexible mind, and languages have 'evolved' in the cultural domain as independent replicators to occupy this newly created cultural niche.

This approach has been supported in a less extreme formulation by a series of realistic computer simulations of the interaction between cultural change and biological evolution. These have indicated that while genetic assimilation can occur for stable aspects of language, once languages are allowed to vary within a cultural timescale the specific aspects of any language become too transient to act as targets for innate programming (see, for example, Munroe and Cangelosi 2002, and for further general discussion of the Baldwin Effect in language evolution see the papers in Weber and Depew 2003). So the Baldwin Effect, if defined as the interaction between behavioural plasticity and genes, could indeed be a powerful underlying factor in the evolution of language, but it does not predict that the result of natural selection will be the creation of a detailed and domain-specific set of language structures. Rather it predicts that, given sufficient environmental consistency and a communicative behaviour that enhances the survival or reproductive success of individuals, structures will evolve to make the communicative behaviour easier to perform or more easily learnt, unless the costs of doing so outweigh the benefits accrued. In this sense, the Baldwin Effect is not one of Dan Dennett's universal sky-hooks but merely a statement of the normal evolutionary interaction of learning and genes. Learned behavioural plasticity can act as an evolutionary 'crane' by permitting individuals to explore aspects of their own niche to determine better survival or reproductive actions, in the absence of specific mutations governing those actions. These successful behaviours can then become the focus of subsequent evolutionary action and become increasingly prevalent in descendant populations, either by the genetic creation of instinctive behaviours or by the genetic creation of better learning mechanisms, either specific to the behaviour or as improvements in general non-specific learning mechanisms. The real extent to which instinctive responses and general or specific learning algorithms become genetically programmed for a particular behaviour will be dependent on the balance of opposing evolutionary factors operating on the individuals concerned.

To return to our own just-so story and the Cognitive Niche hypothesis; amongst early hominins faced by changing climate, environmental variability may have favoured those individuals who were a bit more adventurous and who, for example, consequently tried novel food sources more readily, but who could also remember which foods had made them ill in the past. In times of famine, such individuals would have had a greater chance of being able to recognise and obtain edible foods, effectively extending the niche range that they occupied. Their behavioural flexibility might not only have increased their own survival and reproductive chances, but also have acted as a source of knowledge for survival of other members of the group, who would probably be genetically related to the adventurous individual. Hence, over time, these flexible hominins could give rise to groups of descendants with even more flexible behaviour patterns and social structures, who in turn may have developed larger and more complex brains and group interactions, creating the mental and social niche required for development of many interacting neurological structures that form the foundation of a rich learning environment and communication system (Avital and Jablonka 2000, Dor and Jablonka 2001). Similarly, once group size, or other social pressures, began to favour the development of verbal communication over other forms of group cohesion, genes that favoured **rapid** learning of language will have evolved and become fixed.

We may seem here to be arguing in favour of an evolved and detailed language acquisition device which would itself continue to become more complex over time in order to generate more and more complex speech, eventually giving rise to the biological hard wiring of language universals, and consequent requirement for minimal amounts of training input in the learning of a language. However, as we have outlined above, language evolution happened in an environment involving the balancing of conflicting biological pressures, so that factors such as badges of group membership and the identification of mobile cheaters may well have acted against the development of a single universal language. Once a language-ready brain existed within groups of communicating individuals, languages themselves, as structures showing descent with modification, will have undergone cultural evolutionary change to adapt them to **their** 'niche', without necessitating any significant biological change in the hominid minds forming that environment. The Cognitive Niche hypothesis predicts that the flexibility of behaviour and the rapidity of **cultural** change are the factors that gave hominins their edge. In such a scenario, there are many benefits for keeping much of the 'evolution' of language systems in the cultural domain; and as we saw above, simulations suggest that very simple learning algorithms can generate apparent linguistic universals due to the 'cultural evolutionary' pressure for ease of learning acting on the developing languages themselves when faced with a cultural bottleneck for language acquisition. Under these conditions, most of the structures of individual languages will have evolved in the cultural domain at rates that are too rapid to act as specific targets for genetic evolution. We might predict therefore that many of the structural forms and universals of the world's languages will have a cultural origin and not a genetic underpinning in anything other than the most general way. As Senghas *et al.* (2004: 1782) have said of their 25-year study of a developing sign language amongst successive generations of learners in Nicaragua, 'evolutionary pressures would shape children's language learning (and now, language-building) mechanisms to be analytical and combinatorial. On the other hand, once humans were equipped with analytical, combinatorial learning mechanisms, any subsequently learned languages would be shaped into discrete and hierarchically organized systems.'

This is not to say that brain structures have not been liable to further genetic change after the establishment of modern language patterns. Cultural influences on genetic change are now well established (see Laland, Odling-Smee and Myles 2010 for a recent review), and the effects can be strong, but complicated. Modern language systems are therefore likely to be the product of cultural 'evolution' acting at the level of iterative learning of each language system combined with biological structures – both general, and those specifically evolved to produce and interpret spoken language.

9.8   Summary

We began this chapter with the notion of the 'language-ready brain', and considered the various shapes that have been proposed for protolanguage, and the various routes which would then be required to take us to human language as it is in the historical record and as it remains today. We have returned to a number of themes which have recurred throughout the book, considering arguments for gradual or abrupt transitions to complexity, and a formal versus functional focus on language origins and development. As we saw in Chapter 3, it may not be possible to reconstruct back through attested languages to navigate the passage from historical to evolutionary time, but a new research paradigm is developing in the laboratory to try to bridge that contentious gap. While this approach began with artificial life (or 'ALife') experiments involving computational agents and an experimenter who holds all the cards and can control all the contextual parameters, more recent and innovative approaches work with human subjects who are asked to learn 'alien languages'. Although we can never escape from the fact that these participants already know at least one natural language, it is possible to model gradual inter-generational regularisations of the patterns within the 'toy' systems they are encouraged to learn; and as this research progresses, so we can anticipate achieving more subtle control of individual parameters, and focusing our attention on more specific aspects of language acquisition, structure, evolution and change.

We then turned to motivations for the increasing complexity in human language systems. These arguments are largely (and pleasingly) independent of whether we assume language began with gestures, sounds, or a combination of the two; but in terms of sounds, we have developed a suggestion that perhaps the prosody of modern human languages has deeper roots than its segmental aspects. Perhaps Darwin's inclination towards a musical protolanguage was not so far from the mark; and the prioritisation of prosody, and subsequent development of segmental and syntactic linguistic structure, is currently providing as close to a consensus as we are likely to see in this field.

A further, and pervasive theme of this chapter has been the permeable distinction between biological and cultural evolution. It seems clear from the developing Iterated Learning Paradigm that languages stand or fall on their learnability – at least under laboratory conditions. However, language variation, like many other kinds of human behaviour, can be adopted or retained for social reasons which lie strictly outside their learnability and formal characteristics, and a complete understanding of the relative weighting of social, psychological and genetic factors still lies some way off.

So, here we are, back at the beginning. At the start of this book, we set out a particular approach to evolutionary linguistics, which is built on a tripartite system involving genetic instructions, which build physical and cognitive systems, with observable phenotypic and behavioural consequences. We have shown evidence through the book for all of these levels and for their productive interaction. We do not see the priority from here as a reductive approach, getting rid of some of the mechanisms, or showing language to be solely genetic or solely environmental – we strongly prefer a synthetic and interdisciplinary approach which accepts the contributions of all these levels of understanding and analysis in improving our grasp of what human language is, and where it has come from. And we are not alone: we can do no better than to recommend Kirby's (2009: 678) approach: 'The challenge therefore is to determine for each feature of language that we wish to explain whether natural selection or iterated learning is the right explanatory mechanism.'

Further reading

For a discussion of the conditions that can lead to the evolution of cooperative behaviour, see Nowak 2006. For historical aspects of the Iterated Learning Model see, for example, Kirby 2001, Brighton 2002, Kirby and Hurford 2002, Kirby and Christiansen 2003. The existence, status and meaning of language universals is currently under intensive debate: see Evans and Levinson (2009) and Dunn *et al.* (2011). Hurford (2007) outlines similarities and differences between the concepts of meaning in human and animal minds and presents an evolutionary account of how the human mind could have arisen; in particular, Chapters 8 and 9 provide a more detailed discussion of the evolutionary justification for communication and language evolution, particularly with reference to group dynamics and altruism. The nature of protolanguage in general and the importance of a musical stage in particular is discussed in detail in Fitch (2010: Chs. 12–14). Deutsch (2006) provides a good starting point for exploring the relationship between pitch perception in music and speech. The place of developmental processes in the possession of perfect pitch in adults is discussed from conflicting viewpoints by Saffran and Griepentrog (2001) and Miyazaki and Ogawa (2006). The papers in Weber and Depew (2003) provide a general discussion of the Baldwin Effect in evolution, often from very different standpoints.

One very recent book which seeks to reunify the historical and evolutionary perspectives is Shryock and Smail (2011). This is a multi-authored volume set within the emerging 'big history' framework, which seeks to redefine history, not as what makes us modern, but as what makes us human. There are separate but often linked discussions on how we as a species have developed our current traits and trends, from diet, through cultural goods, body and ecology, to language itself.

Points for discussion

1. Bickerton suggests that deriving negatives and questions in a holistic protolanguage framework would be an insuperable problem. Is it?
2. Learning algorithms in ALife simulations come in for some criticism, as we have seen in this chapter. Look at the content of one simulation in particular, and identify some possible or actual criticisms, and how they might be or have been resolved. You can choose any cases you like – but if you are struggling to choose, you might opt to compare Kenny Smith's neural network approach to any earlier case.
3. In this chapter, we have only discussed experimental investigations of human communication that rely on learning artificial languages invented by the experimenters, or natural languages already spoken by the participants. However, there is also a developing trend towards laboratory investigations of second-language acquisition. What insights might relevant papers bring to the acquisition, use and modification over time of linguistic structures, and how might these be relevant to the debate over the transition from protolanguage to modern language? You might start with, for example, Garrod and Doherty (1994), Tanenhaus et al. (1995), Yang and Givón (1997), Hudson and Newport (1999). You may wish to compare these to the results from the carefully documented development of Nicaraguan Sign Language (Senghas and Coppola 2001, Senghas, Kita and Ozyurek 2004) or Al-Sayyid Bedouin Sign Language (Sandler 2010, Sandler et al. 2005).
4. In Section 9.3 we discussed Bickerton's objections to the holistic protolanguage stage due to the difficulty of a community of protolanguage

speakers agreeing on which of the alternative meanings of formulaic expression to start with when determining segmentation patterns. Discuss these objections in the light of the iterated learning experiments using real people and artificial agents.

5. We have stated above that the Baldwin Effect is really a cover term for the multiple behaviours of genetic systems controlling behavioural flexibility, and hence not a single effect at all. Compare some examples of the discussion about the Baldwin Effect (you could start with papers or references from Weber and Depew 2003) with the discussion of punctuated equilibrium (start with Gould and Eldredge 1993). Do these discussions share any commonalities? Is either really a totally new, creative and explanatory mechanism, or are they both inevitable perceptual consequence of well-known evolutionary principles and consequences? Do we gain anything by using labels like 'the Baldwin Effect' as an alternative mechanism to normal evolution or are we merely creating a potential area of confusion and misunderstanding?

6. Even leaving aside Darwin and Mithen, the idea of a musical or prosodic protolanguage has occurred to several scholars over the years in slightly different guises. Compare the models of, for example, Jespersen (1922), Brown (2000), Dissanayake (2000), Falk (2000, 2004), Fitch (2005), Mithen (2006). Look in particular at their motivations for positing a musical protolanguage stage, and any data that they bring to bear in support of their particular model.

7. If shared evolved universal structures exist for prosodic aspects of speech and music, we might expect to find particular genes that, when damaged, disrupt music and some aspects of language function, but not others. Discuss what you think about this statement in the light of recent genetic studies of speech and music. You might like to start with some of the following papers: Sloboda and Howe (1991), Baharloo et al. (2000), Peretz, Cummings and Dubé (2007), Drayna et al. (2001), Karma (2002), Fisher and Scharff (2009).

Bibliography

Abbeduto, L. and R. Hagerman (1997) 'Language and communication in fragile X syndrome'. *Mental Retardation and Developmental Disabilities Research Reviews* 3: 313–22.

Aikhenvald, Alexandra Y. and R. M. W. Dixon (2001) *Areal Diffusion and Genetic Inheritance: Problems in Comparative Linguistics*. Oxford University Press.

Aitchison, Jean (1983) 'On roots of language'. *Language and Communication* 3: 83–97.

(1987) 'The language lifegame: prediction, explanation and linguistic change'. In Willem Koopman, Frederike van der Leek, Olga Fischer and Roger Eaton (eds.), *Explanation and Language Change*. Amsterdam: Benjamins, 11–32.

(1989a) 'Spaghetti junctions and recurrent routes: some preferred pathways in language evolution'. *Lingua* 77: 151–71.

(1989b) *The Articulate Mammal*. 3rd edition. London: Routledge.

(1996) *The Seeds of Speech*. Cambridge University Press.

Alcock, Katherine J., Richard E. Passingham, Kate E. Watkins and Faraneh Vargha-Khadem (2000) 'Oral dyspraxia in inherited speech and language impairment and acquired dysphasia'. *Brain and Language* 75: 17–33.

Amunts, Katrin, Axel Schleicher and Karl Zillesa (2004) 'Outstanding language competence and cytoarchitecture in Broca's speech region'. *Brain and Language* 89: 346–53.

Andreason, N. C., M. Flaum, V. Swayze, D. S. O'Leary, R. Alliger, G. Cohen, J. Ehrhardt and W. T. Yuh (1993) 'Intelligence and brain structure in normal individuals'. *American Journal of Psychiatry* 150: 130–4.

Andrews, P. and J. E. Cronin (1982) 'The relationship of *Sivapithecus* and *Ramapithecus* and the evolution of the orang-utan'. *Nature* 297: 541–6.

Annett, Marian (1985) *Left, Right, Hand and Brain: The Right Shift Theory*. New Jersey: Erlbaum.

Arbib, Michael A. (2002) 'The mirror system, imitation, and the evolution of language'. In C. Nehaniv and K. Dautenhahn (eds.), *Imitation in Animals and Artifacts*. Cambridge, MA: MIT Press, 229–80.

(2003) 'Schema theory'. In Michael A. Arbib (ed.), *The Handbook of Brain Theory and Neural Networks*. 2nd edition. Cambridge, MA: MIT Press, 993–8.

(2005a) 'From monkey-like action recognition to human language: an evolutionary framework for neurolinguistics'. *Behavioral and Brain Sciences* 28: 105–67.

(2005b) 'The Mirror System Hypothesis: how did protolanguage evolve?' In Tallerman (ed.), 21–47.

Arensburg, B. (1994) 'Middle Paleolithic speech capabilities: a response to Dr Lieberman'. *American Journal of Physical Anthropology* 94: 279–80.

Arensburg, B., A. M. Tillier, B. Vandermeersch, H. Duday, L. A. Schepartz and Y. Rak (1989) 'A middle paleolithic human hyoid bone'. *Nature* 338: 758–60.

Arensburg, B., L. A. Schepartz, A. M. Tillier, B. Vandermeersch and Y. Rak (1990) 'A reappraisal of the anatomical basis for speech in middle Paleolithic hominids'. *American Journal of Physical Anthropology* 83: 137–46.

Argue, Debbie, Denise Donlon, Colin Groves and Richard Wright (2006) '*Homo floresiensis*: Microcephalic, pygmoid, *Australopithecus*, or *Homo?*' *Journal of Human Evolution* 51: 360–74.

Argue, D., M. J. Morwood, T. Sutikna, Jatmiko and E. W. Saptomo (2009) 'Homo floresiensis: A cladistic analysis'. *Journal of Human Evolution* 57: 623–39.

Armstrong, David F. (1999) *Original Signs, Gesture, Sign, and the Sources of Language.* Washington, DC: Gallaudet University Press.

Armstrong, David F., William C. Stokoe and Sherman E. Wilcox (1995) *Gesture and the Nature of Language.* Cambridge University Press.

Ashcraft, Mark H. (1993) 'A personal case history of transient anomia'. *Brain and Language* 44: 47–57.

Ashley-Koch, A., Q. Yang and R. S. Olney (2000) 'Sickle hemoglobin (Hb S) allele and sickle cell disease: A HuGE review'. *American Journal of Epidemiology* 151: 839–45.

Avery, O. T., C. M. MacLeod and M. McCarty (1944) 'Studies on the chemical nature of the substance inducing transformation of Pneumococcal types'. *Journal of Experimental Medicine* 79: 137–58.

Avital, E. and E. Jablonka (2000) *Animal Traditions: Behavioural Inheritance in Evolution.* Cambridge University Press.

Baharloo, S., P. A. Johnston, S. K. Service, J. Gitschier and N. B. Freimer (1998) 'Absolute pitch: an approach for identification of genetic and nongenetic components'. *American Journal of Human Genetics* 62: 224–31.

Baharloo, S., S. K. Service, N. Risch, J. Gitschier and N. B. Freimer (2000) 'Familial aggregation of absolute pitch'. *American Journal of Human Genetics* 67: 755–8.

Baldwin, J. M. (1896) 'A new factor in evolution'. *American Naturalist* 30: 441–51.

Barbujani, Guido, Arianna Magagni, Eric Minch and L. Luca Cavalli-Sforza (1997) 'An apportionment of human DNA diversity'. *Proceedings of the National Academy of Sciences USA* 94: 4516–19.

Barlow, H. B. (1972) 'Single units and sensation: a neuron doctrine for perceptual psychology?' *Perception* 1: 371–94.

Bartlett, C. W., J. F. Flax, M. W. Logue, V. J. Vieland, A. S. Bassett *et al.* (2002) 'A major susceptibility locus for Specific Language Impairment is located on chromosome 13q21'. *American Journal of Human Genetics* 71: 45–55.

Bates, Elizabeth (2004) 'Explaining and interpreting deficits in language development across clinical groups: where do we go from here?' *Brain and Language* 88: 248–53.

Bates, Elizabeth and Jeffrey Elman (1996) 'Learning rediscovered'. *Science* 274: 1849–50.

Bates, Elizabeth and J. Goodman (1997) 'On the inseparability of grammar and the lexicon: evidence from acquisition, aphasia and real-time processing'. In

G. Altmann (ed.), Special issue on the lexicon, *Language and Cognitive Processes* 12(5/6): 507–86.

Bellugi, Ursula, S. Marks, A. Bihrle and H. Sabo (1988) 'Dissociation between language and cognitive functions in Williams syndrome'. In Bishop and Mogford, 177–89.

Bellugi, Ursula, P. P. Wang and T. L. Jernigan (1994) 'Williams syndrome: an unusual neuropsychological profile'. In S. H. Broman and J. Grafman (eds.), *Atypical Cognitive Deficits in Developmental Disorders*. Hillsdale, NJ: Lawrence Erlbaum, 23–56.

Bellugi, Ursula, L. Lichtenberger, W. Jones, Z. Lai and M. St. George (2000) 'The neuro-cognitive profile of Williams syndrome: a complex pattern of strengths and weaknesses'. *Journal of Cognitive Neuroscience* 12: 7–29.

Bellwood, Peter (1987) *The Polynesians*. London: Thames and Hudson.

Bender, Daniel and Xiaoqin Wang (2005) 'The neuronal representation of pitch in primate auditory cortex'. *Nature* 436: 1161–5.

Bengtson, John D. and Merritt Ruhlen (1994) 'Global etymologies'. In Ruhlen (1994a), 277–336.

Bergsland, K. and H. Vogt (1962) 'On the validity of glottochronology'. *Current Anthropology* 3: 115–53.

Bever, T. G. and R. J. Chiarello (1974) 'Cerebral dominance in musicians and nonmusicians'. *Science* 185: 537–9.

Bhattacharyya, Madan, Cathie Martin and Alison Smith (1993) 'The importance of starch biosynthesis in the wrinkled seed shape character of peas studied by Mendel'. *Plant Molecular Biology* 22: 525–31.

Bickerton, Derek (1981) *Roots of Language*. Ann Arbor, MI: Karoma.

(1984) 'The language bioprogram hypothesis'. *Behavioral and Brain Sciences* 7: 173–222.

(1990) *Language and Species*. University of Chicago Press.

(1995) *Language and Human Behavior*. Seattle: University of Washington Press.

(1998) 'Catastrophic evolution: the case for a single step from protolanguage to full human language. In Hurford, Studdert-Kennedy and Knight, 341–58.

(2003) 'Symbol and structure: a comprehensive framework for language evolution'. In Christiansen and Kirby, 77–93.

(2007) 'Language evolution: a brief guide for linguists'. *Lingua* 117: 510–26.

Binder, J. R., J. A. Frost, T. A. Hammeke, S. M. Rao and R. W. Cox (1996) 'Function of the left planum temporale in auditory and linguistic processing'. *Brain* 4: 1239–47.

Binder, J. R., J. A. Frost, T. A. Hammeke, P. S. F. Bellgowan, J. A. Springer, J. N. Kaufman and E. T. Possing (2000) 'Human temporal lobe activation by speech and nonspeech sounds'. *Cerebral Cortex* 10: 512–28.

Bishop, Dorothy V. M. (2001) 'Genetic and environmental risks for specific language impairment in children'. *Philosophical Transactions of the Royal Society of London B: Biological Sciences* 356: 369–80.

(2002) 'Motor immaturity and specific speech and language impairment: evidence for a common genetic basis'. *American Journal of Medical Genetics* 114: 56–63.

Bishop, Dorothy and Kay Mogford (eds.) (1988) *Language Development in Exceptional Circumstances*. London: Longman.

Blevins, Juliette (2004) *Evolutionary Phonology*. Cambridge University Press.

Bloom, L. (1970) *Language Development: Form and Function in Emerging Grammars*. Cambridge, MA: MIT Press.

— (1973) *One Word At a Time: The Use of Single Word Utterances Before Syntax*. The Hague: Mouton.

Boatman, Dana (2004) 'Cortical bases of speech perception: evidence from functional lesion studies'. *Cognition* 92: 47–65.

Boatman, Dana F., Ronald P. Lesser and Barry Gordon (1994) 'Cortical representation of speech perception and production, as revealed by direct cortical electrical interference'. In *ICSLP-1994*: 763–6.

Boatman, D., C. Hall, M. L. Goldstein, R. Lesser and B. Gordon (1997) 'Neuroperceptual differences in consonant and vowel discrimination, as revealed by direct cortical electrical interference'. *Cortex* 33: 83–98.

Boë, L.-J., J.-L. Heim, K. Honda and S. Maeda (2002) 'The potential Neandertal vowel space was as large as that of modern humans'. *Journal of Phonetics* 30: 465–84.

de Boer, Bart (2001) *The Origins of Vowel Systems*. Oxford University Press.

— (2010) 'Modelling vocal anatomy's significant effect on speech'. *Journal of Evolutionary Psychology* 8: 351–66.

Botstein, D. and N. Risch (2003) 'Discovering genotypes underlying human phenotypes: past successes for mendelian disease, future approaches for complex disease'. *Nature Genetics* 33: 228–37.

Boveri, T. H. (1904) *Ergebnisse über die Konstitution der chromatischen Substanz des Zelkerns*. Jena: Fisher.

Brighton, Henry (2002) 'Compositional syntax from cultural transmission'. *Artificial Life* 8: 25–54.

Briscoe, E. J. (1997) 'Co-evolution of language and of the language acquisition device'. In P. R. Cohen and W. Wahlster (eds.), *Proceedings of the Thirty-Fifth Annual Meeting of the Association for Computational Linguistics and Eighth Conference of the European Chapter of the Association for Computational Linguistics*. Somerset, NJ: Association for Computational Linguistics, 418–27.

— (1998) 'Language as a complex adaptive system: co-evolution of language and of the language acquisition device'. *Proceedings of the 8th Computational Linguistics in the Netherlands Meeting, Nijmegen*. URL: http://citeseer.ist.psu.edu/viewdoc/summary?doi=10.1.1.51.7440.

— (2000) 'Grammatical acquisition: inductive bias and coevolution of language and the language acquisition device'. *Language* 76: 245–96.

— (2002a) 'Coevolution of the language faculty and language(s) with decorrelated encodings'. In Tallerman (ed.), 310–33.

— (2002b) 'Grammatical acquisition and linguistic selection'. In E. J. Briscoe (ed.), *Language Acquisition and Linguistic Evolution: Formal and Computational Approaches*. Cambridge University Press.

Broca, Paul (1861) 'Remarques sur le siège de la faculté de langage articulé suivies d'une observation d'aphémie'. *Bull. Soc. Anat. Paris* 6: 330.

— (1865) 'Sur le siège de la faculté de langage articulé'. *Bull. Soc. D'Anthropologie* 6: 337–93.

(1877) 'Rapport sur un mémoire de M. Armand de Fleury intitulé: De l'inégalité dynamique des deux hémisphères cérébraux'. *Bull Acad Médicine* 6: 508–39.

Brodmann, Korbinian (1909) *Vergleichende Lokalisationslehre der Grosshirnrinde in ihren Prinzipien dargestellt auf Grund des Zellenbaues.* Leipzig: Johann Ambrosius Barth Verlag. Available as K. Brodmann and Laurence J. Garey (2010) *Brodmann's Localisation in the Cerebral Cortex.* New York: Springer.

Browman, Catherine P. and Louis Goldstein (1986) 'Towards an articulatory phonology'. *Phonology Yearbook* 3: 219–52.

(1989) 'Articulatory gestures as phonological units'. *Phonology* 6: 201–51.

(1992) 'Articulatory phonology: an overview'. *Phonetica* 49: 155–80.

(2000) 'Competing constraints on intergestural coordination and self-organization of phonological structures'. *Bulletin de la Communication Parlée* 5: 25–34.

Brown, P., T. Sutikna, M. J. Morwood, R. P. Soejono, Jatmiko, E. Wayhu Saptomo and Rokus Awe Due (2004) 'A new small-bodied hominin from the Late Pleistocene of Flores, Indonesia'. *Nature* 431: 1055–61.

Brown, Roger (1973) *A First Language.* Cambridge, MA: Harvard University Press.

Brown, Steven (2000) 'The "Musilanguage" model of music evolution'. In N. L. Wallin, B. Merker and S. Brown (eds.), *The Origins of Music.* Cambridge, MA: MIT Press, 271–300.

Brunet, M., F. Guy, D. Pilbeam, H. Taisso Mackaye, A. Likius, D. Ahounta *et al.* (2002) 'A new hominid from the Upper Miocene of Chad, Central Africa'. *Nature* 418: 145–51.

Buxhoeveden, Daniel P., A. E. Switala, M. Litaker, E. Roy and M. F. Casanova (2001) 'Lateralization in human planum temporale is absent in nonhuman primates'. *Brain and Behavioral Evolution* 57: 349–58.

Buxhoeveden, Daniel P. and Manuel F. Casanova (2002) 'The minicolumn hypothesis in neuroscience'. *Brain* 125: 935–51.

Bybee, J. (2007) *Frequency of Use and the Organization of Language.* Oxford University Press.

Calvin, William H. (1983) 'A stone's throw and its launch window: timing precision and its implications for language and hominid brains'. *Journal of Theoretical Biology* 104: 121–35.

Calvin, William H. and D. Bickerton (2000) *Lingua Ex Machina: Reconciling Darwin with the Human Brain.* Cambridge, MA: MIT Press.

Calvin, William H. and George A. Ojemann (1994) *Conversations with Neil's Brain: The Neural Nature of Thought and Language.* Reading, MA: Addison-Wesley.

Cameron-Faulkner, Thea, E. Lieven and M. Tomasello (2003) 'A construction-based analysis of child-directed speech'. *Cognitive Science* 27: 843–73.

Campbell, Lyle (1988) 'Review of Greenberg (1987)'. *Language* 64: 591–615.

(2003) 'How to show languages are related: methods for distant genetic relationship'. In Joseph and Janda, 262–82.

(2004) *Historical Linguistics.* 2nd edition. Edinburgh University Press.

(forthcoming) 'What can we learn about the earliest human language by comparing languages known today?' To appear in Bernard Laks and Daniel Simeoni (eds.) *Origins and Evolution of Language.*

Cangelosi, A., A. D. M. Smith and K. Smith (eds.) (2006) *Proceedings of Evolang* 6. World Scientific Publishing.

Cansino, S., S. J. Williamson and D. Karron (1994) 'Tonotopic organization of human auditory association cortex'. *Brain Research* 663: 38–50.

Caramazza, Alfonso (1996) 'The brain's dictionary'. *Nature* 380: 485–6.

Carlsson, P. and M. Mahlapuu (2002) 'Forkhead transcription factors: key players in development and metabolism'. *Developmental Biology* 250: 1–23.

Carr, Philip (2000) 'Scientific realism, sociophonetic variation, and innate endowments in phonology'. In Noel Burton-Roberts, Philip Carr and Gerard Docherty (eds.), *Phonological Knowledge: Conceptual and Empirical Issues*. Oxford University Press, 67–104.

Carré, R. (2009) 'Dynamic properties of an acoustic tube: prediction of vowel systems'. *Speech Communication* 51: 26–41.

Catford, Ian (2001) *A Practical Introduction to Phonetics*. 2nd edition. Oxford University Press.

Cavalli-Sforza, L. Luca and Marcus W. Feldman (2003) 'The application of molecular genetic approaches to the study of human evolution'. *Nature Genetics* 33: 266–75.

Changeux J.-P. and A. Danchin (1976) 'Selective stabilization of developing synapses as a mechanism for the specification of neuronal networks'. *Nature* 264: 705–12.

Chao, L. L. and R. T. Knight (1998) 'Contribution of human prefrontal cortex to delay performance'. *Journal of Cognitive Neuroscience* 10(2): 167–77.

Chapman, R. and L. Hesketh (2001) 'Language, cognition and short-term memory in individuals with Down syndrome'. *Down Syndrome: Research and Practice* 7: 1–7.

Chen, F. C. and W. H. Li (2001) 'Genomic divergences between humans and other hominoids and the effective population size and common ancestor of humans and chimpanzees'. *American Journal of Human Genetics* 68: 444–56.

Cheney, Dorothy and Robert Seyfarth (1988) 'Assessment of meaning and the detection of unreliable signals by vervet monkeys'. *Animal Behaviour* 36: 477–86.

Chimp Sequencing and Analysis Consortium (2005) 'Initial sequence of the chimpanzee genome and comparison with the human genome'. *Nature* 437: 69–87.

Chomsky, Noam (1965) *Aspects of the Theory of Syntax*. Cambridge, MA: MIT Press.

(1972) *Language and Mind*. New York: Harcourt Brace Jovanovich.

(1986) *Knowledge of Language: Its Nature, Origin and Use*. Westport, CT: Praeger.

(1988) *Language and Problems of Knowledge: The Managua Lectures*. Cambridge, MA: MIT Press.

Christiansen, Morten H. and Simon Kirby (eds.) (2003) *Language Evolution*. Oxford University Press.

Clackson, James, Peter Forster and Colin Renfrew (eds.) (2005) *Phylogenetic Methods in Historical Linguistics*. Cambridge: McDonald Institute for Archaeological Research.

Clark, Andrew G., Stephen Glanowski, Rasmus Nielsen, Paul D. Thomas, Anish Kejariwal, Melissa A. Todd *et al.* (2003) 'Inferring nonneutral evolution from human-chimp-mouse orthologous gene trios'. *Science* 302: 1960–3.

Clemens, W. A. (1974) 'Purgatorius, an early paromomyid primate (Mammalia)'. *Science* 184: 903–5.

Clutton-Brock, T. H. (2002) 'Breeding together: kin selection and mutualism in cooperative vertebrates'. *Science* 296: 69–72.

Clutton-Brock, T. H. and Paul Harvey (1977) 'Primate ecology and social organization'. *Journal of the Zoological Society of London* 183: 1–39.

(1980) 'Primates, brains and ecology'. *Journal of Zoology* 190: 309–23.

Collins, Francis S., Christian J. Stoeckert, Graham R. Serjeant, Bernard G. Forget and Sherman M. Weissman (1984) '$^{G}\gamma\beta$+ hereditary persistence of fetal haemoglobin: cosmid cloning and identification of a specific mutation 5' to the $^{G}\gamma$ gene'. *Proceedings of the National Academy of Sciences of the United States of America* 81: 4894–98.

Condillac, Étienne Bonnot de (1746) *Essai sur l'origine des connaissances humaines* (Essay on the Origin of Human Knowledge). Electronic edition: http://classiques.uqac.ca/classiques/condillac_etienne_bonnot_de/essai_origine_des_connaissances/origine_des_connaissances_L25.jpg.

Conway, C. M. and M. H. Christiansen (2001) 'Sequential learning in non-human primates'. *Trends in Cognitive Sciences* 5: 539–46.

Corballis, M. C. (1991) *The Lopsided Ape*. Oxford University Press.

(2002) *From Hand to Mouth: The Origin of Language*. Princeton University Press.

(2003) 'From hand to mouth: the gestural origins of language'. In Christiansen and Kirby, 201–18.

Crain, Stephen and Paul Pietroski (2002) 'Why language acquisition is a snap'. In Nancy A. Ritter (ed.), *A Review of 'The Poverty of Stimulus Argument'*, special issue of *The Linguistic Review* 19: 163–83.

Croft, William (1990) *Typology and Universals*. Cambridge University Press.

(2000) *Explaining Language Change: An Evolutionary Approach*. London: Longman.

(2006) 'The relevance of an evolutionary model to historical linguistics'. In Ole Nedergård Thomsen (ed.), *Competing Models of Linguistic Change*, Amsterdam: John Benjamins, 91–132.

Crompton, A. W., R. Z. German, A. J. Thexton (1997) 'Mechanisms of swallowing and airway protection in infant mammals (*Sus domesticus and Macaca. fascicutavis*)'. *Journal of the Zoological Society of London* 241: 89–102.

Crow, T. J. (2000) 'Schizophrenia as the price that homo sapiens pays for language: a resolution of the central paradox in the origin of the species'. *Brain Research Reviews* 31: 118–29.

(2002a) 'ProtocadherinXY: a candidate gene for cerebral asymmetry and language'. In Wray (ed.), 93–112.

Crow, T. J. (ed.) (2002b) *The Speciation of Modern Homo Sapiens*. Oxford University Press.

Crow, T. J. (2002c) 'Introduction'. In Crow (ed.), 1–20.

(2005) 'Who forgot Paul Broca? The origins of language as test case for speciation theory'. *Journal of Linguistics* 41: 133–56.

Cruttenden, Alan (1986) *Intonation*. Cambridge University Press.

Curtiss, Susan (1977) *Genie: A Linguistic Study of a Modern-day 'Wild Child'*. New York: Academic Press.

(1994) 'Language as a cognitive system: its independence and selective vulnerabil-
ity'. In Carlos P. Otero (ed.), *Noam Chomsky: Critical Assessments* Vol. 4.
London: Routledge, 211–55.

Dąbrowska, Ewa (2004) *Language, Mind and Brain*. Edinburgh University Press.

Damasio, H., T. J. Grabowski, D. Tranel, R. D. Hichwa and A. R. Damasio (1996) 'A
neural basis for lexical retrieval'. *Nature* 380: 499–505.

Darwin, Charles (1871) *The Descent of Man and Selection in Relation to Sex*. London:
John Murray.

(1881) *The Formation of Vegetable Mould through the Action of Worms, with
Observations on their Habits*. London: Murray.

Davidson, T. M., J. Sedgh, D. Tran and C. J. Stepnowski (2005) 'The anatomic basis
for the acquisition of speech and obstructive sleep apnea: evidence from
cephalometric analysis supports The Great Leap Forward hypothesis'. *Sleep
Medicine* 6: 497–505.

Dawkins, Richard (1976) *The Selfish Gene*. Oxford University Press.

(1983) *The Extended Phenotype*. Oxford University Press.

(1986) *The Blind Watchmaker*. London: Penguin.

(1995) *River out of Eden*. London: HarperCollins.

(2004) *The Ancestor's Tale*. London: Weidenfeld & Nicolson.

De Boysson-Bardies, B., L. Sagart and C. Durand (1984) 'Discernable differences in the
babbling of infants according to target language'. *Journal of Child Language*
11: 1–15.

De Gusta, David, W. Henry Gilbert and Scott P. Turner (1999) 'Hypoglossal canal size
and hominid speech'. *Proceedings of the National Academy of Science USA*
96: 1800–4.

Deacon, Terrence W. (1990) 'Fallacies of progression in theories of brain-size evolution'.
International Journal of Primatology 11: 193–236.

(1992) 'The human brain'. In S. Jones, R. Martin and D. Pilbeam (eds.),
Cambridge Encyclopedia of Human Evolution. Cambridge University
Press, 115–23.

(1997) *The Symbolic Species: The Co-Evolution of Language and the Brain*.
London: Penguin.

Dediu, Dan and D. Robert Ladd (2007) 'Linguistic tone is related to the population fre-
quency of the adaptive haplogroups of two brain size genes, ASPM and
Microcephalin'. *Proceedings of the National Academy of Sciences of the
USA* 104: 10944–9.

Dennett, Daniel C. (1995) *Darwin's Dangerous Idea: Evolution and the Meanings of
Life*. London: Penguin.

Dennis, M. and B. Kohn (1975) 'Comprehension of syntax in infantile hemiplegics after
cerebral hemidecortification'. *Brain and Language* 2: 475–86.

Deutsch, D. (2006) 'The enigma of absolute pitch'. *Acoustics Today* 2: 11–18.

Deutsch, D., T. Henthorn, E. W. Marvin and H.-S. Xu (2005) 'Perfect pitch in tone language speakers carries over to music'. *Journal of the Acoustical Society of
America* 116: 2580.

Deutscher, Guy (2000) *Syntactic Change in Akkadian: The Evolution of Sentential
Complementation*. Oxford University Press.

Diamond, Jared (1991) *The Rise and Fall of the Third Chimpanzee*. London: Radius.

Dissanayake, Ellen (2000) 'Antecedents of the temporal arts in early mother-infant inter-action'. In N. L. Wallin, B. Merker and S. Brown (eds.), *The Origins of Music*. Cambridge, MA: MIT Press, 389–410.

Dixon, R. M. W. (1997) *The Rise and Fall of Languages*. Cambridge University Press.

Donald, M. (1991) *Origins of the Modern Mind: Three Stages in the Evolution of Culture and Cognition*. Cambridge, MA: Harvard University Press.

(2001) *A Mind so Rare*. New York: W. W. Norton and Co.

Donohue, Mark and Simon Musgrave (2007) 'Typology and the linguistic macrohistory of Island Melanesia'. *Oceanic Linguistics* 46: 348–87.

Dor, A. and E. Jablonka (2001) 'From cultural selection to genetic selection: a framework for the evolution of language'. *Selection* 1: 33–56.

Dorman, C. (1991) 'Microcephaly and intelligence'. *Developmental Medicine and Child Neurology* 33: 267–72.

Doupe, Allison and Patricia Kuhl (1999) 'Birdsong and human speech: common themes and mechanisms'. *Annual Review of Neuroscience* 22: 567–631.

Drachman, David A. (2005) 'Do we have brain to spare?' *Neurology* 64: 2004–5.

Drayna D., A. Manichaikul, M. de Lange, H. Snieder and T. Spector (2001) 'Genetic cor-relates of musical pitch recognition in Humans'. *Science* 291:1969–71.

Dresher, B. Elan and Jonathan D. Kaye (1990) 'A computational learning model for met-rical phonology'. *Cognition* 34: 137–95.

Dronkers, N. F. (1996) 'A new brain region for coordinating speech articulation'. *Nature* 384: 159–60.

Dronkers, N. F., B. B. Redfern and R. T. Knight (2000) 'The neural architecture of lan-guage disorders'. In Gazzaniga, 949 –58.

Dronkers, Nina F., David P. Wilkins, Robert D. Van Valin Jr., Brenda B. Redfern, Jeri J. Jaeger (2004) 'Lesion analysis of the brain areas involved in language com-prehension'. *Cognition* 92: 145–77.

Dunbar, Robin (1992) 'Neocortex size as a constraint on group size in primates'. *Journal of Human Evolution* 22: 469–93.

(1993) 'Coevolution of neocortical size, group size and language in humans'. *Behavioral and Brain Sciences* 16(4): 681–735.

(1997) *Grooming, Gossip and the Evolution of Language*. London: Faber and Faber.

(2004) *The Human Story*. London: Faber and Faber.

Dunn, M., A. Terrill, G. Reesink, R. A. Foley and S. C. Levinson (2005) 'Structural phylogenetics and the reconstruction of ancient language history'. *Science* 309(5743): 2072–5.

Dunn, M., R. Foley, S. C. Levinson, G. Reesink and A. Terrill (2007) 'Statistical reason-ing in the evaluation of typological diversity in Island Melanesia'. *Oceanic Linguistics* 46(2): 388–403.

Dunn, Michael, Simon J. Greenhill, Stephen C. Levinson and Russell D. Gray (2011) 'Evolved structure of language shows lineage-specific trends in word-order universals'. *Nature* 473: 79–82.

Durie, Mark and Malcolm Ross (eds.) (1996) *The Comparative Method Reviewed: Regularity and Irregularity in Language Change*. Oxford University Press.

Earnshaw, W. C., L. M. Martins and S. H. Kaufmann (1999) 'Mammalian caspases: struc-ture, activation, substrates, and functions during apoptosis'. *Annual Review of Biochemistry* 68: 383–424.

Eldredge, N. and S. J. Gould (1972) 'Punctuated equilibria: an alternative to phyletic gradualism'. In T. J. M. Schopf (ed.), *Models in Paleobiology*. San Francisco: Freeman Cooper, 82–115. Available at www.blackwellpublishing.com/ridley/classictexts/eldredge.asp.

Embleton, Sheila (1986) *Statistics in Historical Linguistics*. Bochum: Brockmeyer.

Enard, W., P. Khaitovich, J. Klose, S. Zöllner, F. Heissig, P. Giavalisco *et al.* (2002a) 'Intra-and interspecific variation in primate gene expression patterns'. *Science* 296: 340–3.

Enard, Wolfgang, Molly Przeworski, Simon E. Fisher, Cecilia S. L. Lai, Victor Wiebe, Takashi Kitano *et al.* (2002b) 'Molecular evolution of *FOXP2*, a gene involved in speech and language'. *Nature* 418: 869–72.

Eswaran, Vinayak (2002) 'A diffusion wave out of Africa: the mechanism of the modern human revolution?' *Current Anthropology* 43: 749–74.

Eswaran, Vinayak, Henry Harpending and Alan R. Rogers (2005) 'Genomics refutes an exclusively African origin of humans'. *Journal of Human Evolution* 49: 1–18.

Evans, Nicholas and Stephen C. Levinson (2009) 'The myth of language universals: language diversity and its importance for cognitive science'. *Behavioral and Brain Sciences* 32: 429–92.

Evans, P. D., S. L. Gilbert, N. Mekel-Bobrov, E. J. Vallender, J. R. Anderson *et al.* (2005) 'Microcephalin, a gene regulating brain size, continues to evolve adaptively in humans'. *Science* 309: 1717–20.

Everett, Daniel L. (2005) 'Cultural constraints on grammar and cognition in Pirahã: another look at the design features of human language'. *Current Anthropology* 46: 621–46.

    (2009) 'Pirahã culture and grammar: a response to some criticisms'. *Language* 85: 405–42.

Evett, I. W. and B. S. Weir (1998) *Interpreting DNA Evidence: Statistical Genetics for Forensic Scientists*. Sunderland, MA: Sinauer.

Ewart, A. K., C. A. Morris, D. Atkinson, W. Jin, K. Sternes, P. Spallone *et al.* (1993) 'Hemizygosity at the elastin locus in a developmental disorder, Williams syndrome'. *Nature Genetics* 5: 11–16.

Falconer, Douglas S. and Trudy F. C. Mackay (1996) *Introduction to Quantitative Genetics*. 4th edition. New Jersey: Prentice Hall.

Falk, Dean (1975) 'Comparative anatomy of the larynx in man and the chimpanzee: Implications for language in Neanderthal'. *American Journal of Physical Anthropology* 43: 123–32.

    (1980) 'Language, handedness and primate brains: did the Australopithecines sign?' *American Anthropologist* 82: 72–8.

    (1983) 'Cerebral cortices of East African early hominids'. *Science* 222: 1072–4.

    (1985) 'Apples, oranges, and the lunate sulcus'. *American Journal of Physical Anthropology* 67: 313–15.

    (2000) *Primate Diversity*. New York: Norton.

    (2004) 'Prelinguistic evolution in early hominins: whence motherese?' *Behavioral and Brain Sciences* 27: 491–50.

Fan, W. M., M. Kasahara, J. Gutknecht, D. Klein, W. E. Mayer, M. Jonker and J. Klein (1989) 'Shared class II MHC polymorphisms between humans and chimpanzees'. *Human Immunology* 26: 107–21.

Fedorenko, E., A. D. Patel, D. Casasanto, J. Winawer and E. Gibson (2009) 'Structural integration in language and music: Evidence for a shared system'. *Memory and Cognition* 37: 1–9.

Fernald, Anne (1992) 'Human maternal vocalizations to infants as biological relevant signals: an evolutionary perspective'. in Jerome H. Barkow, Ieda Cosmides and John Tooby *The Adapted Mind: Evolutionary Psychology and the Generation of Culture.* Oxford University Press.

Fernald, A., T. Taeschner, J. Dunn, M. Papousek, B. Boysson-Bardies and I. Fukui (1989) 'A cross-language study of prosodic modifications in mothers' and fathers' speech to preverbal infants'. *Journal of Child Language* 16: 477–501.

Finlay, Barbara L. and Richard B. Darlington (1995) 'Linked regularities in the development and evolution of mammalian brains'. *Science* 268: 1578–84.

Finlay, Barbara L., Richard B. Darlington and Nicholas Nicastro (2001) 'Developmental structure in brain evolution'. *Behavioral and Brain Sciences* 24: 263–308.

Fisher, R. A. (1930) *The Genetical Theory of Natural Selection.* Oxford University Press.

Fisher, Simon E. (2005) 'Dissection of molecular mechanisms underlying speech and language disorders'. *Applied Psycholinguistics* 26: 111–28.

Fisher, Simon E. and Constance Scharff (2009) 'FOXP2 as a molecular window into speech and language'. *Trends in Genetics* 25: 166–77.

Fisher, Simon E., Faraneh Vargha-Khadem, Kate E. Watkins, Anthony P. Monaco and Marcus E. Pembrey (1998) 'Localisation of a gene implicated in a severe speech and language disorder'. *Nature Genetics* 18: 168–70.

Fisher, Simon E., Cecilia Lai and Anthony P. Monaco (2003) 'Deciphering the genetic basis of speech and language disorders'. *Annual Review of Neurosciences* 26: 57–80.

Fitch, W. Tecumseh (2000) 'The evolution of speech: a comparative review'. *Trends in Cognitive Sciences* 4: 258–67.

(2002) 'Comparative vocal production and the evolution of speech: reinterpreting the descent of the larynx'. In Wray (ed.), 21–45.

(2005) 'The evolution of music in comparative perspective'. In G. Avanzini, L. Lopez, S. Koelsch and M. Majno (eds.), *The Neurosciences and Music II: From Perception to Performance*, Vol. 1060. New York Academy of Sciences, 29–49.

(2009) 'Fossil cues to the evolution of speech'. In R. P. Botha and C. Knight (eds.), *The Cradle of Language.* Oxford University Press, 112–34.

(2010) *The Evolution of Language.* Cambridge University Press.

Fitch, W. Tecumseh, Marc D. Hauser and Noam Chomsky (2005) 'The evolution of the language faculty: clarifications and implications'. *Cognition* 97: 179–210.

Fitch, W. Tecumseh and D. Reby (2001) 'The descended larynx is not uniquely human'. *Proceedings of the Royal Society of London (B)* 268: 1669–75.

Fleagle, J. G. (1985) 'Size and adaptation in primates'. In W. L. Jungers (ed.), *Size and Scaling in Primate Biology.* New York: Plenum Press, 1–20.

(1999) *Primate Adaptation and Evolution.* San Diego, CA: Academic Press.

Foley, Robert A. (1987) *Another Unique Species: Patterns in Human Evolutionary Ecology.* London: Longman.

(2001) 'In the shadow of the modern synthesis: alternative perspectives on the last 50 years of palaeoanthropology'. *Evolutionary Anthropology* 10: 5–14.

Foulkes, Paul (1993) *Theoretical Implications of the /p/> /f/> /h/ Change*. Ph.D. dissertation, University of Cambridge.

Fox, Anthony (1995) *Linguistic Reconstruction: An Introduction to Theory and Method*. Oxford University Press.

Fox, P. T., S. Petersen, M. Posner and M. E. Raiche (1988) 'Is Broca's area language-specific?' *Neurology* 38: 172.

Frangiskakis, J. M., A. K. Ewart, C. A. Morris, C. B. Mervis, J. Bertrand, B. F. Robinson *et al.* (1996) 'LIM-kinase1 hemizygosity implicated in impaired visuospatial constructive cognition'. *Cell* 86: 59–69.

Friedrich, Frances J., Robert Egly, Robert D. Rafal and Diane Beck (1998) 'Spatial attention deficits in humans: a comparison of superior parietal and temporal-parietal junction lesions'. *Neuropsychology* 12(2): 193–207.

Gagneux P., C. Wills, U. Gerloff, D. Tautz, P. A. Morin, C.Boesch *et al.* (1999) 'Mitochondrial sequences show diverse evolutionary histories of African hominoids'. *Proceedings National Academy of Sciences USA* 96: 5077–82.

Galantucci, Bruno (2005) 'An experimental study of the emergence of human communication systems'. *Cognitive Science* 29: 737–76.

Gardner, R. A., B. T. Gardner and T. E. van Cantfort (eds.) (1989) *Teaching Sign Language to Chimpanzees*. Albany, NY: State University of New York Press.

Garrod, S. and G. Doherty (1994) 'Conversation, coordination and convention: an empirical investigation of how groups establish linguistic conventions'. *Cognition* 53: 181–215.

Gazzaniga, Michael S. (1995) 'Principles of human brain organization derived from split-brain studies'. *Neuron* 14: 217–28.

Gazzaniga, Michael S. (ed.) (2000) *The New Cognitive Neurosciences*. Cambridge, MA: MIT Press.

Gerken, LouAnn (1994) 'Child phonology: past research, present questions, future directions'. In Morton Ann Gernsbacher (ed.), *The Handbook of Psycholinguistics*. New York: Academic Press, 781–820.

Geschwind, N. (1965) 'Disconnexion syndromes in animals and man'. *Brain* 88: 237–94, 585–644.

Gibson, K. and T. Ingold (1993) *Tools, Language and Cognition in Human Evolution*. Cambridge University Press.

Giedd, J. N., J. W. Snell, N. Lange, J. C. Rajapakse, B. J. Casey, P. L. Kozuch *et al.* (1996) 'Quantitative magnetic resonance imaging of human brain development: ages 4–18'. *Cerebral Cortex* 6: 551–60.

Giegerich, Heinz J. (1992) *English Phonology: An Introduction*. Cambridge University Press.

Gilad, Yoav, Alicia Oshlack, Gordon K. Smyth, Terence P. Speed and Kevin P. White (2006) 'Expression profiling in primates reveals a rapid evolution of human transcription factors'. *Nature* 440: 242–5.

Gilbert, S. F. (2003) *Developmental Biology. 7th edition*. Sunderland, MA: Sinauer.

Givón, T. (1995) *Functionalism and Grammar*. Amsterdam: John Benjamins.

Glezer I. I., P. R. Hof, C. Leranth and P. J. Morgane (1993) 'Calcium-binding protein-containing neuronal populations in mammalian visual cortex: a comparative study in whales, insectivores, bats, rodents, and primates'. *Cerebral Cortex* 3: 249–72.

Goldberg, E. (2001) *The Executive Brain: Frontal Lobes and the Civilised Mind.* Oxford University Press.

Gonick, Larry and Mark Wheelis (1991) *The Cartoon Guide to Genetics.* New York: HarperResource.

Goodall, Jane (1986) *The Chimpanzees of Gombe: Patterns of Behavior.* Boston, MA: Belknap Press of the Harvard University Press.

Gopnik, Myrna (1999) 'Some evidence for impaired grammars'. In R. Jackendoff and P. Bloom (eds.), *Language, Logic and Concepts: Essays in Memory of John Macnamara.* Cambridge, MA: MIT Press. 263–83.

Gopnik, Myrna and M. B. Crago (1991) 'Familial aggregation of a developmental language disorder'. *Cognition* 39: 1–50.

Gould, Stephen J. (1981) *The Mismeasure of Man.* New York: W. W. Norton and Co.
    (2002) *The Structure of Evolutionary Theory.* Cambridge, MA: The Belknap Press of the Harvard University Press.

Gould, Stephen J. and Niles Eldredge (1993) 'Punctuated equilibrium comes of age'. *Nature* 366(6452): 223–7.

Gould, Stephen J. and Richard C. Lewontin (1979) 'The spandrels of San Marco and the Panglossian paradigm: a critique of the adaptationist programme'. *Proceedings of the Royal Society of London, B* 205: 581–98.

Gould, Stephen J. and E. S. Vrba (1982) 'Exaptation – a missing term in the science of form'. *Paleobiology* 8: 4–15.

Grafton, S. T., M. A. Arbib, L. Fadiga and G. Rizzolatti (1996) 'Localization of grasp representations in humans by positron emission tomography. 2. Observation compared with imagination'. *Experimental Brain Research* 112: 103–11.

Green, Richard E., J. Krause, A. W. Briggs, T. Maricic, U. Stenzel, M. Kircher *et al.* (2010) 'A draft sequence of the Neandertal genome'. *Science* 328: 710–22.

Greenberg, Joseph H. (1987) *Language in the Americas.* Stanford University Press.
    (2000) *Indo-European and its Closest Relatives: The Eurasiatic Language Family. Volume 1: Grammar.* Stanford University Press.

Greenberg, S. (2005) 'A multi-tier theoretical framework for understanding spoken language'. In S. Greenberg and W. A. Ainsworth (eds), *Listening to Speech: An Auditory Perspective.* Mahwah, NJ: Erlbaum, 411–33.

Greenblatt, S. H. (1995) 'Phrenology in the science and culture of the 19th century'. *Neurosurgery* 37: 790–805.

Gribaldo, Simonetta and Hervé Philippe (2002) 'Ancient phylogenetic relationships'. *Theoretical Population Biology* 61: 391–408.

Gribbin, John and Jeremy Cherfas (2001) *The First Chimpanzee.* London: Penguin.

Grinnell, J. (1917) 'The niche-relationships of the California Thrasher'. *Auk* 34: 427–33.

Grossman, M. (1980) 'A central processor for hierachically structured material: evidence from Broca's aphasia'. *Neuropsychologia* 18: 299–309.

Haile-Selassie, Yohannes (2001) 'Late Miocene hominids from the Middle Awash, Ethiopia'. *Nature* 412: 178–81.

Hamilton, W. D. (1964) 'The genetical evolution of social behavior'. *Journal of Theoretical Biology* 7: 1–52.

Hamilton, W. D. and M. Zuk (1982) 'Heritable true fitness and bright birds: a role for parasites?' *Science* 218: 384–7.

Harding R. M., S. M. Fullerton, R. C. Griffiths, J. Bond, M. J. Cox, J. A. Schneider *et al.* (1997) 'Archaic African and Asian lineages in the genetic ancestry of modern humans. *American Journal of Human Genetics.* 60: 772–89.

Hardy, G. H. (1908) 'Mendelian proportions in a mixed population'. *Science* 28: 49–50.

Harpending, H. C. and A. R. Rogers (2000) 'A genetic perspectives on human origins and. differentiation'. *Annual Review of Genomics and Human Genetics:* 361–85.

Harris, J., J. Watson and S. Bates (1999) 'Prosody and melody in vowel disorder'. *Journal of Linguistics* 35: 489–525.

Harrison, S. P. (2003) 'On the limits of the Comparative Method'. In Joseph and Janda, 213–43.

Harvey, Paul and Mark Pagel (1991) *The Comparative Method in Evolutionary Biology.* Oxford University Press.

Hauser, Marc D. (1996) *The Evolution of Communication.* Cambridge, MA: MIT Press.

Hauser, Marc D. and C. A. Fowler (1992) 'Fundamental frequency declination is not unique to human speech: evidence from non-human primates'. *Journal of the Acoustical Society of America* 91: 363–9.

Hauser, Marc D. and W. Tecumseh Fitch (2003) 'What are the uniquely human components of the language faculty?' In Christiansen and Kirby, 158–81.

Hauser, Marc D., Noam Chomsky and W. Tecumseh Fitch (2002) 'The faculty of language: what is it, who has it, and how did it evolve?' *Science* 298: 1569–79.

Hawi, Z., R. Segurado, J. Conroy, K. Sheehan, N. Lowe, A. Kirley *et al.* (2005) 'Preferential transmission of paternal alleles at risk genes in attention-deficit/hyperactivity disorder'. *American Journal of Human Genetics* 77: 958–65.

Hawkins, John A. (1988) (ed.) *Explaining Language Universals.* Oxford: Blackwell.

Hayes, K. J. and C. Hayes (1952) 'Imitation in a home-raised chimpanzee'. *Journal of Comparative and Physiological Psychology* 45: 450–9.

Hedges, S. B. and S. Kumar (2003) 'Genomic clocks and evolutionary timescales'. *Trends in Genetics* 19: 200–6.

Heggarty, Paul, Warren Maguire and April McMahon (2010) 'Splits or waves? Trees or webs? How divergence measures and network analysis can unravel language histories'. *Philosophical Transactions of the Royal Society (B)* 365: 3829–43.

Heilbroner, P. L. and R. L. Holloway (1988) 'Anatomical brain asymmetries in New World and Old World monkeys: stages of temporal lobe development in primate Evolution'. *American Journal of Physical Anthropology* 76: 39–48.

Heine, B. and T. Kuteva (2002) 'On the evolution of grammatical forms'. In Wray (ed.), 376–97.

    (2005) *Language Contact and Grammatical Change.* Cambridge University Press.

    (2007) *The Genesis of Grammar: A Reconstruction.* Oxford University Press.

Hendricks, W., J. Leunissen, E. Nevo, H. Bloemendal and W. W. De Jong (1987) 'The lens protein αA-crystallin of the blind mole rat, *Spulux ehrenbergi*: evolutionary change and functional constraints'. *Proceedings of the National Academy of Sciences USA* 84: 5320–4.

Herman, Louis M. (2010) 'What laboratory research has told us about dolphin cognition'. *International Journal of Comparative Psychology* 23: 310–30.

Hewes, G. W. (1973) 'Primate communication and the gestural origin of language'. *Current Anthropology* 14: 5–24.

Hickok, Gregory and David Poeppel (2004) 'Dorsal and ventral streams: a framework for understanding aspects of the functional anatomy of language'. *Cognition* 92: 67–99.

Hildebrand, J. D. (2005) 'Shroom regulates epithelial cell shape via the apical positioning of an actomyosin network'. *Journal of Cell Science* 118: 5191–203.

Hildebrand, J. D. and P. Soriano (1999) 'Shroom, a PDZ domain-containing actin-binding protein, is required for neural tube morphogenesis in mice'. *Cell* 99: 485–97.

Hinton, G. and S. Nowlan (1987) 'How learning can guide evolution'. *Complex Systems* 1: 495–502. Reprinted in R. K. Belew and M. Mitchell (eds.) *Adaptive Individuals in Evolving Populations: Models and Algorithms*. Reading, MA: Addison-Wesley, 447–54.

Hock, Hans Henrich (1991) *Principles of Historical Linguistics*. Berlin/New York: Mouton de Gruyter.

Hockett, Charles F. (1960) 'The origin of speech'. *Scientific American* 203: 88–96.

Hoenigswald, H. M. (1960) *Language Change and Linguistic Reconstruction*. University of Chicago Press.

Höss, M. (2000) 'Neanderthal population genetics'. *Nature* 404: 453–4.

Holloway, R. L. (1970) 'Australopithecine endocast (Taung specimen, 1924): a new volume determination'. *Science* 168: 966–8.

(1979) 'Brain size, allometry and reorganization: toward a synthesis'. In M. E. Hahn, C. Jensen and B. C. Dudek (eds.), *Development and Evolution of Brain Size*. New York: Academic Press, 59–88.

(1983) 'Cerebral brain endocast pattern of *Australopithecus afarensis* hominid'. *Nature* 303: 420–2.

(1991) 'On Falk's 1989 accusation regarding Holloway's study of the Taung endocast: a reply'. *American Journal of Physical Anthropology* 84: 87–8.

(1996) 'Evolution of the human brain'. In A. Lock and C. R. Peters (eds.), *Handbook of Human Symbolic Evolution*. Oxford: Clarendon Press, 74–117.

Holloway, R. L. and M. C. Lacoste-Lareymondie (1982) 'Brain endocast asymmetry in pongids and hominids: some preliminary findings on the paleontology of cerebral dominance'. *American Journal of Physical Anthropology* 58: 101–10.

Hopper, Paul and Elizabeth C. Traugott (1993) *Grammaticalization*. Cambridge University Press.

Houston, Derek M. and Peter W. Jusczyk (2000) 'The role of talker-specific information in word segmentation by infants'. *Journal of Experimental Psychology: Human Perception and Performance* 26: 1570–82.

Hudson, C. L. and E. L. Newport (1999) 'Creolization: could adults really have done it all?' In A. Greenhill *et al.* (eds.), *Proceedings of the 23rd Annual Boston University Conference on Language Development: Vol. 1*. Somerville, MA: Cascadilla Press.

Hull, David (1988) *Science as a Process: an Evolutionary Account of the Social and Conceptual Development of Science*. University of Chicago Press.

(2001) *Science and Selection: Essays on Biological Evolution and the Philosophy of Science*. Cambridge University Press.

Human Genome *Nature* issue: *Nature* 409: 813–958. Available at www.nature.com/genomics/human.

Hurford, James R. (2003) 'The language mosaic and its evolution'. In Christiansen and
     Kirby, 38–57.
     (2007) *The Origins of Meaning*. Oxford University Press.
Hurford, James R., Michael Studdert-Kennedy and Chris Knight (eds.) (1998) *Approaches
     to the Evolution of Language: Social and Cognitive Bases*. Cambridge
     University Press.
Hurst, Jane, M. Baraitser, E. Auger, F. Graham and S. U. Norel (1990) 'An extended fam-
     ily with a dominantly inherited speech disorder'. *Developmental Medicine
     and Clinical Neurology* 32: 352–5.
Husi, Holgar and Seth G. N. Grant (2001) 'Proteomics of the nervous system'. *Trends in
     Neurosciences* 24: 259–66.
Hutsler, J. and R. A. W. Galuske (2003) 'Hemispheric asymmetries in cerebral cortical
     networks'. *Trends in Neurosciences* 26(8): 429–35.
Huttenlocher, P. R. (1990) 'Morphometric study of human cerebral cortex development'.
     Neuropsychologia 28: 517–27.
Huxley, Thomas Henry (1863) *Evidence as to Man's Place in Nature*. London: Williams
     and Norwood.
Ingman M., H. Kaessmann, S. Pääbo and U. Gyllensten (2000) 'Mitochondrial genome
     variation and the origin of modern humans'. *Nature* 408: 708–13
International HapMap Consortium (2005) 'A haplotype map of the human genome'.
     Nature 437: 1299–320.
International Molecular Genetic Study of Autism Consortium (2001) 'Further charac-
     terisation of the autism susceptibility locus AUTS1 on chromosome 7q'.
     Human Molecular Genetics 10: 973–82.
Jackendoff, Ray (1999) 'Possible stages in the evolution of the language capacity'. *Trends
     in Cognitive Sciences* 3: 272–9.
     (2002) *Foundations of Language: Brain, Meaning, Grammar, Evolution*. Oxford
     University Press.
Jackendoff, Ray and Steven Pinker (2005) 'The nature of the language faculty and
     its implications for evolution of language (Reply to Fitch, Hauser and
     Chomsky)'. *Cognition* 97: 211–25.
Jantz, Richard L. (1987) 'Anthropological dermatoglyphic research'. *Annual Review of
     Anthropology* 16: 161–77.
Jeffreys, A. J., V. Wilson and L. S. Thein (1985) 'Individual-specific fingerprints of human
     DNA'. *Nature* 314: 67–73.
Jerison, H. J. (1973) *Evolution of the Brain and Intelligence*. New York: Academic
     Press.
Jespersen, Otto (1922) *Language: Its Nature, Development and Origin*. London: George
     Allen & Unwin.
Jobling, Mark A., Matthew Hurles and Chris Tyler-Smith (eds.) (2004) *Human
     Evolutionary Genetics: Origins, Peoples and Disease*. New York: Garland.
Johanson, Donald C. and Maitland A. Edey (1981) *Lucy: The Beginnings of Humankind*.
     London: Penguin.
Johnson, Keith (2002) *Acoustic and Auditory Phonetics*. 2nd edition. Oxford:
     Blackwell.
Jones, Steve (2000) *The Language of the Genes*. London: Harper Collins (revised edition;
     original 1993).

Jones, Steve, Robert Martin and David Pilbeam (eds.) (1992) *The Cambridge Encyclopedia of Human Evolution*. Cambridge University Press.

Joseph, Brian D. and Richard D. Janda (eds.) (2003) *The Handbook of Historical Linguistics*. Oxford: Blackwell.

Jungers, W. J., A. A. Pokempner, R. F. Kay and M. Cartmill (2003) 'Hypoglossal canal size in living hominoids and the evolution of human speech'. *Human Biology* 75: 473–84.

Jusczyk, P. W., A. D. Friederici, J. Wessels, V. Y. Svenkerud and A. M. Jusczyk (1993) 'Infants' sensitivity to segmental and prosodic characteristics of words in their native language'. *The Journal of Memory and Language* 32: 402–20.

Kaas, J. H. (1995) 'Human visual cortex: progress and puzzles'. *Current Biology* 5: 1126–8.

    (2000) 'Why is brain size so important: design problems and solutions as neocortex gets bigger or smaller'. *Brain and Mind* 1: 7–23.

Kaas, J. H., T. Hackett and M. Tramo (1999) 'Auditory processing in primate cerebral cortex'. *Current Opinions in Neurobiology* 9: 164–70.

Kaessmann, H., V. Wiebe and S. Pääbo (1999) 'Extensive nuclear DNA sequence diversity among chimpanzees'. *Science* 5: 1159–62.

Kaessmann, H., V. Wiebe, G. Weiss, and S. Pääbo (2001) 'Great ape DNA sequences reveal a reduced diversity and an expansion in humans'. *Nature Genetics*. 27: 155–6.

Kark, J. A., D. M. Posey, H. R. Schumacher *et al.* (1987) 'Sickle-cell trait as a risk factor for sudden death in physical training'. *New England Journal of Medicine* 317: 781–87.

Karma, K. (2002) 'Auditory structuring in explaining dyslexia'. In P. McKevitt, S. Ó. Nuallain, and C. Mulvihill (eds.), *Language, Vision and Music*. Amsterdam/Philadelphia: John Benjamins, 221–30.

Karmiloff-Smith, A., L. K. Tyler, K. Voice, K. Sims, O. Udwin, P. Howlins and M. Davies (1997) 'Language and Williams syndrome: how intact is "intact"?' *Child Development* 68: 246–62.

Karn, M. N. and L. S. Penrose (1951) 'Birth weight and gestation time in relation to maternal age, parity, and infant survival'. *Annals of Eugenics* 16: 147–64.

Katz, L. C. and C. J. Shatz (1996) 'Synaptic activity and the construction of the cortical circuits'. *Science* 274: 1133–8.

Kay, R. F., M. Cartmill and M. Balow (1998) 'The hypoglossal canal and the origin of human vocal behavior'. *Proceedings of the National Academy of Sciences USA* 95: 5417–19.

Kegl, Judy, Ann Senghas and Marie Coppola (1999) 'Creation through contact: sign language emergence and sign language change in Nicaragua'. In Michel DeGraff (ed.), *Language Creation and Language Change: Creolization, Diachrony, and Development*. Cambridge, MA: MIT Press, 179–237.

Kendal, J., J. Tehrani and J. Odling-Smee (2011) (eds.) *Transactions of the Royal Society B, 366,* Special Issue: 'Human niche construction'.

Kent, R. D. (1997) *The Speech Sciences*. San Diego, CA: Singular Press.

Kent, R. D. and A. D. Murray (1982) 'Acoustic features of infant vocalic utterances at 3, 6, and 9 months'. *Journal of the Acoustical Society of America* 72: 353–65.

Kertesz, A., D. Lesk and P. McCabe (1977) 'Isotope localization of infarcts in aphasia'. *Archives of Neurology* 34: 590–601.

Khaitovich, Philipp, Bjoern Muetzell, Xinwei She, Michael Lachmann, Ines Hellmann, Janko Dietzsch *et al.* (2004) 'Regional patterns of gene expression in human and chimpanzee brains'. *Genome Research* 14: 1462–73.

Kim, Eunjoon and Morgan Sheng (2004) 'PDZ domain proteins of synapses'. *Nature Reviews Neuroscience* 5: 771–81.

Kimura, M. (1968) 'Evolutionary rate at the molecular level'. *Nature* 217: 624 –6.

   (1983) *The Neutral Theory of Molecular Evolution.* Cambridge University Press.

Kirby, Simon (1999) *Function, Selection and Innateness: The Emergence of Language Universals.* Oxford University Press.

   (2000) 'Syntax without natural selection: how compositionality emerges from vocabulary in a population of learners'. In Knight, Studdert-Kennedy and Hurford, 303–23.

   (2001) 'Spontaneous evolution of linguistic structure: an iterated learning model of the emergence of regularity and irregularity'. *IEEE Journal of Evolutionary Computation* 5(2):102–10.

   (2002) 'Natural language from artificial life'. *Artificial Life* 8: 185–215.

   (2009) 'The evolution of language'. In Robin Dunbar and Louise Barrett (eds.), *The Oxford Handbook of Evolutionary Psychology.* Oxford University Press, 669–81.

Kirby, Simon and Morten H. Christiansen (2003) 'From language learning to language evolution'. In Christiansen and Kirby, 272–94.

Kirby, S. and J. Hurford (2002) 'The emergence of linguistic structure: an overview of the iterated learning model'. In A. Cangelosi and D. Parisi (eds.), *Simulating the Evolution of Language.* London: Springer Verlag, 121–48.

Kirby, Simon, H. Cornish and K. Smith (2008) 'Cumulative cultural evolution in the laboratory: an experimental approach to the origins of structure in human language'. *Proceedings of the National Academy of Sciences, USA* 105: 10681–6.

Kirby, Simon, M. Dowman and T. L. Griffiths (2007) 'Innateness and culture in the evolution of language'. *Proceedings of the National Academy of Sciences, USA* 104: 5241–5.

Knecht, S., B. Drager, M. Deppe, L. Bobe, H. Lohmann, A. Floel, E.-B. Ringelstein and H. Henningsen (2000) 'Handedness and hemispheric language dominance in healthy humans'. *Brain* 123: 2512–18.

Knight, Chris, Michael Studdert-Kennedy and James R. Hurford (eds.) (2000) *The Evolutionary Emergence of Language: Social Function and the Origin of Linguistic Form.* Cambridge University Press.

Knudsen, E. I. and M. Konishi (1978) 'A neural map of auditory space in the owl'. *Science* 200(4343): 795–7.

Koerner, Konrad (ed.) (1983) *Linguistics and Evolutionary Theory: Three Essays by August Schleicher, Ernst Haeckel and Wilhelm Bleek.* Amsterdam: Benjamins.

Krings, M., A. Stone, R. W. Schmitz, H. Krainitzki, M. Stoneking and S. Pääbo (1997) 'Neandertal DNA sequences and the origin of modern humans'. *Cell* 90: 19–30.

Krings, M., H. Geisert, R. W. Schmitz, H. Krainitzki and S. Pääbo (1999) 'DNA sequence of the mitochondrial hypervariable region II from the Neandertal type specimen'. *Proceedings of the National Academy of Sciences, USA* 96: 5581–5.

Krings, M., C. Capelli, F. Tschentscher, H. Geisert, S. Meyer, A. von Haeseler *et al.* (2000) 'A view of Neandertal genetic diversity'. *Nature Genetics* 26: 144–6.

Kroeber, A. L. (1955) 'Linguistic time depth results so far and their meaning'. *International Journal of American Linguistics* 21: 91–104.

Kuhl, P. K. (1979) 'Speech perception in early infancy: perceptual constancy for spectrally dissimilar vowel categories'. *Journal of the Acoustical Society of America* 66: 1668–79.

   (1983) 'Perception of auditory equivalence classes for speech in early infancy'. *Infant Behavior and Development* 6: 263–85.

   (2004) 'Early language acquisition: cracking the speech code'. *Nature Reviews Neuroscience* 5: 831–43.

Ladd, D. Robert (1996) *Intonational Phonology*. Cambridge University Press.

Ladefoged, Peter (1996) *Elements of Acoustic Phonetics*. 2nd edition. University of Chicago Press.

Ladefoged, Peter and Keith Johnson (2010) *A Course in Phonetics*. 6th edition. Boston, MA: Heinle.

Ladefoged, Peter and Ian Maddieson (1996) *The Sounds of the World's Languages*. Oxford: Blackwell.

Lai, Cecilia, Simon E. Fisher, J. A. Hurst, E. R. Levy, S. Hodgson, M. Fox *et al.* (2000) 'The SPCH2 region on human 7q31: genomic characterization of the critical interval and localization of translocations associated with speech and language disorder'. *American Journal of Human Genetics* 67: 357–68.

Lai, Cecilia S. L., Simon E. Fisher, Jane A. Hurst, Faraneh Vargha-Khadem and Anthony P. Monaco (2001) 'A forkhead-domain gene is mutated in a severe speech and language disorder'. *Nature* 413: 519–23.

Laitman, J. T., R. C. Heimbuch and E. C. Crelin (1979) 'The basicranium of fossil hominids as an indicator of their upper respiratory systems'. *American Journal of Physical Anthropology* 51: 15–34.

Laland, Kevin and Gillian Brown (2002) *Sense and Nonsense*. Oxford University Press.

Laland, Kevin, J. Odling-Smee and S. Myles (2010) 'How culture shaped the human genome: bringing genetics and the human sciences together'. *Nature Reviews Genetics* 11: 137–48.

Lass, Roger (1990) 'How to do things with junk: exaptation in language evolution'. *Journal of Linguistics* 26: 79–102.

Laver, John (1994) *Principles of Phonetics*. Cambridge University Press.

Leakey, Louis, Phillip V. Tobias and John Russell Napier (1964) 'A new species of the Genus Homo from Olduvai Gorge'. *Nature* 202: 7–9.

Leakey, M. D. and R. L. Hay (1979) ' Pliocene footprints in the Laetoli Beds at Laetoli, northern Tanzania'. *Nature* 278: 317–23.

Leakey, M. G., C. S. Fiebel, I. McDougall and A. C. Walker (1995) 'New four-million-year-old hominid species from Kanapoi and Allia Bay, Kenya. *Nature* 376: 565–71.

Lenneberg, E. (1967) *Biological Foundations of Language*. New York: Wiley.

Leonard W. R. and M. L. Robertson (1992) 'Nutritional requirements and human evolution: a bioenergetics model'. *American Journal of Human Biology* 4:
179–95.

Levinson, Joshua N. and Alaa El-Husseini (2005) 'New players tip the scales in the balance between excitatory and inhibitory synapses'. *Molecular Pain* 1: 1–6.

Levitt, Pat (2000) 'Molecular determinants of regionalisation of the forebrain and cerebral cortex'. In Gazzaniga, 23–32.

Levitt, Pat, M. F. Barbe and K. E. Eagleson (1997) 'Patterning and specification of the
cerebral cortex'. *Annual Review of Neurosciences* 20: 1–24.

Lewin, Roger (1980) 'Is your brain really necessary?' *Science* 210: 1232–4.

   (1997) *Bones of Contention: Controversies in the Search for Human Origins*. 2nd
   edition. University of Chicago Press.

   (2005) *Human Evolution: An Illustrated Introduction*. 5th edition. Oxford:
   Blackwell.

Lewin, Roger and Robert A. Foley (2004) *Principles of Human Evolution*. Oxford:
Blackwell.

Lewontin, R. C. (1978) 'Adaptation'. *Scientific American* 239: 156–69.

Liberman, A. M., F. S. Cooper, D. P. Shankweiler and M. Studdert-Kennedy (1967)
'Perception of the speech code'. *Psychological Review* 74: 431–61.

Lieberman, D. E. and R. C. McCarthy (1999) 'The ontogeny of cranial base angulation in
humans and chimpanzees and its implications for reconstructing pharyngeal
dimensions'. *Journal of Human Evolution* 36: 487–517.

Lieberman, P. (1975) *On the Origins of Language: An Introduction to the Evolution of
Human Speech*. New York: Macmillan.

   (1984) *The Biology and Evolution of Language*. Cambridge, MA: Harvard
   University Press.

   (1989) 'The origins of some aspects of human language and cognition'. In P. Mellars
   and C. Stringer (eds.), *The Human Revolution: Behavioural and Biological
   Perspectives on the Origins of Modern Humans*. Edinburgh University Press,
   391–414.

   (2003) 'Motor control, speech, and the evolution of human language'. In
   Christiansen and Kirby, 255–71.

   (2007) 'Current views on Neanderthal speech capabilities: a reply to Boë et al.
   (2002)'. *Journal of Phonetics*: 552–63.

Lieberman, P. and E. S. Crelin (1971) 'On the speech of Neanderthal man'. *Linguistic
Inquiry* 2: 203–22.

Lieberman, P., D. H. Klatt and W. H. Wilson (1969) 'Vocal tract limitations on the vowel
repertoires of rhesus monkeys and other nonhuman primates'. *Science* 164:
1185–7.

Liegeois, F., C. S. L. Lai, T. Baldeweg, S. E. Fisher, A. P. Monaco, A. Connelly and
F. Vargha-Khadem (2001) 'Behavioural and neuroimaging correlates of
a chromosome 7q31 deletion containing the *SPCH1* gene'. *Society of
Neuroscience Abstracts* 27: Program No. 529.17.

Liegeois, F., T. Baldeweg, A. Connelly, D. G. Gadian, M. Mishkin and F. Vargha-Khadem
(2003) 'Language fMRI abnormalities associated with FOXP2 gene mutation'. *Nature Neuroscience* 6: 1230–7.

Lightfoot, David (1991) *How to Set Parameters: Arguments from Language Change*.
Cambridge, MA: MIT Press.

Lillien, L. (1998) 'Neural progenitors and stem cells: mechanisms of progenitor hetero-geneity'. *Current Opinions in Neurobiology* 8: 37–44.

Lindesay, Robert, of Pitscottie (1899) *The Histories and Chronicles of Scotland*, ed. by Aeneas J. G. Mackay, Volume I (1437–1542). Edinburgh: Blackwood.

Linnaeus, Carolus (1758) *Systema Naturae*. Holmiae: Laurentius Salvius. Available online at http://gdz.sub.uni-goettingen.de/dms/load/img/?PPN=PPN362053006.

Lohr, Marisa (1999) *Methods for the Genetic Classification of Languages*. Unpublished Ph.D. thesis, University of Cambridge.

Lord, C., M. Rutter and A. LeCouteur (1994) 'Autism diagnostic interview – revised: a revised version of a diagnostic interview for caregivers of individuals with possible pervasive developmental disorders'. *Journal of Autism and Developmental Disorders* 24: 659–85.

Lord, C, S. Risi, L. Lambrecht, E. H. Cook, B. L. Lenventhal, P. S. DiLavore *et al.* (2000) 'The Autism Diagnostic Observation Schedule – Generic: a standard meas-ure of social and communication defects associated with the spectrum of autism'. *Journal of Autism and Developmental Disorders* 30: 205–23.

Loritz, D. (1999) *How the Brain Evolved Language*. Oxford University Press.

Luria, Salvador E., Stephen Jay Gould and Sam Singer (1981) *A View of Life*. Menlo Park, CA: Benjamin/Cummings.

Lynch, Gary and Richard Granger (2008) *Big Brain: The Origins and Future of Human Intelligence*. Basingstoke: Palgrave Macmillan.

Macdermot, K. D., E. Bonora, N. Sykes, A. M. Coupe, C. S. Lai, S. C. Vernes *et al.* (2005) 'Identification of FOXP2 truncation as a novel cause of developmen-tal speech and language deficits'. *American Journal of Human Genetics* 76: 1074–80.

Mackay, Aeneas J. G. (ed.) (1899) *The Historie and Cronicles of Scotland: from the slaughter of King James the First to the ane thousande fyve hundreith thrie scoir fyftein zeir*. 3 vols. Edinburgh: Scottish Text Society.

MacLarnon, A. M. and G. P. Hewitt (1999) 'The evolution of human speech: the role of enhanced breathing control'. *American Journal of Physical Anthropology* 109: 341–63.

Maddox, Brenda (2002) *Rosalind Franklin: The Dark Lady of DNA*. New York: Harper Collins.

Maddieson, I. (1984) *Patterns of Sounds*. Cambridge University Press.

Mann, Fanny, Victoria Zhukareva, Aurea Pimenta, Pat Levitt and Jürgen Bolz (1998) 'Membrane-associated molecules guide limbic and nonlimbic thalamocorti-cal projections'. *The Journal of Neuroscience* 18: 9409–19.

Marcus, G. F. and S. E. Fisher (2003) 'FOXP2 in focus: what can genes tell us about speech and language?'. *Trends in Cognitive Sciences* 7: 257–62.

Markowitsch, Hans J. (2000) 'The anatomical bases of memory'. In Gazzaniga, 781–96.

Marks, Jonathan (2002) *What it Means to be 98% Chimpanzee*. Berkeley, CA: University of California Press.

Marler, Peter (1970) 'A comparative approach to vocal learning: song development in white-crowned sparrows'. *Journal of Comparative Physiological Psychology Monographs* 71: 1–25.

    (1984) 'Song learning: innate species differences in the learning process'. In P. Marler and H. S. Terrace (eds.), *The Biology of Learning*. Berlin: Springer, 289–309.

(1991) 'Differences in behavioural development in closely related species: bird-song'. In P. Bateson (ed.), *The Development and Integration of Behaviour*. Cambridge University Press, 41–70.

Martin, R. D. (1993) 'Primate origins: plugging the gaps'. *Nature* 363: 223–34.

Matisoff, James (1990) 'On megalocomparison'. *Language* 66: 106–20.

Matras, Yaron, April McMahon and Nigel Vincent (eds.) (2005) *Linguistic Areas, Convergence and Language Change*. London: Palgrave.

Mayley, G. (1996) 'The evolutionary cost of learning'. In P. Maes, M. Mataric, J-A. Meyer, J. Pollack and S. Wilson (eds.), *From Animals to Animats: Proceedings of the Fourth International Conference on Simulation of Adaptive Behaviour*. Boston, MA: MIT Press.

Maynard Smith, John (1989) *Evolutionary Genetics*. Oxford University Press.

(1993) *The Theory of Evolution*. Cambridge University Press (Canto edition: first published by Penguin Books, 1958).

McAlpine, David (2005) 'Neural population coding of sound level adapts to stimulus statistics'. *Nature Neuroscience* 8: 1684–9.

McBrearty, Sally and A. S. Brooks (2000) 'The revolution that wasn't: a new interpretation of the origin of modern human behaviour'. *Journal of Human Evolution* 39: 453–563.

McBrearty, Sally and Nina Jablonski (2005) 'First fossil chimpanzee'. *Nature* 437: 105–8.

McHenry, H. M. and K. Coffing (2000) '*Australopithecus* to *Homo*: transformations in body and mind'. *Annual Review of Anthropology* 29: 125–46.

McMahon, April (1994) *Understanding Language Change*. Cambridge University Press.

(2000) *Change, Chance, and Optimality*. Oxford University Press.

(2005a) 'Heads I win, tails you lose'. In Philip Carr, Jacques Durand and Colin Ewen (eds.), *Headhood, Elements, Specification and Contrastivity*. Dordrecht: Kluwer, 255–75.

McMahon, April (ed.) (2005b) *Quantitative Methods in Language Comparison*. Special issue of *Transactions of the Philological Society* 103.

McMahon, April (2007) 'Sounds, brain and evolution: or why phonology is plural'. In Martha C. Pennington (ed.), *Phonology in Context*. Basingstoke: Palgrave Macmillan, 159–85.

McMahon, April and Robert McMahon (1995) 'Linguistics, genetics and archaeology: internal and external evidence in the Amerind controversy'. *Transactions of the Philological Society* 93: 125–225.

(2005) *Language Classification by Numbers*. Oxford University Press.

McManus, Chris (2002) *Right Hand, Left Hand: the Origins of Asymmetry in Brains, Bodies, Atoms and Cultures*. London: Weidenfeld & Nicolson.

McWhorter, John H. (1997) *Towards a New Model of Creole Genesis*. New York: Peter Lang.

(2005) *Defining Creole*. Oxford University Press.

Mellars, Paul (2006) 'Going east: new genetic and archaeological perspectives on the modern human colonization of Eurasia'. *Science* 313: 796–800.

Mehler, J., P. W. Jusczyk, G. Lambertz, N. Halsted, J. Bertoncini and C. Amiel-Tison (1988) 'A precursor of language acquisition in young infants'. *Cognition* 29: 143–78.

Mekel-Bobrov, N., S. L. Gilbert, P. D. Evans, E. J. Vallender, J. R. Anderson *et al.* (2005) 'Ongoing adaptive evolution of ASPM, a brain size determinant in Homo sapiens'. *Science* 309: 1720–22.

Meyer-Lindenberg, Andreas, Carolyn B. Mervis and Karen F. Berman (2006) 'Neural mechanisms in Williams syndrome: a unique window to genetic influences on cognition and behaviour'. *Nature Reviews; Neurosciences* 7: 380–93.

Micelli, G., C. Caltagirone, G. Gainotti and P. Payer-Rigo (1978) 'Discrimination of voice versus place contrasts in aphasia'. *Brain and Language* 6: 47–51.

Miller, G. F. (1999) 'Sexual selection for cultural displays'. In R. Dunbar, C. Knight and C. Power (eds.), *The Evolution of Culture*. Edinburgh University Press, 71–91.

(2001) *The Mating Mind: How Sexual Choice Shaped the Evolution of Human Nature*. Anchor.

Miller, Jim (2006) 'Spoken and written English'. In Bas Aarts and April McMahon (eds.), *The Handbook of English Linguistics*. Oxford: Blackwell, 670–91.

Milner A. D. and M. A. Goodale (1995) *The Visual Brain in Action*. Oxford University Press.

Mithen, Steven (2006) *The Singing Neanderthals: The Origins of Music, Language, Mind and Body*. London: Phoenix.

Mithun, Marianne (2004) '"Unborrowable" areal traits'. Paper presented at the Linguistics Association of Great Britain annual meeting, London, September 2004.

Miyazaki, Ken'ichi and Yoko Ogawa (2006) 'Learning absolute pitch by children: a cross sectional study'. *Music Perception* 24(1): 63–78.

Mogford, Kay and Dorothy Bishop (1988a) 'Language development in unexceptional circumstances'. In Bishop and Mogford, 10–28.

(1988b) 'Five questions about language acquisition considered in the light of exceptional circumstances'. In Bishop and Mogford, 239–60.

Molnar-Szakacs, Istvan and Katie Overy (2006) 'Music and mirror neurons: from motion to 'e'motion'. *SCAN (2006)* 1: 235–41.

Moor, B. C. J. (2003) *An Introduction to the Psychology of Hearing*. London: Academic Press.

Morris, D. (1974) 'Neandertal speech'. *Linguistic Inquiry* 5:144–50.

Morwood, M. J., R. P. Soejono, R. G. Roberts, T. Sutikna, C. S. M. Turney, K. E. Westaway, W. J. Rink, J.-X. Zhao, G. D. van den Bergh, Rokus Awe Due, D. R. Hobbs, M. W. Moore, M. I. Bird and L. K. Fifield (2004) 'Archaeology and age of a new hominin from Flores in eastern Indonesia'. *Nature* 431: 1087–91.

Morwood, M. J., P. Brown, Jatmiko, T. Sutikna, E. Wahyu Saptomo, K. E. Westaway, R. A. Due, R. G. Roberts, T. Maeda, S. Wasisto and T. Djubiantono (2005) 'Further evidence for small-bodied hominins from the Late Pleistocene of Flores, Indonesia'. *Nature* 437: 1012–17.

Mountcastle, V. B. (1997) 'The columnar organization of the neocortex'. *Brain* 120: 701–22.

(2003) 'Untitled-introduction'. *Cerebral Cortex* 13(1): 2–4.

Mufwene, Salikoko (2001) *The Ecology of Language Evolution*. Cambridge University Press.

Müller, R. A. (1996) 'Innateness, autonomy, universality? Neurobiological approaches to language'. *Behavioral and Brain Sciences* 19: 611–75.

Munroe, S. and A. Cangelosi (2002) 'Learning and the evolution of language: the role of cultural variation and learning cost in the Baldwin Effect'. *Artificial Life* 8: 311–39.

Nei, Masatoshi (1987) *Molecular Evolutionary Genetics*. New York: Columbia University Press.

Nettle, Daniel (1999) *Linguistic Diversity*. Oxford University Press.

Nettle, Daniel and Robin Dunbar (1997) 'Social markers and the evolution of reciprocal exchange'. *Current Anthropology* 38: 93–9.

Nevins, Andrew, David Pesetsky and Cilene Rodrigues (2009) 'Pirahã exceptionality: a reassessment'. *Language* 85: 355–404.

Newbury, D. F. and A. P. Monaco (2010) 'Genetic advances in the study of speech and language disorders'. *Neuron* 68: 309–20.

Newmeyer, Frederick J. (1991) 'Functional explanation in linguistics and the origins of language'. *Language and Communication* 11: 3–28, 97–108.

(1998) *Language Form and Language Function*. Cambridge, MA: MIT Press.

(2003) 'What can the field of linguistics tell us about the origins of language?' In Christiansen and Kirby, 58–76.

Newport, Elissa (1990) 'Maturational constraints on language learning'. *Cognitive Science* 14: 11–28.

Newport, Elissa, Henry Gleitman and Lila R. Gleitman (1977) 'Mother, I'd rather do it myself: some effects and non-effects of maternal speech style'. In C. Snow and C. Ferguson (eds.), *Talking to Children: Language Input and Acquisition*. Cambridge University Press.

Nichols, Johanna (1992) *Linguistic Diversity in Space and Time*. University of Chicago Press.

(1995) 'Diachronically stable structural features'. In Henning Andersen (ed.), *Historical Linguistics 1993: Selected Papers from the 11th International Conference on Historical Linguistics*. Amsterdam: John Benjamins, 337–56.

(1998) 'The origins and dispersal of languages: linguistic evidence'. In Nina Jablonski and Leslie Aiello (eds.), *The Origin and Diversification of Language*. San Francisco: California Academy of Sciences, 127–70.

Nilsson, Dan-E. and Susanne Pelger (1994) 'A pessimistic estimate of the time required for an eye to evolve'. *Proceedings of the Royal Society of London, Series B*, 256: 53–8. Reprinted as pp. 293–301 of Mark Ridley (ed.) *Evolution*. Oxford Readers: Oxford University Press.

Nishimura, T. (2005) 'Developmental changes in the shape of the supralaryngeal vocal tract in chimpanzees'. *American Journal of Physical Anthropology* 126: 193–204.

Nishimura, T., A. Mikami, J. Suzuki and T. Matsuzawa (2003) 'Descent of the larynx in chimpanzee infants'. *Proceedings of the National Academy of Sciences USA* 100: 6930–3.

Nolte, John (2002) *The Human Brain: An Introduction To Its Functional Anatomy*. 5th edition. St Louis, MO: Mosby.

North, Michael (2005) 'Sick puppies, chief nipple twisters and the king of gall wasps'. *The Times Higher Education Supplement*, 30 September 2005.

Nowak, Martin A. (2006) 'Five rules for the evolution of cooperation'. *Science* 314(5805): 1560–3.

Obler, Loraine K. and Kris Gjerlow (1999) *Language and the Brain*. Cambridge University Press.

Odling-Smee, J. J., K. N. Laland and M. W. Feldman (2003) *Niche Construction: The Neglected Process in Evolution*. Princeton University Press.

Ojemann, G. (1983) 'Brain organization for language from the perspective of electrical stimulation mapping'. *Behavioural and Brain Sciences* 6: 189–230.

Ojemann, G. and C. Mateer (1979) 'Human language cortex: localization of memory, syntax, and sequential motor-phoneme identification systems'. *Science* 205: 1401–3.

Oppenheimer, Stephen (2003) *Out of Eden: The Peopling of the World*. London: Robinson.

Ovchinnikov, I. V., A. Götherström, G. P. Romanova, V. M. Kharitonov, K. Lidén and W. Goodwin (2000) 'Molecular analysis of Neanderthal DNA from the northern Caucasus'. *Nature* 404: 490–3.

Pääbo, S. (2003) 'The mosaic that is our genome'. *Nature* 421: 409–12.

Page, R. D. M. and E. C. Holmes (1998) *Molecular Evolution: A Phylogenetic Approach*. Oxford: Blackwell.

Pagel, Mark and Andrew Meade (2005) 'Estimating rates of meaning evolution on phylogenetic trees of languages'. In Clackson, Forster and Renfrew.

Paley, William (1847) [1802] *Natural Theology – Or Evidences of the Existence and Attributes of the Deity Collected from the Appearances of Nature*. 2nd edn; New York: Harper and Brothers.

Panger, Melissa A., Alison S. Brooks, Brian G. Richmond and Bernard Wood (2002) 'Older than the Oldowan? Rethinking the emergence of Hominin tool use'. *Evolutionary Anthropology* 11: 235–45.

Pardis C. Sabeti, Patrick Varilly, Ben Fry, Jason Lohmueller, Elizabeth Hostetter, Chris Cotsapas, Xiaohui Xie, Elizabeth H. Byrne, Steven A. McCarroll, Rachelle Gaudet, Stephen F. Schaffner, Eric S. Lander and The International HapMap Consortium (2007) 'Genome-wide detection and characterization of positive selection in human populations'. *Nature* 449: 913–18.

Parker, Andrew J., Bruce G. Cumming and Jon V. Dodd (2000) 'Binocular neurons and the perception of depth'. In Gazzaniga, 263–77.

Parker, Andrew J. and W. T. Newsome (1998) 'Sense and the single neuron: probing the physiology of perception'. *Annual Review of Neurosciences* 21: 227–77.

Parker, Anna (2006) 'Evolving the narrow language faculty: was recursion the pivotal step?' In A. Cangelosi, A. D. M. Smith and K. Smith (eds.), *The Evolution of Language: Proceedings of the 6th International Conference on the Evolution of Language*. World Scientific Press, 239–46.

Patel, A. D. (2003) 'Language, music, syntax, and the brain'. *Nature Neuroscience* 6: 674–81.

(2010) *Music, Language and the Brain*. New York: Oxford University Press.

Paterson, S. J., J. H. Brown, M. K. Gsödl, M. H. Johnson and A. Karmiloff-Smith (1999) 'A cognitive modularity and genetic disorders'. *Science* 286: 2355–8.

Paus, Tomáš (2005) 'Mapping brain maturation and cognitive development during adolescence'. *Trends in Cognitive Sciences* 9: 60–8.

Pearson, Helen (2006) 'What is a gene?' *Nature* 441: 398–401.

Penagos, Hector, Jennifer R. Melcher and Andrew J. Oxenham (2004) 'A neural representa-
    tion of pitch salience in nonprimary human auditory cortex revealed with func-
    tional magnetic resonance imaging'. *Journal of Neuroscience* 24: 6810–15.
Penfield, W. and T. Rasmussen (1950) *The Cerebral Cortex of Man.* New York:
    Macmillan.
Penfield, W. and L. Roberts (1959) *Speech and Brain-Mechanisms.* Princeton University
    Press.
Pepperberg, Irene (2005) 'An avian perspective on language evolution: implications of
    simultaneous development of vocal and physical object combinations by a
    Grey Parrot (Psittacus erithacus)'. In Tallerman, 239–61.
    (2010) 'Vocal learning in Grey Parrots: a brief review of perception, production
    and cross-species comparison'. *Brain and Language* 115: 81–91.
Peretz, I., S. Cummings and M.-P., Dubé (2007) 'The genetics of congenital amusia
    (tone deafness): a family-aggregation study'. *American Journal of Human
    Genetics* 81: 582–8.
Pickford, M. and B. Senut (2001) '"Millennium ancestor", a 6-million-year-old bipedal
    hominid from Kenya'. *South African Journal of Science* 97: 1–22.
Pilbeam, David (1982) 'New hominoid skull material from the Miocene of Pakistan'.
    Nature 295: 232–4.
Pinker, Steven (1994) *The Language Instinct.* London: Penguin.
    (1999) *How the Mind Works,* London: Penguin Press Science.
Pinker, S. and P. Bloom (1990) 'Natural language and natural selection'. *Behavioural and
    Brain Sciences* 13(4): 707–84.
Pinker, Steven and Ray Jackendoff (2005) 'The faculty of language: what's special about
    it?' *Cognition* 95: 201–36.
Poeppel, D. and G. Hickok (eds.) (2004) *Towards a New Functional Anatomy of Language.*
    Special issue of *Cognition* 92: 1–270.
Poeppel, D., W. Idsardi and V. van Wassenhove (2008) 'Speech perception at the interface
    of neurobiology and linguistics'. *Philosophical Transactions of the Royal
    Society London B* 363: 1071–86.
Polaszek, Andrew, Miguel A. Alonso-Zarazaga, Philippe Bouchet, Denis J. Brothers, Neal
    L. Evenhuis, Frank T. Krell, Christopher H. C. Lyal, Alessandro Minelli,
    Richard L. Pyle, Nigel Robinson, F. C. Thompson and J. van Tol (2005)
    'ZooBank: the open-access register for zoological taxonomy: Technical
    Discussion Paper'. *Bulletin of Zoological Nomenclature* 62: 210–20.
Polechová, J. and D. Storch (2008) 'Ecological niche'. In S. E. Jorgensen and B. Fath
    (eds.), *Encyclopedia of Ecology.* Oxford: Elsevier, 1088–97.
Ponting, Chris P. (2006) 'A novel domain suggests a ciliary function for ASPM, a brain
    size determining gene'. *Bioinformatics* 22: 1031–5.
Preuss, T. M. (2000) 'What's human about the human brain?' In Gazzaniga, 1219–34.
Provine, Robert (1996) 'Laughter'. *American Scientist* 84: 38–47.
Pullum, Geoffrey K. and Barbara C. Scholz (2002) 'Empirical assessment of stimulus
    poverty arguments'. In Nancy A. Ritter (ed.), *A Review of 'The Poverty of
    Stimulus Argument',* special issue of *The Linguistic Review* 19: 9–50.
Rakic, Pasko (1988) 'Specification of cerebral cortical areas'. *Science* 241: 170–6.
    (2000) 'Setting the stage for cognition: genesis of the primate cerebral cortex'. In
    Gazzaniga, 7–21.
Rankin, Robert L. (2003) 'The Comparative Method'. In Joseph and Janda, 183–212.

Raphael, Lawrence J., Gloria J. Borden and Katherine S. Harris (2006) *Speech Science Primer: Physiology, Acoustics and Perception of Speech.* 5th edition. Baltimore, MD: Lippincott, Williams & Wilkins.

Rauschecker, J. P. (1998) 'Cortical processing of complex sounds'. *Current Opinions in Neurobiology* 8: 516–21.

Relethford J. H. (2001) *Genetics and the Search for Modern Human Origins.* New York: Wiley-Liss.

Renfrew, Colin (2000) 'The problem of time depth'. In Colin Renfrew, April McMahon and Larry Trask (eds.), *Time Depth in Historical Linguistics*: Volume I, ix–xiv. Cambridge: McDonald Institute for Archaeological Research.

Renfrew, Colin and Paul G. Bahn (2004) *Archaeology: Theories, Methods and Practice.* London, Thames and Hudson.

Renfrew, Colin, April McMahon and Larry Trask (eds.) (2000) *Time Depth in Historical Linguistics*: 2 Volumes. Cambridge: McDonald Institute for Archaeological Research.

Renfrew, Colin and Daniel Nettle (1999) *Nostratic: Examining a Linguistic Macrofamily.* Cambridge: McDonald Institute for Archaeological Research.

de Renzi, E. and L. A. Vignolo (1962) 'The Token Test: a sensitive test to detect receptive disturbances in aphasics'. *Brain* 85: 665–78.

Rice, Mabel L. and Steven F. Warren (eds.) (2004) *Developmental Language Disorders: From Phenotype to Etiology.* Mahwah, NJ: Erlbaum.

Rice, Mabel L., Steven F. Warren and Stacy K. Betz (2005) 'Language symptoms of developmental language disorders: an overview of autism, Down syndrome, fragile X, specific language impairment, and Williams syndrome'. *Applied Psycholinguistics* 26: 7–27.

Richards, M. and V. Macaulay (2000) 'Genetic data and the colonization of Europe: genealogies and founders'. In Colin Renfrew and K. Boyle (eds.), *Archaeogenetics: DNA and the Population Prehistory of Europe.* Cambridge: McDonald Institute for Archaeological Research, 139–51.

Richards Robert J. (1987) *Darwin and the Emergence of Evolutionary Theories of Mind and Behavior.* University of Chicago Press.

Ridley, Mark (ed.) (1997) *Evolution.* Oxford Readers: Oxford University Press.

Ridley, Matt (1994) *The Red Queen: Sex and the Evolution of Human Nature.* London: Penguin Books.

(1999) *Genome: The Autobiography of a Species in 23 Chapters.* London: Fourth Estate.

(2003) *Nature via Nurture.* London: HarperCollins.

Ringe, Don (1992) '*On calculating the factor of chance in language comparison*'. *Transactions of the American Philosophical Society* 82. Philadelphia: American Philosophical Society.

(1996) 'The mathematics of "Amerind"'. *Diachronica* 13: 135–54.

(1999) 'How hard is it to match CVC roots?' *Transactions of the Philological Society* 97: 213–44.

(2003) 'Internal reconstruction'. In Joseph and Janda, 244–61.

Ritter, Nancy A. (ed.) (2002) *A Review of 'The Poverty of Stimulus Argument'*, special issue of *The Linguistic Review* 19.

Rizzolatti, G. and Michael A. Arbib (1998) 'Language within our grasp'. *Trends in Neurosciences* 21: 188–94.

Rizzolatti G., L. Fadiga, M. Matelli, V. Bettinardi, E. Paulesu, D. Perani and F. Fazio (1996) 'Localization of grasp representations in humans by PET: 1. Observation versus execution'. *Experimental Brain Research* 111: 246–52.

Roberts, G. W., N. Leigh and D. R. Weinberger (1993) *Neuropsychiatric Disorder.* London: Wolfe.

Roberts, Gareth (2008) 'Language and the freerider problem: an experimental paradigm'. *Biological Theory* 3(2): 174–83.

(2010) 'An experimental study of social selection and frequency of interaction in linguistic diversity'. *Interaction Studies* 11(1): 138–59.

Rogalsky, Corianne, Feng Rong, Kourosh Saberi and Gregory Hickok (2011) 'Functional anatomy of language and music perception: temporal and structural factors investigated using functional magnetic resonance imaging'. *The Journal of Neuroscience* 31: 3843–52.

Romaine, Suzanne (1988) *Pidgin and Creole Languages.* London: Longman.

Romani, G. L., S. J. Williamson and L. Kaufman (1982) 'Tonotopic organization of the human auditory cortex'. *Science* 18: 1339–40.

Rosch, E. H., C. B. Mervis, W. D. Gray, D. M. Johnson and P. Boyes-Braem (1976) 'Basic objects in natural categories'. *Cognitive Psychology* 8(3): 382–439.

Ross, Elliot D. (2010) 'Cerebral localization of functions and the neurology of language: fact versus fiction or is it something else? *The Neuroscientist* 16(3): 222–43.

Rubens, A. B., M. W. Mahowlad and J. T. Hutton (1976) 'Asymmetry of the lateral (sylvian) fissures in man'. *Neurology* 26: 620–4.

Ruhlen, Merritt (1994a) *On the Origin of Languages: Studies in Linguistic Taxonomy.* Stanford University Press.

(1994b) *The Origin of Language: Tracing the Evolution of the Mother Tongue.* New York: John Wiley and Sons.

Ryan, Ciara A. and Guy S. Salvesen (2003) 'Caspases and neurodevelopment'. *Biological Chemistry* 384: 855–61.

Sabeti, P. C., S. F. Schaffner, B. Fry, J. Lohmueller, P. Varilly, O. Shamovsky *et al.* (2006) 'Positive natural selection in human lineages'. *Science* 312: 1614–20.

Sabeti, P. C., P. Varilly *et al.* (2007) 'Genome-wide detection and characterization of positive selection in human populations'. *Nature* 449: 913–8.

Saffran, J. R. and G. J. Griepentrog (2001) 'Absolute pitch in infant auditory learning: evidence for developmental reorganization'. *Developmental Psychology* 37(1): 74–85.

Salmons, Joe (1992) 'A look at the data for a global etymology: *tik*, "finger"'. In Garry W. Davis and Gregory K. Iverson (eds.), *Explanation in Historical Linguistics.* Amsterdam: Benjamins, 208–28.

Sampson, Geoffrey (1999) *Educating Eve: The 'Language Instinct' Debate.* London/ New York: Continuum.

(2002) 'Exploring the richness of the stimulus'. In Nancy A. Ritter (ed.), *A Review of 'The Poverty of Stimulus Argument'*, special issue of *The Linguistic Review* 19: 73–104.

Samuels, Michael (1972) *Linguistic Evolution: With Special Reference to English.* Cambridge University Press.

Sandler, Wendy (2010) 'The uniformity and diversity of language: evidence from sign language'. *Lingua* 120: 2727–32.

Sandler, Wendy, Irit Meir, Carol Padden and Mark Aronoff (2005) 'The emergence of grammar: systematic structure in a new language'. *PNAS.* 102(7): 2661–5.

Sapir, Edward (1921) *Language.* New York: Harcourt, Brace and Co.

Sarich, Vincent M. and Allan C. Wilson (1967) 'Immunological time scale for hominid evolution'. *Science* 158: 1200–3.

Satz, P., E. Strauss and H. Whitaker (1990) 'The ontogeny of hemispheric specialization: some old hypotheses revisited. *Brain and Language* 38: 596–614.

Savage-Rumbaugh, Sue and Roger Lewin (1994) *Kanzi: The Ape At the Brink of the Human Mind.* New York: John Wiley and Sons.

Savage-Rumbaugh, Sue, Stuart Shanker and Talbot Taylor (1998) *Apes, Language, and the Human Mind.* Oxford University Press.

Scharff, C. and S. A. White (2004) 'Genetic components of vocal learning'. *Annals of the New York Academy of Sciences* 1016: 325–47.

Schmitz R. W., D. Serre, G. Bonani, S. Feine, F. Hillgruber, H. Krainitzki *et al.* (2002) 'The Neandertal type site revisited: interdisciplinary investigations of skeletal remains from the Neander Valley, Germany'. *Proceedings of the National Academy of Sciences, USA* 99: 13342–7.

Schwartz, Jean-Luc, Louis-Jean Boë, Nathalie Vallée and Christian Abry (1997) 'Major trends in vowel system inventories'. *Journal of Phonetics* 25: 233–53

Scott-Phillips, T. C. and S. Kirby (2010) 'Language evolution in the laboratory'. *Trends in Cognitive Sciences* 14: 411–17.

Scott-Phillips, T. C., S. Kirby and G. R. S. Ritchie (2009) 'Signalling signalhood and the emergence of communication'. *Cognition* 113(2): 226–33.

Seldon, H. L. (1981a) 'Structure of human auditory cortex. I. Cytoarchitectonics and dendritic distributions'. *Brain Research* 229: 277–94.

  (1981b) 'Structure of human auditory cortex. II. Axon distributions and morphological correlates of speech perception'. *Brain Research* 229: 295–310.

Senghas, A. and M. Coppola (2001) 'Children creating language: how Nicaraguan Sign Language acquired a spatial grammar'. *Psychological Science* 12: 323–8.

Senghas, A., S. Kita and A. Ozyurek (2004) 'Children creating core properties of language: evidence from an emerging sign language in Nicaragua'. *Science* 305: 1779–82.

Senut, Brigitte, Martin Pickford, Dominique Gommery, Pierre Mein, Kiptalam Cheboi and Yves Coppens (2001) 'First hominid from the Miocene (Lukeino Formation, Kenya)'. *Comptes Rendus de l'Academie des Sciences, Series IIA – Earth and Planetary Science* 332: 137–44.

Seyfarth, Robert and Dorothy Cheney (1986) 'Vocal development in vervet monkeys'. *Animal Behaviour* 34: 1640–58.

  (1990) 'The assessment by vervet monkeys of their own and another species' alarm calls'. *Animal Behaviour* 40: 754–64.

Seyfarth, Robert, Dorothy Cheney and Peter Marler (1980) 'Monkey responses to three different alarm calls: evidence for predator classification and semantic communication'. *Science* 210: 801–3.

Sharp, S. P., A. McGowan, M. J. Wood and B. J. Hatchwell (2005) 'Learned kin recognition cues in a social bird'. *Nature 434*: 1127–30.

Shaw, P., D. Greenstein, J. Lerch, L. Casen, R. Lenroot, N. Gogtay *et al.* (2006) 'Intellectual ability and cortical development in children and adolescents'. *Nature* 440: 676–9.

Shipster, C., D. Hearst, J. E. Dockrell, E. Kilby and R. Hayward (2002) 'Speech and language skills and cognitive functioning in children with Apert syndrome: a pilot study'. *International Journal of Language and Communication Disorders* 37: 325–43.

Shoshani, J., C. P. Groves, E. L. Simons and G. F. Gunnell (1996) 'Primate phylogeny: morphological versus molecular results'. *Molecular Phylogenetics and Evolution* 5: 102–54.

Shreeve, J. (1995.) *The Neanderthal Enigma. Solving the Mystery of Modern Human Origins*. New York, William Morrow.

Shryock, Andrew and Daniel Lord Smail (eds.) (2011) *Deep History: The Architecture of Past and Present*. Berkeley: University of California Press.

Shu, W., H. Yang, L. Zhang, M. M. Lu and E. E. Morrisey (2001) 'Characterization of a new subfamily of winged-helix/forkhead (Fox) genes that are expressed in the lung and act as transcriptional repressors'. *Journal of Biological Chemistry* 276: 27488–97.

Sibley, C. G. and J. E. Ahlquist (1984) 'The phylogeny of the hominoid primates, as indicated by DNA-DNA hybridization'. *Journal of Molecular Evolution* 20: 2–15.

Singer, W. and C. M. Gray (1995) 'Visual feature integration and the temporal correlation hypothesis'. *Annual Review of Neurosciences* 18: 555–86.

Singh, Ishtla (2000) *Pidgins and Creoles: An Introduction*. London: Arnold.

Skuse, D. H. (1988) 'Extreme deprivation in early childhood'. In Bishop and Mogford, 29–46.

Slaska, Natalia (2006) *Meaning Lists in Historical Linguistics – A Critical Appraisal and Case Study*. Ph.D. dissertation, University of Sheffield.

SLI Consortium (2002) 'A genome-wide scan identifies two novel loci involved in specific language impairment (SLI)'. *American Journal of Human Genetics* 70: 384–98.

   (2004) 'Highly significant linkage to SLI1 locus in an expanded sample of individuals affected by Specific Language Impairment (SLI)'. *American Journal of Human Genetics* 74: 1225–38.

Sloboda, J. A. and M. J. A. Howe (1991) 'Biographical precursors of musical excellence: an interview study'. *Psychology of Music* 19: 3–21.

Smith, Andrew D. M. (2008) 'Protolanguage reconstructed'. *Interaction Studies* 9(1): 98–114. Also reprinted as Smith, A. D. M. (2010) 'Protolanguage reconstructed'. In M. A. Arbib and D. Bickerton (eds.), *The Emergence of Protolanguage: Holophrasis vs Compositionality*. Amsterdam: John Benjamins, 99–115.

Smith, A. D. M., K. Smith and R. Ferrer i Cancho (eds.) (2008) *The Evolution of Language: Proceedings of the 7th International Conference (EVOLANG7)*. Singapore: World Scientific Press.

Smith, A. D. M., M. Schouwstra, B. de Boer and K. Smith (eds) (2010) *The Evolution of Language (EVOLANG8)*. Singapore: World Scientific Press.

Smith, K. (2006) 'The protolanguage debate: bridging the gap?' In A. Cangelosi, A. D. M. Smith and K. Smith, *The Evolution of Language (EVOLANG 6)*. Singapore, World Scientific Publishing, 315–22.

   (2008) 'Is a holistic protolanguage a plausible precursor to language? A test case for a modern evolutionary linguistics'. *Interaction Studies* 9: 1–17.

Smith, K., H. Brighton and S. Kirby (2003) 'Complex systems in language evolution: the cultural emergence of compositional structure'. *Advances in Complex Systems* 6(4): 537–58.

Smith, N. V. (1973) *The Acquisition of Phonology*. Cambridge University Press.

Smith, N. V. and Ianthi Tsimpli (1995) *The Mind of a Savant: Language Learning and Modularity*. Oxford: Blackwell.

Snow, C. and C. Ferguson (eds.) (1977) *Talking to Children: Language Input and Acquisition*. Cambridge University Press.

Snow, D. (1994) 'Phrase-final syllable lengthening and intonation in early child speech'. *Journal of Communication Disorders* 23: 325–36.

Solecki, R. S. (1975) 'Shanidar IV, a Neanderthal flower burial in northern Iraq'. *Science* 190: 880.

Somerville M. J., C. B. Mervis, E. J. Young, E. J. Seo, M. del Campo, S. Bamforth *et al.* (2005) 'Severe expressive-language delay related to duplication of the Williams-Beuren locus'. *New England Journal of Medicine* 353: 1694–701.

Spencer, John P., Mark S. Blumberg, Bob McMurray, Scott R. Robinson, Larissa K. Samuelson and J. Bruce Tomblin (2009) 'Short arms and talking eggs: why we should no longer abide the nativist–empiricist debate'. *Child Development Perspectives* 3(2): 79–87.

Springer, Mark S., Michael J. Stanhope, Ole Madsen and Wilfried W. de Jong (2004) 'Molecules consolidate the placental mammal tree'. *Trends in Ecology and Evolution* 19: 430–8.

Stark, R. E. (1980) 'Stages of speech development in the first year of life'. In G. H. Yeni-Komshian, J. F. Kavanagh and G. A. Ferguson (eds.), *Child Phonology: Vol. I*. New York: Academic Press, 73–92.

Starostin, Sergei (2000) 'Comparative-historical linguistics and lexicostatistics'. In Colin Renfrew, April McMahon and Larry Trask (eds.), *Time Depth in Historical Linguistics*. Cambridge: McDonald Institute for Archaeological Research, 223–66.

Stauffer, R. L., A. Walker, O. Ryder, M. Lyons-Weiler, and S. B. Hedges (2001) 'Human and ape molecular clocks and constraints on paleontological hypotheses'. *Journal of Heredity* 92: 469–74.

Steels, L. (2003) 'Evolving grounded communication for robots'. *Trends in Cognitive Sciences* 7: 308–12.

Steinbeis, N. and S. Koelsch (2008) 'Shared neural resources between music and language indicate semantic processing of musical tension-resolution patterns'. *Cerebral Corte* 18: 1169–78.

Stojanovik, Vesna, M. Perkins and S. Howard (2004) 'Williams syndrome and specific language impairment do not support claims for developmental double dissociations and innate modularity'. *Journal of Neurolinguistics* 17: 403–24.

Strachan, Tom and Andrew P. Read (2004) *Human Molecular Genetics*. Third Edition. London/New York: Garland Science.

Streeter, Lynn (1976) 'Language perception in 2-month-old infants shows effects of both innate mechanisms and experience'. *Nature* 259: 39–41.

Stringer, C. and P. Andrews (2005) *The Complete World of Human Evolution*. London: Thames and Hudson.

Stringer, C. and C. Gamble (1993) *In Search of the Neanderthals*. London: Thames and Hudson.

Stringer, C. and R. McKie (1996) *African Exodus: The Origins of Modern Humanity*. New York: Henry Holt.

Stromswold, K. (2000) 'The cognitive neuroscience of language acquisition'. In M. S. Gazzaniga (ed.), *The New Cognitive Neurosciences*. Cambridge, MA: MIT Press, 909–32.

Studdert-Kennedy, M. (1998) 'The particulate origins of language generativity: from syllable to gesture'. In J. R. Hurford, M. Studdert-Kennedy, and C. Knight (eds.), *Approaches to the Evolution of Language: Social and Cognitive Bases*. Cambridge University Press.

(2000) 'Evolutionary implications of the particulate principle: Imitation and the dissociation of phonetic form from semantic function'. In C. Knight, J. R. Hurford and M. Studdert-Kennedy (eds.), *The Evolutionary Emergence of Language: Social Function and the Origins of Linguistic Form*. Cambridge University Press.

(2005) 'How did language go discrete?' In M. Tallerman, *Language Origins: Perspectives on Evolution*. Oxford University Press, 48–67.

Studdert-Kennedy, M. and L. Goldstein (2003) 'Launching language: the gestural origins of discrete infinity'. In Christiansen and Kirby, 235–54.

Sulek, A. (1989) 'The experiment of Psammetichus: fact, fiction, and model to follow'. *Journal of the History of Ideas* 50: 645–51.

Sutton, W. S. (1903) 'The chromosomes in heredity'. *Biological Bulletin* 4: 231–51.

Sykes, Bryan (2001) *The Seven Daughters of Eve*. New York: W. W. Norton and Co.

(2003) *Adam's Curse: A Future Without Men*. London: Corgi.

(2006) *Blood of the Isles: Exploring the Genetic Roots of our Tribal History*. London: Bantam Press.

Számadó, S. and E. Szathmáry (2006) 'Competing selective scenarios for the emergence of natural language'. *Trends in Ecology and Evolution* 21: 555–61.

Tallerman, Maggie (ed.) (2005) *Language Origins: Perspectives on Evolution*. Oxford University Press.

Tallerman, Maggie (2007) 'Did our ancestors speak a holistic protolanguage? In A. Carstairs-McCarthy (ed.), *The Evolution of Language*. Special issue of *Lingua* 117: 579–604.

Tanenhaus, M. K., M. J. Spivey-Knowlton, K. M. Eberhard and J. C. Sedivy (1995) 'Integration of visual and linguistic information in spoken language comprehension'. *Science* 268: 1632–4.

Tanner, Joanne E. and Richard W. Byrne (1996) 'Representation of action through iconic gesture in a captive lowland gorilla'. *Current Anthropology* 37(1): 162–73.

Tanner, Joanne E., Francine G. Patterson and Richard W. Byrne (2006) 'The development of spontaneous gestures in zoo-living gorillas and sign-taught gorillas: from action and location to object representation'. *Journal of Developmental Processes* 1: 69–103.

Taylor, John (1995) *Linguistic Categorization: Prototypes in Linguistic Theory*. 2nd edition. Oxford University Press.

Templeton, Alan (2002) 'Out of Africa again and again'. *Nature* 416: 45–51.

Terrace, Herbert (1979) *Nim*. New York: Knopf.

Thomason, Sarah G. (2001) *Language Contact: An Introduction*. Edinburgh University Press.

Thomason, Sarah G. and Terrence Kaufman (1988) *Language Contact, Creolization, and Genetic Linguistics*. Berkeley, CA: University of California Press.

Thomson R., J. K. Pritchard, P. Shen, P. J. Oefner and M. W. Feldman (2000) 'Recent common ancestry of human Y chromosomes: Evidence from DNA sequence data'. *Proceedings of the National Academy of Sciences* 97:7360–5.

Tobias, P. V. (1995) 'The brains of the first hominids'. In J.-P. Changeux and J. Chavaillon (eds.), *Origins of the Human Brain*. Oxford University Press, 61–82.

Tomasello, M. (2003) *Constructing a Language: A Usage-Based Theory of Child Language Acquisition*. Cambridge, MA: Harvard University Press.

Tomasello, M., J. Call, J. Warren, T. Frost, M. Carpenter and K. Nagell (1997) 'The ontogeny of chimpanzee gestural signals'. In S. Wilcox, B. King and L. Steels (eds.), *Evolution of Communication*. Amsterdam: John Benjamins, 224–59.

Tooby, J. and I. DeVore (1987) 'The reconstruction of hominid behavioral evolution through strategic modeling'. In Warren G. Kinzey (ed.), *The Evolution of Human Behavior: Primate Models*. Albany, NY: SUNY Press, 186–213.

Toro, R. and Y. Burnod (2005) 'Morphogenetic model for the development of cortical convolutions'. *Cerebral Cortex* 15: 1900–13.

Trask, R. L. (1996) *Historical Linguistics: An Introduction*. London: Arnold.

Trinkaus, Erik and Pat Shipman (1993) *The Neandertals: Changing the Image of Mankind*. London: Pimlico.

Trivers, R. L. (1971) 'The evolution of reciprocal altruism'. *Quarterly Review of Biology* 46: 35–57.

Troth, N. (1985) 'Archaeological evidence for preferential right-handedness in the lower and middle Pleistocene, and its possible implications'. *Journal of Human Evolution* 14: 607–14.

Trudgill, Peter (1992) 'Dialect typology and social structure'. In Ernst Hakon Jahr (ed.), *Language Contact*. New York: Mouton de Gruyter, 195–211.

Tulving, E. (2002) 'Episodic memory: from mind to brain'. *Annual Review of Psychology* 53: 1–25.

Tyler-Smith, C. (2002) 'What can the Y chromosome tell us about the origin of modern humans?' In T. J. Crow (ed.), *The Speciation of Modern Homo Sapiens*. Oxford University Press, 217–30.

Ullman, M. and M. Gopnik (1999) 'The production of inflectional morphology in hereditary specific language impairment'. *Applied Psycholinguistics* 20: 51–117.

Ultan, R. (1978) 'Some general characteristics of interrogative systems'. In J. H. Greenberg, C. A. Ferguson and E. A. Moravcsik (eds.), *Universals of Human Language. Vol. II: Syntax*. Stanford University Press, 211–48.

Vargha-Khadem, Faraneh, K. Watkins, K. Alcock, P. Fletcher and R. Passingham (1995) 'Praxic and nonverbal cognitive deficits in a large family with a genetically transmitted speech and language disorder'. *Proceedings of the National Academy of Sciences of the USA* 92: 930–3.

Vihman, Marilyn May (1996) *Phonological Development: The Origins of Language in the Child*. Oxford: Blackwell.

Vincent, Nigel (1978) 'Is sound change teleological?' In J. Fisiak (ed.), *Recent Developments in Historical Phonology*. The Hague: Mouton, 409–29.

Voight, B. F., S. Kudaravalli, X. Wen and J. K. Pritchard (2006) 'A map of recent positive selection in the human genome'. *PLoS Biol* 4(3): e72.doi:10.1371/journal. pbio.0040072.

von Frisch, Karl (1967) *A Biologist Remembers*. Oxford: Pergamon Press.

Wada, Juhn A. (1997) 'Clinical experimental observations of carotid artery injections of sodium amytal'. *Brain and Cognition* 33: 11–13. (Translation of J. Wada (1949) *Igaku to Seibutsuqaku 14*: 221–2).

Wada, Juhn A. and T. Rasmussen (1960) 'Intracarotid injection of sodium amytal for the lateralization of speech dominance'. *Journal of Neurosurgery* 17: 266–82.

Waddington, C. H. (1942) 'Canalisation of development and the inheritance of acquired characters'. *Nature* 150: 563–5.

(1975) *The Evolution of an Evolutionist*. Edinburgh University Press.

Walker, A. and M. F. Teaford (1988) 'The Kaswanga Primate Site: An early Miocene hominoid site on Rusinga Island, Kenya'. *Journal of Human Evolution* 17: 539–44.

Wallman, Joel (1992) *Aping Language*. Cambridge University Press.

Ward, C. (2002) 'Interpreting the posture and locomotion of *Australopithecus afarensis*: where do we stand?' *Yearbook of Physical Anthropology Supplement* 35: 185–215.

Watkins, K. E., F. Vargha-Khadem, J. Ashburner, R. E. Passingham, K. J. Friston, A. Connelly *et al.* (2002) 'MRI analysis of an inherited speech and language disorder: structural brain abnormalities'. *Brain* 125: 465–78.

Watson, J. D. and F. H. C. Crick (1953) 'Molecular structure of nucleic acids. A structure for deoxyribose nucleic acid'. *Nature* 171: 737–8.

Weber, Bruce H. and David J. Depew (eds.) (2003) *Evolution and Learning: The Baldwin Effect Reconsidered*. Cambridge, MA: MIT Press.

Weinberg, W. (1908) 'Über den Nachweis der Vererbung beim Menschen'. *Jahreshefte des Vereins für vaterländische Naturkunde in Württemberg* 64: 368–82.

Wells, J. C. (1983) *Accents of English*. 3 volumes. Cambridge University Press.

Werker, J. F. and C. E. Lalonde (1988) 'Cross-language speech perception: initial capabilities and developmental change'. *Developmental Psychology* 24: 672–83.

Wernicke, C. (1874) *Der aphasische Symptomencomplex*. Breslau: M. Cohn und Weigert.

West, S. A., A. S. Griffin and A. Gardner (2007) 'Evolutionary explanations for cooperation'. *Current Biology* 17: 661–72.

Westbury C. F., R. J. Zatorre and A. C. Evans (1999) 'Quantifying variability in the planum temporale: a probability map'. *Cerebral Cortex* 9: 392– 405.

Westergaard, Gregory C. and Stephen J. Suomi (1994) 'A simple stone tool technology in monkeys'. *Journal of Human Evolution* 27: 399–404.

(1996a) 'Hand preference for a bimanual task in tufted capuchins (Cebus apella) and rhesus macaques (Macaca mulatta)'. *Journal of Comparative Psychology* 110: 406–11.

(1996b) 'Hand preference for stone artefact production and tool use by monkeys: possible implications for the evolution of right-handedness in hominids'. *Journal of Human Evolution* 30: 291–98.

White, D., G. Suwa and B. Asfaw (1994) 'Australopithecus ramidus, a new species of early hominid from Aramis, Ethiopia'. *Nature* 371: 306.

Wichmann, Søren and Arpiar Saunders (2007) 'How to use typological databases in historical linguistic research'. *Diachronica* 24(2): 373–404.

Wilson, Allan C. and Vincent M. Sarich (1969) 'A molecular time scale for human evolution'. *PNAS* 63: 1088–93.

Wolpoff, M. H. and R. Caspari (1997) 'What does it mean to be modern?' In G. A. Clark and C. M. Willermet (eds.), *Conceptual Issues in Modern Human Origins Research*. New York: Aldine de Gruyter, 28–44.

Wolpoff, M. H., Wu Xinzhi and A. G. Thorne (1984) 'Modern *Homo sapiens* origins: a general theory of Hominid evolution involving the fossil evidence from East Asia'. In F. H. Smith and F. Spencer (eds.), *The Origins of Modern Humans: A World Survey of the Fossil Evidence*. New York: Liss, 411–83.

Wood, B. and M. Collard (1999) 'The human genus'. *Science* 284: 65–71.

Wray, Alison (1998) 'Protolanguage as a holistic system for social interaction' *Language & Communication* 18: 47–67.

       (2000) 'Holistic utterances in protolanguage: the link from primates to humans'. In Knight, Studdert-Kennedy and Hurford, 285–302.

       (2002a) *Formulaic Language and the Lexicon*. Cambridge University Press.

Wray, Alison (ed.) (2002b) *The Transition to Language*. Oxford University Press.

Wray, Alison (2008) *Formulaic Language: Pushing the Boundaries*. Oxford University Press.

Wright, Sewell (1931) 'Evolution in Mendelian populations'. *Genetics* 16: 97–159.

Yang L. R. and T. Givón (1997) 'Benefits and drawbacks of controlled laboratory studies of second language acquisition: the Keck second language learning project'. *Studies in Second Language Acquisition* 19: 173–94.

Yeh, Edward, Taizo Kawano, Robby M. Weimer, Jean-Louis Bessereau and Mei Zhen (2005) 'Identification of genes involved in synaptogenesis using a fluorescent active zone marker in Caenorhabditis elegans'. *The Journal of Neuroscience* 25: 3833–41.

Yeni-Komishian, G. H. and D. A. Benson (1976) 'Anatomical study of cerebral asymmetry in the temporal lobe of humans, chimpanzees and rhesus monkeys'. *Science* 192: 387–9.

Yunis, J. J. and O. Prakash (1982) 'The origin of man: a chromosomal pictorial legacy'. *Science* 215: 1525–30.

Zatorre, R. J., P. Belin and V. B. Penhune (2002) 'Structure and function of auditory cortex: music and speech'. *Trends in Cognitive Sciences* 6: 37–46.

Zollikofer, Christoph P. E., Marcia S. Ponce de León, Daniel E. Lieberman, Franck Guy, David Pilbeam, Andossa Likius, Hassane T. Mackaye, Patrick Vignaud and Michel Brunet (2005) 'Virtual cranial reconstruction of *Sahelanthropus tchadensis*'. *Nature* 434: 755–9.

Zubrow, Ezra (1989) 'The demographic modeling of neanderthal extinction'. In P. Mellars and C. Stringer (eds.), *The Human Revolution: Behavioral and Biological Perspectives on the Origins of Modern Humans*. Princeton University Press, 212–31.

Index